Contributions to Environmental Sciences & Innovative Business Technology

Contributions to Environmental Sciences & Innovative Business Technology (CESIBT) is an interdisciplinary series of peer-reviewed books dedicated to addressing emerging research trends relevant to the interplay between Environmental Sciences, Innovation, and Business Technology in their broadest sense. This series constitutes a comprehensive up-to-date interdisciplinary reference that develops integrated concepts for sustainability and discusses the emerging trends and practices that will define the future of these disciplines.

This series publishes the latest developments and research in the various areas of Environmental Sciences, Innovation, and Business Technology, combined with scientific quality and timeliness. It encompasses the theoretical, practical, and methodological aspects of all branches of these scientific disciplines embedded in the fields of Environmental Sciences, Innovation, and Business Technology.

The series also draws on the best research papers from EuroMid Academy of Business and Technology (EMABT) and other international conferences to foster the creation and development of sustainable solutions for local and international organizations worldwide. By including interdisciplinary contributions, this series introduces innovative tools that can best support and shape both the economical and sustainability agenda for the welfare of all countries, through better use of data, a more effective organization, and global, local, and individual work. The series can also present new case studies in real-world settings offering solid examples of recent innovations and business technology with special consideration for resolving environmental issues in different regions of the world.

The series can be beneficial to researchers, instructors, practitioners, consultants, and industrial experts, in addition to governments from around the world. Published in collaboration with EMABT, the Springer CESIBT series will bring together the latest research that addresses key challenges and issues in the domain of Environmental Sciences & Innovative Business Technology for sustainable development.

Ahmad Rafiki
Editor

Digitalization in Halal Management

Springer

Editor
Ahmad Rafiki
Universitas Medan Area
Medan, Indonesia

ISSN 2731-8303 ISSN 2731-8311 (electronic)
Contributions to Environmental Sciences & Innovative Business Technology
ISBN 978-981-99-5145-1 ISBN 978-981-99-5146-8 (eBook)
https://doi.org/10.1007/978-981-99-5146-8

This Springer imprint is published by the registered company Springer Nature Singapore Pte Ltd.
The registered company address is: 152 Beach Road, #21-01/04 Gateway East, Singapore 189721, Singapore

Paper in this product is recyclable.

Contents

Halal Logo as Consumer Alternative Solution for Nutraceutical Products Issues

1

Siti Nurunnajwa Shamsudin, Elistina Abu Bakar, Nusaibah Anuar Musaddad, and Nuradli Ridzwan Shah Mohd Dali

Abstract

Nutraceutical is defined as any food or food component that provides medical or health benefits, including the treatment or prevention of disease. Nutraceuticals have sparked considerable interest recently due to their nutritional values and therapeutic benefits. However, a few issues related to this product could affect consumers, especially Muslims. This article examines the nutraceutical product issues regarding halal status, safety, and legality. Consumers need to protect themselves by choosing the best nutraceutical product in the market. Therefore, this study suggests that consumers choose nutraceutical products with a halal logo because they can be the benchmark for halal, safe, and good-quality products.

Keywords

Halal • Nutraceuticals • Halal logo • Food safety • Consumer protection

1.1 Introduction

Consistent with the old Arabic proverb, "prevention is better than cure," more people nowadays take dietary supplements, including nutraceutical products, to maintain health. Health is the greatest gift from God to all human beings. The Prophet (p.b.u.h.) has highlighted the importance of good health, "No blessing

S. N. Shamsudin (✉) · E. Abu Bakar · N. Anuar Musaddad
Halal Product Research Institute, Universiti Putra Malaysia, 43400 Serdang, Selangor, Malaysia
e-mail: najwashams@gmail.com

N. R. S. Mohd Dali
Faculty of Economic and Muamalat, Universiti Sains Islam Malaysia, 71800 Nilai, Negeri Sembilan, Malaysia

A. Rafiki (ed.), *Digitalization in Halal Management*, Contributions to Environmental Sciences & Innovative Business Technology,
https://doi.org/10.1007/978-981-99-5146-8_1

other than faith is better than well-being" (Sahih al-Bukhari). The hadeeth advised Muslims to take care of their health to thank God for this great blessing. As people began to be concerned about their physical health by consuming nutraceutical products, their demand increased. With a healthy growth rate of 4% over the forecast period of 2022–2030, the global demand for nutraceutical products is projected to grow from USD 429.5 billion in 2021 to USD 652.6 billion by 2030 (Market Watch 2022).

People have consumed numerous nutraceutical products throughout recent years to improve their health. These items enhance bodily functions, mental performance, and weight loss. Khan et al. (2011) discovered that Malaysians commonly use nutraceutical products. This statement is supported by National Health and Morbidity Report in 2017; two in five adolescents in Malaysia consume food supplements (Institute for Public Health (IPH) 2017). According to Harian (2018), about 80% of Malaysians were believed to consume nutraceutical products at least once in their lifetimes. Herbal-based nutraceutical products have been growing in the Malaysian market, and the numbers of consumers keep increasing. Some of the well-known nutraceuticals producers in Malaysia are D'Herbs, V'Asia, Vida Beauty, Seri Ayu, Mustika Ratu, Orang Kampung, Nona Roguy, Dunia Herbs, White Pearl and Jamu Ratu Malaya.

Several issues concerning nutraceutical products have recently arisen, which may potentially harm consumers. Furthermore, Muslim consumers need to be extra careful when choosing nutraceutical products, as these products can contain non-halal ingredients. As a result, consumers, particularly Muslims, must become more self-protective to avoid purchasing and consuming harmful or non-halal products (Shamsudin et al. 2022). Consumers should choose halal-certified nutraceutical products by looking for a specific halal logo on the product's packaging because they are guaranteed halal, safe, and high quality. Hence, applying the halal logo is a lifesaver for all consumers, not limited to Muslims. Therefore, this study aims to analyze the nutraceutical products issues and the halal logo as the consumers' solution to protect themselves from non-halal and unsafe nutraceutical products.

1.2 Literature Review

1.2.1 Overview of Halal Nutraceuticals

Stephen De Felice first introduced the term "nutraceutical" in 1989. It is a combination of nutrition and pharmaceuticals. According to De Felice (1997), a nutraceutical is "a food, or parts of a food, that provide medical or health benefits including the prevention and treatment of disease." The development and connection between "nutrition" and "health" led to the nutraceutical concepts' emergence. Even though the nutraceutical term has been used for nearly 30 years, there is no exact definition for the products or materials under this term (Télessy 2018). Generally, the term "nutraceutical" is defined differently in different countries (Nasri et al. 2014). Every country has its own terminologies in various jurisdictions to regulate nutraceuticals. However, Malaysia has no specific definition and regulation related to nutraceuticals. Therefore, nutraceutical products in Malaysia may be subjected to drugs or food depending on the classification given by the manufacturers. Nutraceutical products have four categories: dietary supplements, functional foods, therapeutic foods, and pharmaceuticals (Nawsheen Bee and Rahmanpoor 2019; Sreeramya et al. 2018). In recent times, various issues have arisen that may harm consumers. According to Shamsudin et al. (2023), nutraceuticals have attracted substantial interest due to their prospective nutritional, safety, and therapeutic effects, especially post-pandemic when people search for more natural supplements and treatments, compared to "chemical medication." Moreover, in comparison to ordinary dietary supplements, the health benefits provided by nutraceutical supplements extend beyond their primary nutritional value by supplementing the diet and retaining therapeutic functions (Daud et al. 2017).

According to Islamic law, consuming halal is an absolute obligation for Muslims (Mohd Dali et al. 2019). Thus, ensuring halal in nutraceuticals is pertinent, especially in majority Muslim countries like Malaysia. Halal nutraceuticals should refer to nutraceuticals that adhere to Islamic dietary laws and are permitted for consumption by Muslims (Alzeer and Hadeed 2020). So far, there has been little discussion about halal nutraceuticals. This research topic was started in 2008 by Ismail and Ehsan (2008), who listed the potential ingredients and types of nutraceutical products that may be using non-halal sources. Moreover, Alzeer and Hadeed (2020) used a similar approach by discussing the potential non-halal ingredients in food and pharmaceuticals and also introduced the concept of halalopathy (a holistic approach to disease prevention and treatment that integrates naturopathy, allopathy, and halal guidelines). Alzeer and Hadeed (2020); Regenstein (2019) conducted studies that examined the concept of halal nutraceuticals from Islamic perspectives. Similarly, Mohamad et al. (2018) also conducted a review of halal nutraceuticals but focused exclusively on those mentioned in the Quran and hadeeth, including myrtle, figs, olive oil, black seed, dates, honey and vinegar. A study by Umar et al. (2016) also discussed the issues and challenges of nutraceuticals but only focused on prophetic nutraceuticals. Despite the fact that several studies have been conducted in this area, it is clear that such studies are still required to discuss current issues and possible solutions.

1.3 Research Methodology

This paper adopted a qualitative content analysis method to analyze existing literature concerning nutraceutical products and halal logos available across the databases. Mayring (2014) defined content analysis as an empirical and methodological approach for analyzing texts within their context of understanding, following analytical rules of content analysis and step-by-step models, without rash quantification. The documents for this study are collected from the available databases, including Scopus, ResearchGate, Elsevier, and Dimensions. Data gathered from these documents are analyzed and deduced to emulate the objective of this paper, which is to provide insight into how the halal logo can act as an alternative solution for nutraceutical product issues. The result and discussion will be as follows.

1.4 Discussion

1.4.1 Nutraceutical Products Issues

Halal Issue

Typically, nutraceutical products are made of botanical and plant extracts. Numerous herbal and botanical extracts, including lingzhi, long jack, and turmeric, have been developed as nutraceutical products. Formulation of nutraceutical products using animal-derived ingredients should be avoided. However, other ingredients used as flavors, stabilizers, or coloring agents may be derived from non-halal ingredients. Some sources of non-halal ingredients include pork, animals that were not appropriately slaughtered, genetically modified, extracted with ethanol, or derived from blood.

Some nutraceuticals use gelatin to produce soft and hard shell capsules, as well as for tableting, granulation, tablet coating, encapsulation, and microencapsulation (Mariod and Adam 2013). Gelatin prevents oxidation, facilitates swallowing, and masks unpleasant flavors to make the product more palatable (Uddin et al. 2021). It is typically produced through the selective hydrolysis of collagen extracted from the animal's skin, hides, white connective tissues, and bones (Abdullah Amqizal et al. 2017). Gelatin may be derived from cattle bone and skin. It could be considered halal if the Islamic slaughtered procedure was applied to slaughter the cattle. Recently, Muslim consumers have alternative halal gelatin derived from seaweed (Adam et al. 2020) and fish (Karim and Bhat 2009).

However, most currently available gelatin is not acceptable for Muslim consumers because it might be derived from pork. Demirhan et al. (2012) found that about 80% of Europe's edible gelatin is made from pigskin. Erwanto et al. (2018) also claimed that about 90% of gelatin used in food and pharmaceutical products, including nutraceuticals, has a porcine origin. In the meantime, Sultana et al. (2020) found that out of 35 halal-branded supplements, two tested positive for porcine. Therefore, the halal status of gelatin is highly critical as it may be derived from a non-halal source (Alzeer and Hadeed 2020).

Other ingredients such as glycerin, stearates, and tweens can be considered to be potential non-halal ingredients in most capsule and tablet formulations. For example, glycerin is used as a carrier in many medicines and as a plasticizer in gelatin capsules (Stout and McKessor 2012). Meanwhile, tweens were used for coating and polishing tablets (Riaz et al. 2019) and stearates as a lubricant to improve tablet brittleness (Paul and Sun 2017). All these three ingredients are usually obtained from plant and animal sources. However, animal-based ingredients should be prevented because they can come from non-halal sources.

Furthermore, alcohol is one of the ingredients with a highly critical halal status used in the nutraceutical industry. Numerous liquid formulations use ethyl alcohol (ethanol) as a preservative or solvent (Ismail and Ehsan 2008). The halal status of ethanol is a highly debated topic, and the majority of Muslim scholars hold the opinion that it cannot be considered halal under any circumstances, regardless of how it was manufactured. The halal status of ethanol is contingent on the source as well as the concentration of the alcohol. According to Alzeer and Abou Hadeed (2016), any ethanol produced through anaerobic fermentation with a concentration ranging from 1 to 15% was regarded as non-halal. On the other hand, ethanol produced organically through fermentation and has a concentration of less than 1% which is recognized as a preservative. Its usage is allowed in Islam because it is regarded as halal.

A current study by Qomariah et al. (2021) discovered that out of 18 herbal products sold in the Indonesian market, six of them can be considered non-halal due to their high alcohol content. All 18 products were categorized as beverages. As a result, no clinical evaluation was required before they could be sold in the market. Therefore, liquid formulation products in the nutraceutical industry, especially herbal products, were exposed to high alcohol content and could be considered non-halal.

Some flavors and colorants used in nutraceutical products also might have hidden alcohol, such as artificial colors and flavors, vanilla, and vanillin extract (Chaudry et al. 2000). However, industrial ethanol can be used to extract flavors, but the amount should be less than 0.5% (Riaz et al. 2019). Other colorants and flavorings such as oleoresins, glutamic acid, civet oil, ambergris, and castoreum may be derived from haram animal origin (Chaudry et al. 2000; Riaz et al. 2019). In creating new flavors, it is best to avoid using any ingredients derived from animals unless those ingredients have been confirmed to be halal. Therefore, consumers should be aware that nutraceutical products that use flavorings and colorants may be exposed to non-halal ingredients and should be avoided. Table 1.1 summarizes the potential non-halal ingredients used in nutraceutical products.

Safety Issues

Another issue that previous studies have highlighted is the safety of nutraceutical products. In a study, Nasri et al. (2014) pointed out that people had consumed thousands of species of plants to meet their basic nutritional needs for many years. However, very few of them have received substantial safety studies. Since most nutraceuticals are made from plant extracts, they will likely contain pesticides, mycotoxins, metals (such as arsenic, cadmium, lead, and mercury), and other plant alkaloids, such as pyrrolizidine alkaloids.

Many nutraceuticals, such as Aloe vera, St. John's wort, bitter melon, goldenseal, green tea extract, Ginkgo biloba, Ma Huang, and Garcinia cambogia, have the potential to be harmful. You can find many toxic contaminants in foods and nutraceuticals, such as metals, phytotoxicants, mycotoxins, radiation, pesticides, therapeutic drugs, and illegal drugs. When these contaminants are present in higher amounts, they can cause serious health problems or even death. In addition, it is

Table 1.1 Potential non-halal ingredients in nutraceutical products

Ingredients	Comments
Ethanol	Used to maximize the therapeutic properties of the herbal extracts and as preservatives to keep a longer shelf-life. Any ethanol produced through anaerobic fermentation with a concentration between 1 and 5% is haram. However, ethanol produced through natural fermentation and containing less than 1% is permitted and halal
Colorants and flavors	In the formulations, there may be hidden alcohol or haram animal-derived ingredients, such as civet oil, ambergris, and castoreum
Gelatin	Used in the production of capsules, both soft gel and two-piece hard shells, and typically derived from animal sources, primarily pig. Halal gelatin, derived from seaweed or fish, can be substituted for porcine gelatin
Glycerin	Used as a stabilizer and emulsifier in dietary supplements to protect the ingredients from drying. It may be derived from both vegetable and animal fats
Stearates	It has lubricating properties that make it suitable as a flow agent in powders and tableting aids in tablets. For halal products, manufacturers may use plant-based stearates rather than those derived from animals
Tweens	Utilize for polishing and coating tablets. In halal products, tweens derived from plants should replace those derived from animals

Source Author's adapted concept from Alzeer and Hadeed (2020), Riaz et al. (2019)

important to note that children, pregnant women, and the elderly are more at risk if these contaminants are higher than what is considered safe (Gupta et al. 2018).

Another major issue that needs to be concerned about is the adulteration of nutraceuticals with illegal drugs. Nounou et al. (2021) have brought attention to the fact that male-enhancement nutraceutical products that claim to be 100% natural have been found to contain harmful ingredients. The products contain undeclared adulterants (sildenafil citrate) and are highly toxic and potentially harmful to consumers. In addition, drug interactions with food, herbs, and dietary supplements can have severe consequences. Interactions between prescription medications and herbal supplements significantly threaten patients' safety, particularly in medications with a narrow therapeutic index. This interaction can result in fatalities (Anadón et al. 2021).

According to a report by Harian (2018), dozens of chemical compounds in herbal-based health supplement products sold on the market have not undergone clinical testing. As a result, they have not been approved by the Ministry of Health (MOH). Various chemical compounds are used in the production of nutritional supplements. However, it is not tested directly by the National Pharmaceutical Regulatory Agency (NPRA) because some were sold as food. In 2017, MOH Malaysia issued a list of 162 adulterated products that contained illegal substances and scheduled poisons, of which 78 were classified as nutraceuticals (Aziz et al. 2020).

Legal Issue

Several issues concerning the regulation of nutraceutical products in Malaysia have been highlighted by previous research. One of the most significant issues related to nutraceuticals' regulation in Malaysia is the nutraceutical term under Malaysian law. Daud et al. (2017) claimed that this nutraceuticals' term has yet to be appropriately defined. Another study by Ismail et al. (2020) also pointed out that nutraceuticals are loosely categorized between food and drugs. Therefore, the regulation of nutraceutical products in Malaysia's legal system remains unclear (Safuan et al. 2020). As a result, nutraceutical products may fall under the jurisdiction of distinct regulatory agencies, either Food Safety and Quality Division (FSQD) or NPRA.

NPRA and FSQD are departments under the MOH. NPRA was established to implement quality control on pharmaceutical products. Meanwhile, FSQD was formed to coordinate food quality control activities. Most nutraceutical products in Malaysia are classified as food products under the control of FSQD. When a product is subjected to FSQD, it is only required to fulfill and comply with the provision of the Food Act 1983. If nutraceutical products are classified as food, they are exempted from strict requirements imposed by the Control of Drugs and Cosmetics Regulation 1984. Consequently, the manufacturer can escape the strict clinical evaluation (Safuan et al. 2020), and the product may be freely sold on the open market.

Furthermore, most nutraceutical products claim that they can treat, reduce, or even prevent serious diseases such as hypertension, asthma, and heart attack (Daud et al. 2017). Following subsection 17(1) of the Food Act of 1983, these health claims made on food products are illegal. Unlike pharmaceutical products, food-registered nutraceuticals are not required to undergo clinical testing to prove that the products are effective. Thus, the health claims made by sellers could not be true. Moreover, if the products contain illegal substances or poisons, they may harm the body and cause death. In 2017, D'Herbs Holdings (M) Sdn Bhd, a well-known seller of nutraceutical products in Malaysia, was fined for promoting two products through deceptive advertisements. The company claimed its products which could enhance immune and blood circulation. However, this assertion contradicts Section 17(1) of the Food Act of 1983, and the Magistrate's Court fined the company RM11,000 for this violation (Bernama, 2017).

The Importance of Halal Logo for Consumer

To overcome the previous issues, the consumer should play an important role by choosing the best nutraceutical products to be consumed. Then, using the rule of "caveat emptor" or "let the buyer beware," the consumers are responsible for verifying that the products they want to purchase are safe and halal (Dahlan and Abdullah 2017). This responsibility was also called self-protection. Self-protection is how a person defends himself, his property, and his well-being against any harm, and it is a key element outlined in the Malaysian National Consumer Policy (NCP) (Jing Yi and Afida Mastura 2021).

Therefore, consumers should choose nutraceutical products with the halal logo. This is because the halal logo signifies that a product can be relied upon in terms of halal, safety, and quality (Ambali and Bakar 2014). In Malaysia, the halal logo is protected by the Trade Description Act 2011. This act provides legal protection and stricter enforcement against halal fraud. In addition, specific government agencies have been authorized by law to issue halal certification and conduct halal surveillance in this country (Mohd Dali et al. 2009; Zakaria and Ismail 2014). In Malaysia, the government bodies that can issue halal certification is the Department of Islamic Development Malaysia (JAKIM), the State Islamic Religious Council (MAIN), and the State Islamic Religious Department (JAIN).

The halal certification is granted following the fundamental principles and procedures stipulating that the items must be halal, safe, high quality, and suitable for human consumption (Noordin et al. 2009). The procedure adopted by JAKIM to determine halal products, premises, and services is based on Manual Procedure for Malaysia Halal Certification (Domestic) 2020. It specifies the essential principles that all halal certificate holders in Malaysia must adhere to, and it is used as a guideline by JAKIM to coordinate the implementation of halal certification activities in Malaysia. This manual covers the process of applying for halal certification and monitoring, inspection, and enforcement of halal products (Arif and Sidek 2015). Its content and requirement must be read with the Malaysian Halal Standards, Malaysian Halal Management System (MHMS) 2020, fatwa decisions, relevant regulatory documents, rules, and circulars enforced in the halal certification procedures (JAKIM 2020). The MHMS 2020 is used to ensure halal integrity and practices. This document explains the requirements for developing, implementing, and maintaining halal assurance to fulfill and comply with the Malaysian Halal Certification procedure.

Meanwhile, the Malaysian Halal Standards provide practical guidelines for preparing and handling halal products in Malaysia. It assures the manufacturing and production of halal-compliant products. Specific standards have been established to provide general guidelines for halal products and services. Table 1.2 shows a list of Malaysian Halal Standards. However, there are no specific standards related to nutraceuticals. Now, halal nutraceutical products should follow either MS1500:2019 for halal food or MS2424:2019 for halal pharmaceuticals to fulfill the halal requirements. Apart from standards, there are also a few acts and regulations that need to be adhered to, such as the Food Act of 1983, Food Hygiene Regulation of 2009, Food Regulation of 1985, Poisons Act of 1952, Sales of Drugs Act of 1952, and other related guidelines and regulations (JAKIM 2020).

Ensure Product Halalness

All JAKIM-certified products must have halal status in terms of ingredients, packaging, processing, and storage area, as well as the logistics process. According to Manual Procedure for Malaysia Halal Certification (Domestic) 2020, all the raw materials and ingredients must be halal-certified. All uncertified halal raw materials and ingredients must be accompanied by detailed specifications (indicating material composition, flow chart, and its raw material sources). In addition, they

Table 1.2 List of Malaysia Halal standards

No.	MS No.	Title	Status
1	MS1500:2019	Halal food—general requirements (third revision)	Third revision
2	MS1500:2009	Halal food—production, preparation, handling, and storage—general guidelines (second revision)	Second revision
3	MS1900:2014	Shariah-based quality management system—requirement with guidance (first revision)	First revision
4	MS2200:Part 1:2008	Islamic consumer goods—part 1: cosmetic and personal care-general guidelines	Original
5	MS2200-2:2013	Islamic consumer goods—part 2: usage of animal bone, skin and hair—general guidelines	Original
6	MS2393:2013	Islamic and halal principles—definitions and interpretation of terminology	Original
7	MS2400-1:2019	Halal supply chain management system—part 1: transportation—general requirements (first revision)	First revision
8	MS2400-2:2019	Halal supply chain management system—part 2: warehousing—general requirements (first revision)	First revision
9	MS2400-3:2019	Halal supply chain management system—part 3: retailing—general requirements (first revision)	First1st revision
10	MS2424:2019	Halal pharmaceutical—general requirements (first revision)	First revision
11	MS2565:2014	Halal packaging—general guidelines	Original
12	MS2594:2015	Halal chemical for use in portable water treatment—general guidelines	Original
13	MS2610:2015	Muslim-friendly hospitality services—requirements	Original
14	MS2634:2019	Halal cosmetics—general requirements (first revision)	First revision
15	MS2636:2019	Halal medical device—general requirements	Original
16	MS2691:2021	Halal profession—general requirements	Original

Source Author's compilation from Malaysian Standards Online (2023)

must not be prohibited by the Shariah law and fatwa. These requirements must be met in order to be considered halal. MS1500:2019 and MS2424:2019 have explained the sources of halal ingredients, including plants, animals, minerals, chemicals, microorganisms, and genetically modified organisms (GMOs). Products with questionable materials or critical ingredients must perform a laboratory analysis in the specific lab that JAKIM appointed. Usually, the analysis will be related to the DNA, alcohol concentration, protein, oils, and fats' source.

Furthermore, non-halal materials should not be used or stored in the halal-certified premise to prevent contamination. When preparing halal products, the equipment, utensils, machines, and other processing aids used cannot be made of or contain any substances deemed to be najis according to Shariah law. The product's packaging also must not be made of non-halal materials, and the information on the label must not be deceptive or go against the principles of Shariah law.

It is forbidden to give halal product names similar to non-halal products, such as bacon, rum, beer, or ham, because doing so could lead to confusion. Halal logistics is required so that raw materials and finished products do not become contaminated during transportation or distribution. These regulations and procedures ensure that Muslim consumers only receive and consume halal tayyiban products (Ahmad et al. 2014).

Ensure Product Safety and Quality

The Shariah and technical auditors will evaluate the premises to ensure that the halal-certified products are safe and high quality. JAKIM auditors use a few standards for this purpose, such as MS1514:2022 Good Manufacturing Practice (GMP) for Food, MS1480:2019 Food safety according to Hazard Analysis and Critical Control Point (HACCP) system, MS1900:2014, MS1500:2019, MS2424:2019, MPPH 2020, and MHMS 2020 (Arif and Sidek 2015). The GMP standards outline the necessary hygiene and food safety conditions that must be implemented in the food manufacturing process to guarantee the production of safe and suitable food for consumption by consumers. In the meantime, the HACCP standards outline the requirements for food safety according to the HACCP system. These requirements are intended to ensure that food is not contaminated during any stage of the food chain, including preparation, processing, manufacturing, packaging, storage, transportation, distribution, offering for sale, or supply.

Every premise that produces halal-certified products must also follow the requirements of the MHMS 2020. Therefore, all premises should have their internal halal control and assurance systems to ensure the products' quality. The preparation of halal products must comply with certain conditions, including hygiene, sanitation, and the safety of the food itself. It includes the many facets of personal hygiene, clothing, devices, utensils, machines, and processing aids, as well as product processing, manufacturing, and storage facilities. All the procedures and regulations in the halal certification process guarantee that all halal products are safe and of superior quality. The halal certificate can be withdrawn if the premises or factory does not comply with the guidelines, regulations, and standards. Therefore, the halal logo represents the symbol of halal, safety, and superior quality of nutraceutical products. Hence, it should be an important determinant for consumers to be considered when consuming nutraceutical products.

1.5 Conclusion and Recommendation

In summary, this paper highlights the issues of nutraceutical products in terms of their halal status, safety, and legal issues. Not all nutraceutical products are halal and safe to be consumed. In Malaysia, nutraceuticals' terms have yet to be appropriately defined. As a result, some nutraceutical products have been treated like food and frequently make false health claims, like being able to treat, reduce, or even prevent serious illnesses. As a result, consumers are exposed to non-halal and dangerous products when these products are sold in the open market.

Hence, consumers need to practice self-protection by choosing the best nutraceutical products. This paper proposes the halal logo as the benchmark for safe and high-quality products. This is because for a product to use the halal logo on the product label, that products need to fulfill specific procedures and requirements based on Manual Procedure for Malaysia Halal Certification (Domestic) 2020, MHMS 2020, Malaysian Halal Standards, fatwa decisions, and relevant regulations and guidelines related to the halal certificate. This procedure ensures the halalness of the products and makes the halal-certified products safe and high in quality. If the demand for halal-certified products increases, nutraceutical manufacturers will make more efforts to apply for halal certification, which will benefit the consumers.

Acknowledgement This research was supported by Graduate Initiative Putra Grant, Universiti Putra Malaysia, entitled "Halalan Tayyiban Purchase Behavior toward Nutraceutical Products among Muslim Consumers in Malaysia," GP-IPS/2022/9724400.

References

Abdullah Amqizal HI, Al-Kahtani HA, Ismail EA, Hayat K, Jaswir I (2017) Identification and verification of porcine DNA in commercial gelatin and gelatin containing processed foods. Food Control 78:297–303. https://doi.org/10.1016/j.foodcont.2017.02.024

Adam F, Jamaludin J, Abu Bakar SH, Abdul Rasid R, Hassan Z (2020) Evaluation of hard capsule application from seaweed: gum Arabic-Kappa carrageenan biocomposite films. Cogent Eng 7(1). https://doi.org/10.1080/23311916.2020.1765682

Ahmad H, Hirawaty N, Abd I (2014) Factors influencing readiness towards halal logistics among food- based logistics players in Malaysia. UMK Procedia 1:42–49. https://doi.org/10.1016/j.umkpro.2014.07.006

Alzeer J, Abou Hadeed K (2016) Ethanol and its Halal status in food industries. Trends Food Sci Technol 58:14–20. https://doi.org/10.1016/j.tifs.2016.10.018

Alzeer J, Hadeed KA (2020) Halal certification of food, nutraceuticals, and pharmaceuticals in the Arab world. In: Handbook of healthcare in the Arab world, pp 1–22. https://doi.org/10.1007/978-3-319-74365-3_36-1

Ambali AR, Bakar AN (2014) People's awareness on halal foods and products: potential issues for policy-makers. Procedia Social Behav Sci 121:3–25. https://doi.org/10.1016/j.sbspro.2014.01.1104

Anadón A, Ares I, Martínez-Larrañaga MR, Martínez MA (2021) Interactions between nutraceuticals/nutrients and nutrients and therapeutic drugs. In: Nutraceuticals: efficacy, safety and toxicity, pp 1175–1197. https://doi.org/10.1016/B978-0-12-821038-3.00070-7

Arif S, Sidek S (2015) Application of Halalan Tayyiban in the standard reference for determining Malaysian halal food. Asian Soc Sci 11(17):116–129. https://doi.org/10.5539/ass.v11n17p116

Aziz A, Bakar NA, Zakaria Z (2020) Unsafe ingredients included in Malaysian food drug interphase (FDI) products: Toyyiban perspective. Malaysian J Halal Res 3(2):63–68. https://doi.org/10.2478/mjhr-2020-0011

Harian B (2018, Sept 13) Rakyat kita taasub ambil suplemen. Berita Harian. https://www.bharian.com.my/bhbongkar/2018/09/473195/rakyat-kita-taasub-ambil-suplemen

Bernama (2017, Feb 21) D'Herbs company fined RM11,000 for misleading advertisements. The Sunday Daily. https://www.thesundaily.my/archive/2170152-HTARCH429147

Chaudry MM, Jackson MA, Hussaini MM, Riaz MN (2000) Halal industrial production standards. J&M Food Products Company Illinois.

Dahlan HA, Abdullah N (2017) Comparison and challenges in the implementation of halal food laws in Malaysia, the Netherlands and United States of America. Jurnal Undang-Undang Dan Masyarakat 21(1):53–62. https://doi.org/10.17576/juum-2017-2101-06

Daud M, Jalil JA, Azmi IMAG, Ismail SF, Safuan S (2017) "Unsafe" nutraceuticals products on the internet: the need for stricter regulation in Malaysia. In: 2017 5th international conference on cyber and IT service management, CITSM 2017, Aug. https://doi.org/10.1109/CITSM.2017.8089269

De Felice SL (1997) The nutraceutical revolution: what is it? Clin Res Regul Affairs 14(1):1–4. https://doi.org/10.3109/10601339709019434

Demirhan Y, Ulca P, Senyuva HZ (2012) Detection of porcine DNA in gelatine and gelatine-containing processed food products-Halal/Kosher authentication. Meat Sci 90(3):686–689. https://doi.org/10.1016/j.meatsci.2011.10.014

Erwanto Y, Rohman A, Arsyanti L, Pranoto Y (2018) Identification of pig DNA in food products using polymerase chain reaction (PCR) for halal authentication—a review. Int Food Res J 25(4):1322–1331

Gupta RC, Srivastava A, Lall R (2018) Toxicity potential of nutraceuticals. Methods Mole Biol 1800:367–394. https://doi.org/10.1007/978-1-4939-7899-1_18

Ibn Battal A-H, Ali KA-M (n.d.) Syarah Sahih al-Bukhari li Ibn Battal, vol 6. Maktabah al-Rushd

Institute for Public Health (IPH) (2017) National health and morbidity survey 2017: adolescent nutrition survey

Ismail SF, Daud M, Ida Madieha AGA, Juriah AJ (2020) Enhancing regulation of nutraceutical products in Malaysia: lessons from Japan. IIUM Law J 28(S1):351–372. https://doi.org/10.31436/iiumlj.v28i(S1).589

Ismail Z, Ehsan AH (2008) Halal nutraceutical market: issues and challenges. In: 2008 Asian forum on business education (AFBE) conference, 573–597

JAKIM (2020) Manual Prosedur Pensijilan Halal Malaysia (Domestik) 2020. Forth 7

Jing Yi T, Afida Mastura MA (2021) The relationsip between self-efficacy, awareness and subjective norms with online sopping self-protection practices among Chinese consumers in Seremban, Negeri Sembilan. Jurnal Pengguna Malaysia 36:81–92

Karim AA, Bhat R (2009) Fish gelatin: properties, challenges, and prospects as an alternative to mammalian gelatins. Food Hydrocolloids 23(3):563–576. https://doi.org/10.1016/j.foodhyd.2008.07.002

Khan TM, Hassali MA, Al-Haddad MSM (2011) Nutraceuticals use among the inhabitants of Penang, Malaysia. Int J Collab Res Internal Med Publ Health 3(5):402–414

Malaysian Standards Online (2023). https://mysol.jsm.gov.my/

Mariod AA, Adam HF (2013) Review: gelatin, source, extraction and industrial applications. Acta Scientiarum Polonorum, Technologia Alimentaria 12(2):135–147

Market Watch (2022) Nutraceuticals market size, demand, share, key players, growth, and forecast 2030. Market Watch. https://www.marketwatch.com/press-release/nutraceuticals-market-size-demand-share-key-players-growth-and-forecast-2030-2022-12-14?mod=search_headline

Mayring P (2014) Qualitative content analysis: theoretical foundation, basic procedures and software solution

Mohamad AA, Baharuddin AS, Ruskam A (2018) Healthy lifestyle education from halal nutraceutical concept. UMRAN Int J Islamic Civil Stud 5(1):58–73. https://doi.org/10.11113/umran2018.5n1.182

Mohd Dali NRS, Nooh MN, Nawai N, Mohammad H (2009) Is halal products are more expensive as perceived by the consumers? Muslimpreneurs challenges and opportunities in establishing a blue ocean playing field. J Manage Muamalah 2:39–62

Mohd Dali NRS, Yousafzai S, Abdul Hamid H (2019) Religiosity scale development. J Islamic Mark 10(1):227–248. https://doi.org/10.1108/JIMA-11-2016-0087

Nasri H, Baradaran A, Shirzad H, Kopaei MR (2014) New concepts in nutraceuticals as alternative for pharmaceuticals. Int J Preventive Med 5(12):1487–1499

Nawsheen Bee T, Rahmanpoor HA (2019) Nutraceuticals: nutritionally functional foods—an overview. Biomed J Sci Techn Res 15(4):11480–11482. https://doi.org/10.26717/bjstr.2019.15.002728
Noordin N, Md Noor NL, Hashim M, Samicho Z (2009) Value chain of halal certification system: a case of the Malaysia halal industry. In: Proceedings of the European and Mediterranean conference on information systems, EMCIS 2009, 2008, 1–14
Nounou MI, Eassa HA, Helal NA, AboulFotouh K, Mansoor I, Latz IK, Zheng C, Eassa HA, Mohamed D, Huynh DM, Wiss AR, Sweeney J, Oakes MT, Mikhail MM, Amine N, Kaur H, Echeverry N, Orzechowski K, Szollosi D (2021) The safety, efficacy and pharmaceutical quality of male enhancement nutraceuticals bought online: Truth versus claim. J Integr Med 19(3):232–242. https://doi.org/10.1016/j.joim.2021.02.003
Paul S, Sun CC (2017) Lubrication with magnesium stearate increases tablet brittleness. Powder Technol 309:126–132. https://doi.org/10.1016/j.powtec.2016.12.012
Qomariah R, Anna Priangani R, Dedy S (2021) Analysis of alcohol content in a herbal medicine of noni using gas chromatography method. Int J Halal Res 3(1):1–7. https://doi.org/10.18517/ijhr.3.1.1-7.2021
Regenstein JM (2019) Kosher and halal regulations for nutraceutical and functional foods. In: Nutraceutical and functional food regulations in the united states and around the world, 3rd edn. Elsevier Inc. https://doi.org/10.1016/B978-0-12-816467-9.00036-8
Riaz MN, Chaudry MM, Riaz MN, Abbas S, Chaudry MM (2019) Halal production requirements for nutritional food supplements. In: Handbook of halal food production. Taylor & Francis Group, pp 221–224. https://doi.org/10.1201/9781315119564-31
Safuan S, Ismail SF, Abdul Ghani Azmi IM, Daud M (2020) Penetapan standard bagi industri nutraseutikal: Melindungi hak pengguna. Kanun: Jurnal Undang-Undang Malaysia 32(2):177–194
Shamsudin SN, Bakar EA, Osman S, Mohd Dali NRS (2023) Konsep halalan tayyiban bagi produk nutraseutikal. J Fatwa Manage Res 28(1):69–84. https://doi.org/10.33102/jfatwa.vol28no1.460
Shamsudin SN, Bakar EA, Osman S, Mohd Dali NRS, Anuar Musaddad N (2022) Research trend in halal studies: a bibliometric analysis using the Scopus database. J Halal Industr Serv 5(4). https://doi.org/10.36877/jhis.a0000300
Sreeramya G, Nalini CN, Ramalakshmi N, Sahini K, Lakshmi AS (2018) A new era in medicine—neutraceuticals. Res J Pharm Technol 11(8):3572–3576. https://doi.org/10.5958/0974-360X.2018.00657.1
Stout EI, McKessor A (2012) Glycerin-based hydrogel for infection control. Adv Wound Care 1(1):48–51. https://doi.org/10.1089/wound.2011.0288
Sultana S, Hossain MAM, Azlan A, Johan MR, Chowdhury ZZ, Ali ME (2020) TaqMan probe based multiplex quantitative PCR assay for determination of bovine, porcine and fish DNA in gelatin admixture, food products and dietary supplements. Food Chem325.https://doi.org/10.1016/j.foodchem.2020.126756
Télessy IG (2018) Nutraceuticals. In: The role of functional food security in global health. Academic Press, pp 409–421. https://doi.org/10.1016/B978-0-12-813148-0.00024-4
Uddin SMK, Hossain MAM, Sagadevan S, Al Amin M, Johan MR (2021) Halal and kosher gelatin: applications as well as detection approaches with challenges and prospects. Food Biosci 44. https://doi.org/10.1016/j.fbio.2021.101422
Umar NS, Rani MDM, Aripin KNN, Rahman TSA, Salleh NM (2016) A review of prophetic nutraceutical foods: Issues and challenges. Adv Sci Lett 22(9):2147–2151. https://doi.org/10.1166/asl.2016.7572
Zakaria Z, Ismail SZ (2014) The trade description act 2011: regulating 'halal' in Malaysia. In: International conference on law, management and humanities (ICLMH'14), 8–10. https://doi.org/10.15242/icehm.ed0614020

Toward the Global Halal MSMEs Hub Through the Digitally Enabled Society: An Institutional Arrangement with Islamic N-Tuple Helix

2

Sutan Emir Hidayat and Khairunnisa Musari

Abstract

This chapter develops the institutional arrangement concept by adopting the Islamic Helix approach to guide the Indonesian halal micro, small, and medium enterprises (MSMEs) to transform into a digitally enabled society toward a global halal MSMEs' hub. This institutional arrangement also can be adopted by other emerging countries with Muslim-majority populations to manage Islamic social finance to fund entrepreneurship and MSMEs and through a digitally enabled society as an enabler to open the business innovation. By using a literature review of the empirical studies, then coupled with the argument for presenting the model of Islamic N-Tuple Helix theoretically as the approach to develop institutional arrangement, this chapter can be categorized as a Conceptual Paper with Models research design as a methodological approach. This chapter promotes the integration between Islamic Triple Helix and information and communications technology (ICT) or technologies toward Islamic Quadruple Helix to create a digitally enabled society. Then, to construct the framework of the Islamic N-Tuple Helix, this chapter proposes additional components, namely University and Environment, to develop the concept of the Islamic Quintuple Helix and Islamic Sextuple Helix.

S. E. Hidayat (✉)
Gunadarma University and National Committee for Islamic Economy and Finance (KNEKS), Jakarta, Indonesia
e-mail: sutan.emir@kneks.go.id

K. Musari
Kiai Haji Achmad Siddiq State Islamic University (UIN KHAS), Jember, Indonesia

A. Rafiki (ed.), *Digitalization in Halal Management*, Contributions to Environmental Sciences & Innovative Business Technology,
https://doi.org/10.1007/978-981-99-5146-8_2

Keywords

Digitalization • Digital transformation • Islamic triple helix • Islamic quadruple helix • Islamic quintuple helix • Islamic sextuple helix • Islamic partnership model • Islamic social finance

2.1 Introduction

Indonesia is a portrayal of emerging market economies that engage financial inclusion as a structural constituent of development to realize inclusive growth through financial technology (fintech) lending and platform of philanthropic (Musari 2022). Philanthropy in Indonesia has always been driven primarily by religion and traditional faith. Faith-based philanthropy contributes remarkably to this field, particularly through Islamic philanthropy which refers to the concept of *zakat* where most of the giving is intended for religious or social welfare purposes (Hartnell 2020). Apart from *zakat,* Islamic philanthropy also has *infaq, sadaqa*, and *waqf*, the four of which are often called ZISWa. Currently, all four in Indonesia are managed with a productive activity orientation, one of which is to support entrepreneurship through financing and micro, small, and medium enterprises' (MSMEs) empowerment (Musari and Fathorrazi 2021).

Entrepreneurship and MSMEs in Indonesia are the growth engine of the national economy. The number of MSMEs in the country with the fourth-largest population in the world reaches 99.99% and as many as 98.67% are micro-enterprises. For this reason, supporting the micro-enterprises is actually also supporting the Indonesian economy. The Coordinating Ministry for Economic Affairs of the Republic of Indonesia (2021) noted that the contribution of MSMEs has reached 61% of the gross domestic product (GDP) with the ability to absorb labor reaching 97% of the total national employment. Total investment in the MSME sector has reached 60% of the total national investment. By 2024, MSME loans, which are currently at 20%, are targeted to reach 24%.

As a country with a great number of Islamic financial institutions in the world (Ismal 2012), Indonesia has a lot of potential to become a major player in the development of Islamic philanthropy which is often also mentioned as Islamic social finance. Islamic social finance has an immense capacity to contribute to economic recovery. During the pandemic COVID-19, Islamic social finance became the necessary buffer to safeguard lives, households, financial institutions, and businesses as well as to restore economic activities. It has also huge readiness to bestow in achieving economic development goals along with social and environmental goals, including the Sustainable Development Goals (SDGs) (Bashori 2020; Juwaini 2020; Ascarya 2020).

A study by Musari and Sayah (2022) in Indonesia and Algeria strengthened the previous finding of Musari and Fathorrazi (2021) in Indonesia that a base arrangement in realizing financial inclusion by using Islamic social finance as the funding source for financing MSMEs was Islamic Triple Helix which composes 'Islamic (Government–Civil Society–Industry)'. As a social finance institution partnership model, this composition initiates the Theory of the Islamic Helix which turned out to be dissimilar from the (Conventional) Triple Helix which has been notable to be structured into 'University-Industry-Government'. There are three assumptions for Islamic Triple Helix within those studies: (1) leastwise single partner institution is embraced according to sharia; (2) leastwise two partner institutions have collaborated; (3) the proceed of Islamic Triple Helix is the total proceeds of Islamic (Government–Civil Society–Industry).

Starting from those two studies, this chapter develops the institutional arrangement concept by adopting the Islamic Helix approach to guide the Indonesian halal MSMEs to transform into a digitally enabled society toward a global halal MSMEs hub. Considering its existence as the largest Muslim-majority country in the world with the domination of MSMEs in the national economy, Indonesian MSMEs have the opportunity to position themselves as a global halal MSMEs' hub. This chapter promotes the integration between Islamic Triple Helix as well as information and communications technology (ICT) or technologies toward Islamic Quadruple Helix to create a digitally enabled society.

By using a literature review of the empirical studies, then coupled with the argument for presenting the model of Islamic N-Tuple Helix theoretically as the approach to develop institutional arrangement, this chapter can be categorized as a Conceptual Paper with Models research design as a methodological approach. Overall, these institutional arrangements can be adopted by other emerging countries with Muslim-majority populations to manage Islamic social finance for financing entrepreneurship and MSMEs and through a digitally enabled society as an enabler to open the business innovation through Islamic Quadruple Helix. To construct the framework of the other Islamic N-Tuple Helixes, this chapter proposes additional components, namely University and Environment, to develop the concept of the Islamic Quintuple Helix and Islamic Sextuple Helix.

2.2 Global Halal MSMEs Hub

Based on the State of the Global Islamic Economy (GIE) Report, Indonesia is one of the four member countries of the Association of Southeast Asian Nations (ASEAN) which is always included in the top 15 countries in the world in leading the global halal economy. Table 2.1 shows Indonesia's ranking in the GIE Indicator. GIE Indicator is a comprehensive picture of countries with best position in mastering halal industry. Exception to the year of 2013, the State of the GIE Report at that time still used the term 'Top Centers of the Islamic Economy' to establish the top 15 countries with the highest scores through a method which today mentioned as GIE Indicators.

In addition, as shown in Table 2.2, Indonesia has always been in the top ten Organization of Islamic Cooperation (OIC) Destinations of the Global Muslim Travel Index (GMTI). Since Mastercard-CrescentRating launched GMTI in 2015, Indonesia has been included in the top six OIC Destination Countries. In 2022,

Table 2.1 Rank of Indonesia in GIE indicators (2013–2022)

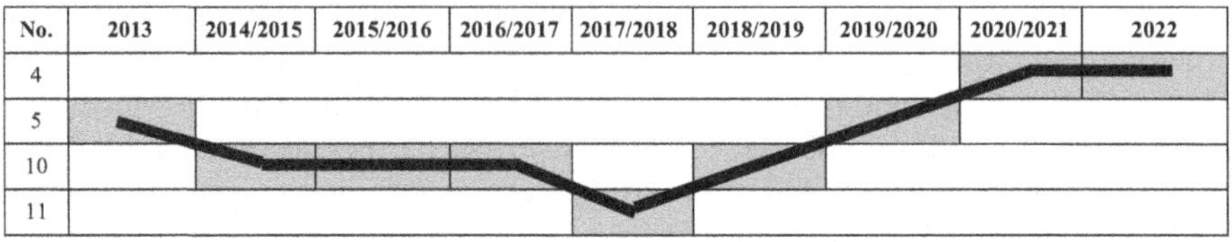

Source Thomson Reuters (2013), Hidayat and Musari (2022), DinarStandard (2020, 2022), Modified

among 138 countries listed in this report, Indonesia was ranked second. This indicates that Indonesia is at the forefront of the Muslim-Friendly Travel sector.

In order to carry out the mandate of the Halal Product Assurance Act (UU JPH) as well as to realize its mission of Indonesia toward a global halal hub, Indonesian MSMEs have the opportunity to set their position as a barometer of global halal MSMEs. MSMEs in Indonesia present highly to the national economy. They are a driving force of growth and work not only to strengthen the economic structure but also for absorbing employment and playing a function in the distribution of goods and services. Its existence has straightened out to diverse parts of the country and brings a way to the broad community's potency to reach a better livelihood.

No doubt, the halal symbol today has become the global standard that controls the quality of products. Therefore, halal currently gains a universal happening that is appreciated by various states, both Muslim and non-Muslim. Very great occasions for the halal market have been taken on by a number of countries in the world. Likewise, Muslim-majority countries have a large chance to take part in this global halal trade. Moreover, Indonesia which has huge measurable potential is reflected in its achievements in GIE Indicators and GMTI. In the context of global halal hub development, the role of the ICT sector is crucial to create a digitally enabled society as an enabler for transforming Indonesian halal MSMEs toward a global halal MSMEs hub.

2.3 Digitally Enabled Society

National Committee for Islamic Economy and Finance [KNEKS] (2021) reported that only around 25.6% of MSMEs or around 16.4 million business actors are utilizing digital technology. The role of MSMEs in the digital sector has not yet shown significant results, compared to the growth rate of the MSME market in Indonesia. However, referring to Statistics Indonesia (2022), development in the ICT sector shows a positive trend for internet penetration in Indonesia during 2017–2021. This positive trend is further driven by the COVID-19 pandemic which has shifted the behavior of people to eliminate physical contact with other people but still has to carry out daily activities online through various digital platforms.

Table 2.3 shows that there is a positive trend in several indicators of ICT in Indonesia during 2017–2021, except for the number of fixed telephone subscribers

Table 2.2 Rank of Indonesia in top OIC destinations of GMTI (2015–2022)

No.	2015	2016	2017	2018	2019	2020	2021	2022
1								
2								
3						Not Available		
4								
6								

Source Mastercard-CrescentRating (2015, 2016, 2017, 2018, 2019, 2021, 2022), Modified

which has decreased since 2019. In 2021, there are only three to four fixed telephone subscribers per 100 residents. In contrast, the number of mobile phone subscribers per 100 residents has increased since 2019 and reached 134.18 subscribers per 100 residents in 2021. A value greater than 100 indicates that there is one resident who subscribes to more than one type of mobile phone. Other ICT that shows a positive trend is indicators related to internet usage.

Then, the number of fixed-broadband subscribers per 100 residents is always increasing from year to year during 2017–2021. There were 4.58 fixed-broadband subscribers per 100 residents in 2021. The number of mobile broadband subscribers per 100 residents has increased since 2019 to reach 112.45 subscribers per 100 residents in 2021. It means that there is resident who subscribes to more than one type of mobile broadband. The escalation of internet subscribers, both fixed-broadband and mobile broadband, reflects the phenomenon of internet usage in society as an impact of the COVID-19 pandemic, to support activities that are increasingly being carried out online. Overall, the performance of the ICT sector in Indonesia during 2017–2021 is also shown in Fig. 2.1.

Undoubtedly, the COVID-19 pandemic has become a momentum for the Indonesian Government to accelerate digital transformation. Digitalization must be utilized by MSMEs because digital infrastructure continues to be developed

Table 2.3 Development of ICT in Indonesia (2017–2021)

ICT Indicators	2017	2018	2019	2020	2021
Fixed telephone subscribers per 100 residents	4.22	4.23	3.54	3.40	3.31
Mobile phone subscribers per 100 residents	166.17	121.04	128.70	131.66	134.18
Fixed-broadband subscribers per 100 residents	2.37	3.31	3.51	3.96	4.58
Mobile broadband subscribers per 100 residents	96.79	88.12	92.02	104.00	112.45
Percentage of the internet users	32.34	39.90	47.69	53.73	62.10

Source Statistics Indonesia (2022)

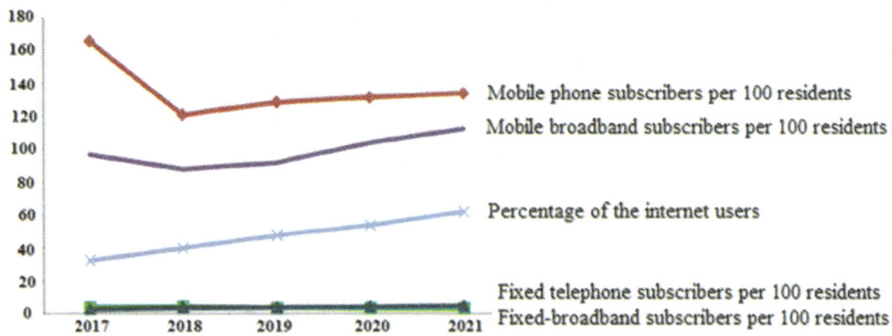

Fig. 2.1 Performance of ICT sector in Indonesia (2017–2021). *Source* Statistics Indonesia (2022)

massively and people's digital literacy and connectivity between regions are getting better. E-commerce can be an instrument of economic equity (Coordinating Ministry for the Economy of the Republic of Indonesia 2020a; b). Therefore, ICT sector/technologies are enabler to transform the Indonesian halal MSMEs into a digitally enabled society for realizing the mission as a global halal MSMEs' hub. To actualize this mission, the institutional arrangement must be implemented to support the achievement.

2.4 Institutional Arrangement with Islamic Helix Approach

To generate knowledge, a Triple Helix Model emphasizes three interweave 'helices' in creating an innovation system that includes: Academia/Universities, Industries, and Governments (Etzkowitz and Leydesdorff 1995, 1998, 2000). The concept of Triple Helix comes from the analysis of Double Helix (University–Industry) which is then followed by the awareness that Government is a necessary piece of the innovation equation (Etzkowitz 2002, 2008; Etzkowitz et al. 2007; Carayannis and Campbell 2011; Carayannis et al. 2012, 2018; Sunina and Rivza 2016).

As a core design for innovation gained from the interaction between University–Industry–Government, Triple Helix in its progress then develops into a wider innovation model, namely Quadruple Helix which consists of University, Industry, Government, and Civil Society by arguing that the Triple Helix is not sufficient for long-term innovative growth (Leydesdorff 2012). Through Quadruple Helix, this model generates new knowledge, technology, as well as products and services to meet the need of society (Afonso et al. 2012).

Regarding the institutional arrangement concept by adopting the Islamic Helix approach to guide the Indonesian halal MSMEs to transform into a digitally enabled society toward a global halal MSMEs hub, the phases are as follows.

2.4.1 Islamic Double Helix

There are two components to the institutional arrangements at the Islamic Helix at this simplest level. This helix has three variants (Musari and Fathorrazi 2021; Musari and Sayah 2022), namely: (1) Islamic (Government–Industry); (2) Islamic (Government–Civil Society); and (3) Islamic (Industry–Civil Society) as shown in Fig. 2.2.

As the assumption of the Islamic Triple Helix, at least one partner institution is sharia compliance and leastwise two collaborated partner institutions, so the Islamic Double Helix consists of two institutions with at least one sharia-affiliated institution. (Islamic) Industry in this model is more interpreted as financial institutions such as Conventional/Islamic Banks; Conventional/Islamic Microfinance or Nanofinance Institutions; Cooperation; *Baitul Mal wat Tamwil; Venture* Capital Firms; Sovereign Wealth Funds; Islamic Financial Industry Associations.

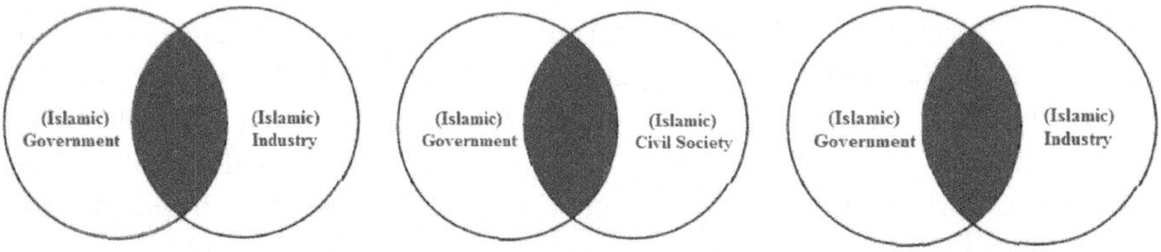

Fig. 2.2 Model of Islamic double helix. *Source* Authors' Proposed Model

Then, (Islamic) Government in this model is more interpreted as Legal or Regulatory such as Regional/National/International Sharia Authorities; Regional/National/International Sharia Industry Organization; or Regional/National/International Sharia Multilateral Institutions. Regarding the (Islamic) Civil Society, it is more interpreted as Advisory Regional/National Sharia Scholars Communities; Advisory Regional/National Sharia Scholars Organizations; Philanthropic Institutions; Companion Volunteers; Consultants; Non-governmental Organizations (NGOs).

2.4.2 Islamic Triple Helix

There are three components to the institutional arrangements at the Islamic Helix at this second level (Musari and Fathorrazi 2021; Musari and Sayah 2022), namely: Islamic (Government–Civil Society–Industry), as shown in Fig. 2.3.

As the assumption of the Islamic Triple Helix, the proceed of this model is the total proceeds of Islamic (Government–Civil Society–Industry) with leastwise two collaborated partner institutions and at least one partner institution is sharia

Fig. 2.3 Model of Islamic triple helix. *Source* Authors' Proposed Model

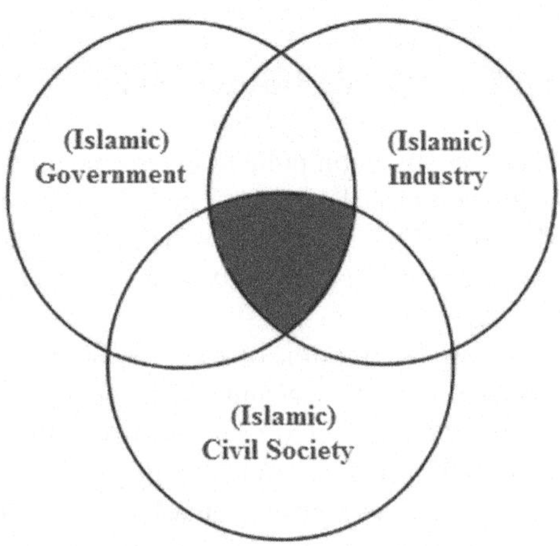

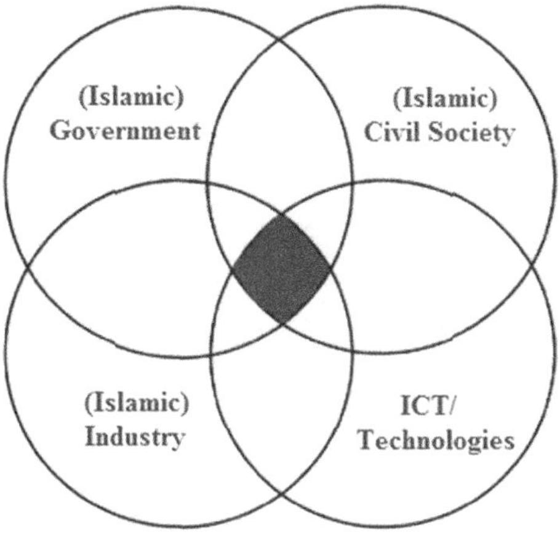

Fig. 2.4 Model of Islamic quadruple helix. *Source* Authors' Proposed Model

compliance. In other words, the Islamic Triple Helix is a combination of three variants of the Islamic Double Helix.

2.4.3 Islamic Quadruple Helix

There are four components to the institutional arrangements at the Islamic Helix at this third level, Islamic (Government–Civil Society–Industry) and ICT sector/technologies, as shown in Fig. 2.4. In other words, the Islamic Quadruple Helix is a combination of Islamic Triple Helix with ICT/technologies.

In this phase, to open the business innovation of MSMEs, the output of institutional arrangement through Islamic Quadruple Helix results a digitally enabled society. ICT sector/technologies include technology provider to Artificial Intelligence (Big Data, Data and Analytics, Machine Learning); Security (Biometrics, Cybersecurity, Identity Verification); Cloud and Software; Distributed Ledger Technology (Blockchain, Crypto, Tokenization); and Technology Company (Fintech Firms, Fintech Associations).

2.4.4 Islamic Quintuple Helix

There are five components to the institutional arrangements at the Islamic Helix at this fourth level. This helix has two variants, namely: (1) Islamic (Government–Civil Society–Industry)–ICT sector/technologies–University and (2) Islamic (Government–Civil Society–Industry)–ICT sector/technologies–Environment, as shown in Fig. 2.5. In other words, the Islamic Quintuple Helix is a combination of Islamic Quadruple Helix with Universities or Islamic Quadruple Helix with Environment.

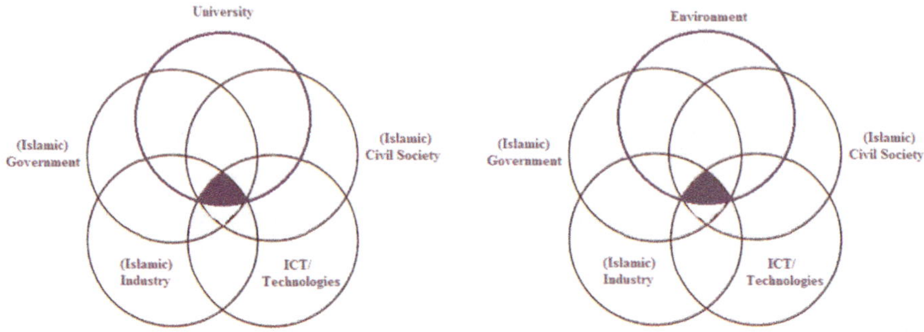

Fig. 2.5 Model of Islamic quintuple helix. *Source* Authors' Proposed Model

The proposed models in this phase consider the (Conventional) Triple Helix which presents the University as one of the main institutions for creating innovation. Traditionally, the University has been noticed as a bolstering construction for innovation, supplying skilled people, generating research and development (R&D), as well as serving knowledge to industry. Nowadays, the University has more and more become involved in the company's formation, frequently referring to the developed technologies initiated from academic research (Etzkowitz 2003). The University in this model is more interpreted as Higher Education Institution; Education and Training Institutes; R&D Institutions.

Likewise, the existence of the Environment as another variant in this phase is driven by the urgency so that MSMEs can become subjects in the green economy, including in tackling climate change, circular economy, renewable energy, and other eligible green sectors. Moreover, there is a very close relationship between the halal economy and the obligation to preserve the Environment. The institutional arrangements based on Islamic Helix must serve the enforcement tool for *hifdzul al-bi'ah/riayatu al-bi'ah* and *maqasid al-shari'a.*

2.4.5 Islamic Sextuple Helix

There are six components to the institutional arrangements at the Islamic Helix at this fifth level, namely: Islamic (Government–Civil Society–Industry)–ICT sector/ technologies–University–Environment, as shown in Fig. 2.6. In other words, the Islamic Sextuple Helix is a combination of two variants of the Islamic Quintuple Helix.

Another figure of the Islamic Sextuple Helix also can be seen in Fig. 2.7. Overall, this model is the development of the results of empirical studies, then coupled with the argument for presenting the model of Islamic N-Tuple Helix theoretically as the approach to develop institutional arrangement. If Islamic Double Helix and Islamic Triple Helix are developed through empirical studies, so the next Islamic

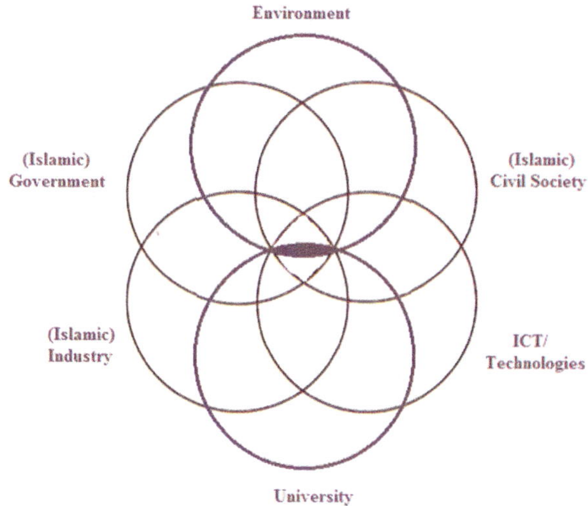

Fig. 2.6 Model of Islamic sextuple helix. *Source* Authors' Proposed Model

N-Tuple Helix—Islamic Quadruple Helix, Islamic Quintuple Helix, Islamic Sextuple Helix—is developed by arguing and inspired by the (Conventional) Triple Helix.

Last but not least, at a more complicated level, such as the mission to become a global halal MSME hub, the Islamic Triple Helix may not be effective for Indonesian halal MSMEs to achieve this mission. As digital transformation accelerates throughout the financial ecosystem around the world, Indonesian halal MSMEs have the opportunity to lead the global halal MSME hub by presenting the ICT sector/technologies to open innovation. The Islamic Triple Helix must involve the

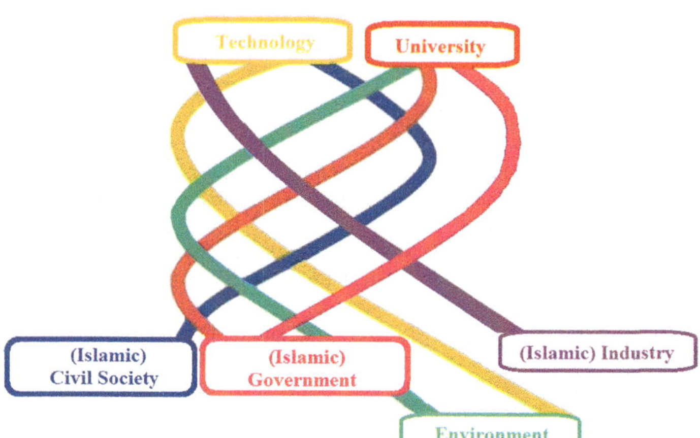

Fig. 2.7 Another model of the Islamic sextuple helix. *Source* Authors' Proposed Model

ICT sector/technologies also for developing itself to become the Islamic Quadruple Helix. If the presence of Generation Z as the 'true digital natives' who make up the majority of Indonesia's population is believed can be an enabler in strengthening the Islamic fintech (Musari and Hidayat 2022), the presence of Generation Z also can be a supporting system to strengthen the Islamic Quadruple Helix in the future.

Then, along with the greater need to develop and strengthen MSMEs, in the next phase, the Islamic Quadruple Helix may no longer be effective enough to answer future challenges. At this level, University and Environment are needed because the Islamic Quadruple Helix is no longer effective enough. Islamic Quintuple Helix and Islamic Sextuple Helix exist to answer bigger challenges, not only to open doors for innovation, but also to become subjects of active business actors to solve the crucial problems of the world, particularly the climate change.

2.5 Conclusion

The institutional arrangement concept by adopting the Islamic Helix approach can guide the Indonesian halal MSMEs to transform into a digitally enabled society toward a global halal MSMEs hub. The existence of the ICT sector/technologies covers the weaknesses of the Islamic Triple Helix by forming the Islamic Quadruple Helix. The integration between Islamic Triple Helix and the ICT sector/technologies will create a digitally enabled society. This is the real output of institutional arrangement through the Islamic Quadruple Helix. Then, to construct the framework of the Islamic N-Tuple Helix, this chapter recommends the additional components, namely University and Environment, to develop the concept of the Islamic Quintuple Helix and Islamic Sextuple Helix.

References

Afonso O, Monteiro S, Thompson M (2012) A growth model for the quadruple helix. J Bus Econ Manag 13(5):849–865. https://doi.org/10.3846/16111699.2011.626438

Ascarya (2020) The role of islamic social finance to improve well-being. A presentation material in Webinar "Akselerasi Peran Islamic Social Finance di Tengah Pandemi" held by STIE Perbanas Surabaya, 20 Sept

Bashori MA (2020) Optimalisasi Islamic Social Finance dalam Mendukung Pemulihan Ekonomi Daerah. A presentation material for Festival Ekonomi Syariah, Departemen Ekonomi dan Keuangan Syariah Bank Indonesia (DEKS-BI), 5 Oct

Carayannis EG, Barth TD, Campbell DFJ (2012) The quintuple helix innovation model: Global warming as a challenge and driver for innovation. J Innov Entrepreneur 1(2). https://doi.org/10.1186/2192-5372-1-2

Carayannis EG, Campbell DFJ (2011) Open innovation diplomacy and a 21st century fractal research, education and innovation (FREIE) ecosystem: Building on the quadruple and quintuple helix innovation concepts and the "mode 3" knowledge production system. J Knowl Econ 2:327–372

Carayannis EG, Grigoroudis E, Campbell DFJ, Meissner D, Stamati D (2018) The ecosystem as helix: an exploratory theory-building study of regional co-opetitive entrepreneurial ecosystems as quadruple/quintuple helix innovation models. R & D Manage 48(1):148–162

Coordinating Ministry for Economic Affairs of the Republic of Indonesia (2020a) Digital transformation accelerates the development of Indonesian MSMEs. Press Release Number HM.4.6/69/SET.M.EKON.2.3/06/2020a, 30 June

Coordinating Ministry for Economic Affairs of the Republic of Indonesia (2020b) National coordination meeting for inflation 2020b: digital transformation of MSMEs helps maintain price stability and supports economic recovery. Press Release Number HM.4.6/152/SET.M.EKON.2.3/10/2020b, 22 Oct

Coordinating Ministry for Economic Affairs of the Republic of Indonesia (2021) Launching the book of MSMEs' financing, coordinating minister Airlangga conveys the government's support for MSMEs. Press Release Number HM.4.6/391/SET.M.EKON.3/10/2021, 11 Nov

DinarStandard (2022) State of the global Islamic economy report 2022. DinarStandard, Dubai Islamic Economic Development Center, and Salaam Gateway, Dubai

DinarStandard (2020) State of the global Islamic economy report 2020/2021. DinarStandard, Dubai Islamic Economic Development Center, and Salaam Gateway, Dubai

Etkowitz H (2002) The triple helix of university—industry—government implications for policy and evaluation. Working Paper 11 Institutet för Studier av Utbildning och Forskning

Etzkowitz H (2003) Innovation in innovation: the triple helix of university-industry-government relations. Soc Sci Inf 42(3):293–337. https://doi.org/10.1177/05390184030423002

Etzkowitz H (2008) The triple helix: university-industry-government innovation in action. Routledge, London

Etzkowitz H, Dzisah J, Ranga M, Zhou C (2007) The triple helix model of innovation: university-industry-government interaction. In: Asia-Pacific Tech Monitor, 14–23 Jan–Feb

Etzkowitz H, Leydesdorff L (1995) The triple helix---university-industry-government relations: a laboratory for knowledge based economic development. Euro Assoc Study Sci Technol (EASST) Rev 14(1):14–19

Etzkowitz H, Leydesdorff L (1998) The endless transition: a "triple helix" of university-industry-government relations. Minerva 36(3):203–208

Etzkowitz H, Leydesdorff L (2000) The dynamics of innovation: from national systems and 'mode 2' to a triple helix of university-industry-government relations. Res Policy 29(2):109–123

Hartnell C (2020) Philanthropy in Indonesia. A working paper of philanthropy for social justice and peace in association with alliance, Filantropi Indonesia, Indonesia for Humanity and WINGS. Retrieved from https://globalfundcommunityfoundations.org/wp-content/uploads/2020/02/Philanthropy-in-Indonesia-Feb-2020.pdf

Hidayat SE, Musari K (2022) ASEAN towards a global halal logistics through the digitally-enabled community. Int J Asian Bus Inf Manage (IJABIM) 13(2):1–15. https://doi.org/10.4018/IJABIM.20220701.oa1

Ismal R (2012) The Indonesian Islamic banking: performance and outlook. A presentation material for Panel Discussion held by Bakrie University, Jakarta, 6 Feb

Juwaini A (2020) Optimalisasi Islamic Social Finance dalam Mendukung Pemulihan Ekonomi Daerah. A Keynote Speech for Festival Ekonomi Syariah (FESyar) Jawa 2020, 5 Oct

KNEKS (2021) Halal MSMEs and Indonesia's economic resilience. *INSIGHT*. Islamic Economy Bulletin, 13th edn. 1–6 Dec

Leydesdorff L (2012) The triple helix, quadruple helix, ..., and an n-tuple of helices: explanatory models for analyzing the knowledge-based economy? J Knowl Econ 3(1):25–35

Mastercard-CrescentRating (2015) Global Muslim travel index (GMTI) 2015. Retrieved from https://www.crescentrating.com/halal-muslim-travel-market-reports.html

Mastercard-CrescentRating (2016) Global Muslim travel index (GMTI) 2016. Retrieved from https://www.crescentrating.com/as-it-happened/gmti-2016-launch.html

Mastercard-CrescentRating (2017) Global Muslim travel index (GMTI) 2017. Retrieved from https://www.crescentrating.com/as-it-happened/gmti-2017-launch.html

Mastercard-CrescentRating (2018) Global Muslim travel index (GMTI) 2018. CrescentRating and Mastercard Asia/Pacific, Singapore

Mastercard-CrescentRating (2019) Global Muslim travel index (GMTI) 2019. CrescentRating and Mastercard Asia/Pacific, Singapore

Mastercard-CrescentRating (2021) Global Muslim travel index (GMTI) 2021. CrescentRating and Mastercard Asia/Pacific, Singapore

Mastercard-CrescentRating (2022) Global Muslim travel index (GMTI) 2022. CrescentRating and Mastercard Asia/Pacific, Singapore

Musari K (2022) P2P lending & philanthropy platform: a new face of asian digital financial inclusion (evidence from Indonesia). In: Almunawar MN, Islam Z, Ordóñez de Pablos P (eds) Digital transformation management: challenges and futures in the asian digital economy. Informa UK Limited, Taylor & Francis/Routledge, New York, NY, pp 185–206. https://doi.org/10.4324/9781003224532

Musari K, Fathorrazi M (2021) Islamic helix approach, the Islamic social finance partnership models for MSMEs: lesson learned from Indonesia. In Azid T, Mukhlisin M, Akbar N, Tahir M (eds) Monetary policy, Islamic finance, and Islamic corporate governance: An international overview. Emerald Publishing Limited, Bingley, UK, pp. 303–322. https://doi.org/10.1108/978-1-80043-786-920211017

Musari K, Hidayat SE (2022) Digital financial knowledge and behavior of generation Z in Indonesia: a survey of Islamic fintech literacy toward digital financial inclusion. In: Anshari M, Almunawar MN, Masri M (eds) Fintech development for financial inclusiveness. IGI Global, Hershey, PA, pp 96–117. https://doi.org/10.4018/978-1-7998-8447-7.ch007

Musari K, Sayah F (2022) Finding the Islamic social finance partnership models for micro entrepreneurship through Islamic helix approach: case study in Indonesia and Algeria. A paper was presented at entrepreneurship, sustainable development, and halal economy conference: concepts, framework and practice held by College of Islamic Studies (CIS), Hamad Bin Khalifa University (HBKU), Doha, Qatar, 27–28 Mar

Statistics Indonesia (2022) Information and communication technology development index 2021. Statistics Indonesia (BPS), Jakarta

Sunina L, Rivza B (2016) The quintuple helix model: Way of regional development centres in Latvia to smart public administration. Res Rural Dev 2:135–142

Thomson Reuters (2013) State of the global Islamic economy report 2013. Thomson Reuters, DinarStandard, and Dubai Islamic Economic Development Center, Dubai

Harnessing Internet of Everything (IoE) for Sustainability of Halal Cosmetics Ecosystem

3

Adlin Masood, Alina Abdul Rahim, and Umi Hamidaton Mohd Soffian Lee

Abstract

Driven by the Islamic faith-inspired ethical consumption needs of affluent digital-savvy Muslims, comprising a third of the world's population, *halal* spending on food, pharmaceuticals, cosmetics, fashion, travel, and media/ recreation, reached US$2 trillion in 2021, reflecting an 8.9% growth compared to 2020 (Elasrag, 2016; GIER 2022). Consistent with this expansion, the *halal* cosmetics industry is projected to grow at a 4-year CAGR of 7.4% and reach US$93 billion in 2025 (GIER 2022). The increased interest in *halal* is brought on by the perspective that *halal* is a new paradigm and solution to global issues such as food safety, ethical and sustainable practices, environmental awareness, economic resilience, and animal welfare. Integrating the different dimensions of the *halal* economy for sustainability and expansion thus needs an innovative approach through the application of the Internet of Everything (IoE). The IoE concept is based on the notion of multifaceted integration, intelligence, and cognition whereby people, things, data, and processes are connected in a collective unified system, the objective of which is to enhance stakeholders' practices and decision-making. This study seeks to analyze the feasibility of implementing IoE into the *halal* cosmetics ecosystem by examining the state of readiness of related agencies, institutions, industry players, and consumers, by examining current publications via a systematic literature review. In most parts, all stakeholders within the *halal* cosmetics ecosystem are ready to embrace IoE, a condition facilitated by COVID-19 pandemic restrictions that pushed the entire world to go online and adopt Industry Revolution 4.0 technology to continue their lives. Assuredly, the *halal* cosmetics ecosystem that is fortified with

A. Masood (✉) · U. H. M. S. Lee
Faculty of Economics and Muamalat, Universiti Sains Islam Malaysia, Nilai, Malaysia
e-mail: adlin.masood@usim.edu.my

A. A. Rahim
Faculty of Science and Technology, Universiti Sains Islam Malaysia, Nilai, Malaysia

A. Rafiki (ed.), *Digitalization in Halal Management*, Contributions to Environmental Sciences & Innovative Business Technology,
https://doi.org/10.1007/978-981-99-5146-8_3

the implementation of IoE will ensure that *halal* stakeholders are abreast, with present and future situations, and enabled to make timely and more effective decisions thereby ensuring a sustainable trade.

Keywords

Halal • Internet of everything • IoE • Halal cosmetics industry • IR4.0

3.1 Introduction

Once a marginal sector, the vibrant *halal* industry now influences every aspect of the global supply chain. No other significant market segment has this much potential for growth and worldwide value. Halal economy has continued its rate of accelerated growth despite the paralyzing disruption caused by the COVID-19 pandemic and its constricting limitations. This is attested when the State of the Global Islamic Economy Report 2022 (GIER 2022) reported that the global consumption of *halal* goods and services, comprising food and beverage, pharmaceutical, cosmetics, fashion, travel, media, and recreation, is forecasted to grow by 9.1% by 2025 to touch US$2.8 trillion (GIER 2022). The report also published that Malaysia is expected to continue to lead in the coming years, as it had done so for the past nine consecutive years (GIER 2022). Mr. Abdul Rasheed Ghaffour, Deputy Governor of Bank Negara Malaysia, when addressing the World Halal Business Conference 2021, stated that the rise in *halal* economy is driven by both, the 1.9 billion affluent digital-savvy Muslim population, who practice a faith-based lifestyle, and the rising trend in ethical consumerism of non-Muslims (BIS 2021).

The Muslim consumer market is rapidly expanding, fueled by a growing population that is more geographically, ethnically, and economically diverse than ever before. Thereby, to capture Muslim consumers, businesses must ensure that their product offerings meet criteria set by Shariah laws. Halal brand is recognized as the global symbol of quality assurance, underlined by ethical international transactions, and addresses sustainable concerns, qualities which have subsequently led to high acceptance by non-Muslims (Islam and Chandrasekaran 2013; Ab Talib et al. 2013). Halal certification, when applied together with Good Manufacturing Practices (GMP) and/or Hazard Analysis of Critical Control Points (HACCP), commands greater value, trust, and acceptance from Muslims and non-Muslims.

Halal industry represents a dynamic lucrative opportunity that is substantial and sustainable for the worldwide economy. Reflecting on the enormous economic possibilities that *halal* economy has to offer, especially in increasing their GDP growth rate, business, and trade activities, as well as providing better inclusive employment prospects, many nations have had to re-strategize their focus to achieve supremacy in the global *halal* marketplace. Some are even rushing to invest millions in upgrading the technology aspect directly or indirectly by aiding the SMEs involved in the *halal* industry. However, as in any economic ecosystem, *halal* ecosystem also has its highlights and woes caused by the need to ensure halal integrity and meet the standard requirements of importing countries.

The social and economic turmoil that resulted from the COVID-19 pandemic and the disruption to the supply chain has sped the utilization of Fourth Industrial Revolution technology (IR4.0), including the implementation of Enterprise Resource Planning (ERP), Artificial Intelligence, Blockchain Technology (BC), Internet of Things (IoT), Big Data Analytics (BD), radio-frequency identification system (RFID), and deployment of drones to assess *halal* integrity standards compliance. Humankind is more at ease with the use of IR4.0 technology as a means for communications, transactions, and marketing unlike the era before. Sectors seen to have benefitted from IR4.0 technology are the manufacturing and financial sectors. The most recent technological ecosystem model proposed is Internet of Everything (IoE). IoE is an information technology term that combines functionalities such as sensing, computation, information extraction, and communication in a smart device. This technological ecosystem allows different electronic devices with different capabilities to sense the environment and to communicate for data exchange, subsequently empowering stakeholders to make better informed decisions and gain fruitful experiences, once given accurate data, at the right time and place.

This is a concept paper, where all the discussions are based on secondary data and information on *halal* development and IoE literature review. Information is extracted from bibliographic databases, online journal publications, conference proceedings, and data from institutional reports and publications. It seeks to do a preliminary survey on the possibility of implementing IoE in the *halal* ecosystem thereby ensuring greater transparency on the *halal* production process, and logistic activities, while maintaining *halal* integrity and matching *halal* requirements of various nations. Ultimately if the possibility of such implementation is possible, then IoE can be harnessed and will eventually be a means for *halal* trade sustainability and sustainability. However, since the global *halal* ecosystem is extremely large and diverse, the study will focus on the *halal* cosmetics ecosystem in Malaysia, the leading nation in *halal* ecosystem for the past nine years (GIER 2022).

3.2 Research Methodology

3.2.1 Methodology

As outlined in the introduction, the objective of the study is to do a preliminary survey on the possibility of implementing IoE in the *halal* cosmetics ecosystem thereby ensuring greater transparency on the *halal* cosmetics production process, and logistic activities, while maintaining *halal* integrity and matching *halal* requirements in Malaysia. Ultimately if the possibility of such implementation is conceivable, then IoE can be harnessed, to capture customers trust and loyalty, and will eventually be a means for *the halal* cosmetics industry sustainability and *halal* trade transactions between Malaysia and other nations.

First, the study seeks a systematic review of understanding different means of communication and sharing of information about *halal* cosmetics ecosystem between various stakeholders. It will also look into the state of readiness of the current system and thus the possibility to integrate Internet of Everything framework into the system. All these information will be retrieved from bibliographic databases to reveal the focus of earlier research on *the halal* cosmetics ecosystem and IoE application in an industry setting.

The remaining study will discuss the findings and put forth recommendations and actions to be taken by producers and owners to capture, maintain, and grow their customer base as well as ways to expand overseas, in addition to the direction of future research studies.

3.3 Literature Review

The increased interest in halal is brought on by the holistic philosophy behind halal production. Specifically, halal is a new paradigm for global issues such as food safety, environmental awareness, economic sustainability, and animal welfare. Concurrently, the Sustainable Development Goals of the UN Global Compact have a lot in common with halal values. While SDG12 (sustainable consumption and production patterns) is addressed and aligned with the philosophy of halal-tayyib that propagates good, pure, and wholesome elements in consumption and upholds responsible production processes, SDG8 objective is to stimulate sustained, inclusive, and viable economic growth, full and productive employment, and decent work for all (United Nations Sustainable Development Goals). These two SDGs can be attained through a halal economy that has seen continued growth, despite the global crisis, and has the potential to narrow the income gap between different societal classes, so doing result in inclusive development for all. Halal stakeholders should seek an alternative solution to address prevailing yet persistent halal issues affecting halal integrity, and lack of standardized global standards, and consider the opportunity provided by the application of the Internet of Everything (IoE) to the universal halal ecosystem. Such innovative efforts will also future-proof the halal ecosystem against disruptive worldwide crises and ensure sustainability and development of the halal economy.

3.3.1 Halal Economy

Rising demand for *halal* products and services among Muslim millennials has been marked by a sizeable growth in over a decade for *halal* sales with Muslim expenditure forecasted to grow by 3.1% per annum to reach USD2.4 trillion by 2024 (GIER 2020, 2021). Despite the catastrophic chaos brought by the COVID-19 pandemic that also resulted in the disruption of global *halal* supply chains,

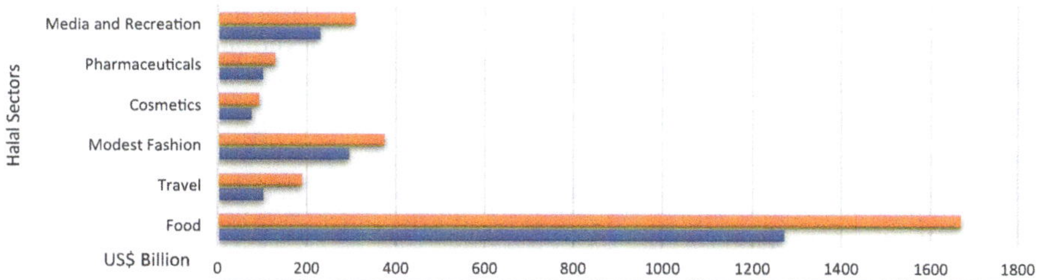

	Food	Travel	Modest Fashion	Cosmetics	Pharmaceuticals	Media and Recreation
2025	1670	189	375	93	129	308
2021	1270	102	295	75	100	231

Fig. 3.1 Global halal economy (2021 and forecasted 2025). *Source* Adapted from GIER(2022)

the Islamic economy has continued to grow, led by an acceleration in the digital revolution and increased government focus on *halal* products' and services' security-related investments.

The drivers of the global Islamic economy continue to be reinforced by a large and growing well-to-do 1.9 billion Muslim population and increasing adherence to Islamic ethical values impacting consumption. This is combined with an upsurge in demand for *halal* products by non-Muslims, practicing ethical consumerism (Pacific 2010). Other influential contributing factors comprise rise in digital connectivity, national strategies dedicated to *halal* products and service development, and businesses pressured to maintain growth momentum through new markets and diversification (GIER 2019, 2020; Azam and Abdullah 2020).

While countries continue to build more robust Islamic economy ecosystems, based on the infrastructure, support, and policies implemented, Malaysia is currently leading the overall Global Islamic Economy Indicator (GIEI) ranking for the ninth consecutive year, followed by Saudi Arabia, UAE, and Indonesia (GIER 2022). According to the Halal Development Corporation (HDC), Malaysia's *halal* exports are expected to reach US$9.6 billion in 2022, with exports in the first quarter up to double in 2021, signaling robust growth ahead. In 2021, *halal* exports rose by 19% to US$8.3 billion (SalaamGateway 2022).

Figure 3.1 demonstrates the growth in revenues, from 2021 and forecast 2025, for each of the various halal business sectors after considering the negativity of COVID-19 pandemic and the war between Ukraine and Russia.

3.3.2 Halal Cosmetics Ecosystem and Stakeholders

DinarStandard, announced in its GIER 2022 that Muslim expenditure on *halal* cosmetics and skincare products, is estimated to reach US$93 billion by 2025 in comparison to US$70 billion in 2021, a situation that bodes well for *halal* cosmetics companies and investors.

In a recent report published by Modor Intelligence, the top three countries with the highest *halal* cosmetics spending are India (US$6 bil), Indonesia (US$5.5 bil), and Turkey (US$4.5 bil), followed by Russia (US$4.2 bil) and Malaysia (US$4 bil). This demonstrates clearly that the first and fourth largest *halal* cosmetics consumer markets are from Muslim minority countries. This may be attributed to the revival of Islam and pious values among young educated Muslim millennials and Gen Z in emerging countries. It is them that has driven the demand for faith-based products and services (Özlem Sandikci 2011) and promoted the increase in *halal* consumption. Despite the impact of the COVID-19 pandemic and the ongoing war between Ukraine and Russia, the long-term growth of *the halal* cosmetics industry is still expected to multiply even though slower than forecasted earlier (Gerstell et al. 2020).

It is believed that attainment of SDG 8 (economic growth and decent work), SDG9 (industry and innovation), and *12* (sustainable consumption and production) will create greater economic security with better employment opportunities for citizens of developing nations while maintaining sustainable development for the future generation (United Nations website). Malaysia can target to expand on SDG8, SDG9, and SDG12 by providing *halal* cosmetics SMEs a more supportive and favorable *halal* cosmetics ecosystem, as economic enablers, so that they can prosper, both domestically and internationally.

The Halal Ecosystem is defined as "A network of components involved in the growth cycle and delivery of *hala*l products or services and contributes to the overall *halal* socio-economy". Each component in the ecosystem has its unique activities that are interrelated with each other, creating a constantly evolving relationship toward sustainability (msiahalaldirectory.com). The Malaysian government showed its stance when *the halal* industry was constantly highlighted in all its three Malaysia Plans (RMK9, 10, and 11) and instituted substantial measures including creating a governance structure for its *halal* industry and its ecosystem. Figure 3.3 displays the diverse mechanisms within the dynamic *halal* ecosystem in Malaysia. The *halal* ecosystem comprises interconnected components that are in perpetual progressing assembly with the main goal of attaining sustainability. The ecosystem includes the consumers, industry, supporting infrastructure, and impactor.

Bom (2019) in his study affirmed that the "cosmetic industry needs to adapt and innovate to design products and processes that can improve the sector sustainability, acting across the entire value chain" … "it is essential to understand which factors to consider when aiming for sustainability". He believed that a sustainable approach, specific to the cosmetics industry, can be conducted by identification and management of multiple environmental, economic, social, and safety aspects.

Various government agencies, government-linked agencies, and private enterprises manage Malaysia's halal cosmetics industry by incorporating standards, regulations, enviable infrastructure, and logistics with human capital development programs to ensure that it remains sustainable and able to expand domestically and internationally (Table 3.1).

Table 3.1 List of agencies involved in cosmetics halal governance in Malaysia

Agencies	Responsibility
Malaysian Islamic Development Department (JAKIM)	Manages issues regarding halal certification, administration Sharia laws including monitoring and enforcing *halal* certification integrity both locally and globally
Standards Malaysia	Establishes *halal* standards and acts as the national accreditation body
Halal Industry Development Corporation (HDC)	To position Malaysia as a global *halal* hub and accelerate the development of the halal industry worldwide in particular on the development of *halal* standards and capacity building for halal products and services
Malaysia Halal Council (MHC)	Consolidates, reconciles, and resolves issues concerning JAKIM, Standards Malaysia and HDC and develop *halal* in a more integrated and holistic manner NOTE: members includes JAKIM, Standards Malaysia, HDC as well as Ministries of Health; Agriculture and Agro-based Industry; Domestic Trade and Consumers Affairs; and Education
SME Corporation Malaysia (SME Corp)	Is the central coordinating agency (CCA) under the Ministry of Entrepreneur Development and Cooperatives (MEDAC) that coordinates the implementation of development programs for small and medium enterprises (SMEs) across all related ministries and agencies including acts as the central point of reference for research and data dissemination on SMEs and entrepreneurs, as well as provides business advisory services for SMEs
Center for Cosmetics, National Pharmaceutical Regulatory Agency (NPRA)	To safeguard the nation's health by determining quality, efficacy, and safety of cosmetic products through continuous monitoring and dissemination of information
Ministry of Domestic Trade and Consumers Affairs (MDTCC)	Conducts enforcement and monitoring program to protect Halal Integrity Halal logo and consumer's interest
Ministry of International Trade and Industry (MITI)	Formulates strategies and provides incentives to encourage trade and investment in *halal* products and services at the domestic level and promote halal standards as a benchmark tool at the international level

(continued)

Table 3.1 (continued)

Agencies	Responsibility
Malaysia External Trade Development Corporation (MATRADE)	To provide an avenue for businesses to develop and expand their exports of *halal* products overseas. Organizes trade events such the Malaysian International Halal Showcase (MIHAS) annually which is considered the world's largest *halal* trade fair
Cosmetic, Toiletry and Fragrance Association (CTFA)	Works to promote trust and confidence in the personal care industry and to ensure that regulations are competitive and internationally harmonized
Malaysian Industrial Development Authority (MIDA)	Promotes investments in the manufacturing and services sector as well as coordination of industrial development in Malaysia
Malaysia Productivity Corporation (MPC)	Identifies potentials in *halal* industry as well provides training and consultancy to SMEs and industries to meet *halal* standards and requirements

Source Adapted from Masood and Zaidi (2021)

Table 3.1 demonstrates the role of the various government bodies, government-linked corporations, and other institutions and agencies. These entities are operated by people carrying out different activities (processes) whose output is in the form of data and information, which can be relayed to each unit and be used for decision-making and forecasting.

3.3.3 Challenges and Issues of Halal Cosmetics Industry

Halal cosmetics can be defined as cosmetics that do not contain any alcohol or any ingredients from animals or parts of animals including insects and reptiles that are forbidden for Muslims or that are not slaughtered according to Islamic law. Also, it must not include ingredients such as coloring agents derived from insects, collagen, elastin, or hyaluronic acid-derived animal and instead substitute them with alternatives. More recently, the formulation of alcohol-free products, e.g., perfumes and hydroalcoholic gels used during the pandemic has ignited debate in countries such as Malaysia and Indonesia (Aliénor 2020). Consequently, producers of *halal* cosmetics need to be innovative to apply new technologies in the areas of research and development, formulation, production, and marketing. From a business standpoint, producers also need to be aware of any changes in policy in the domestic market, especially during an economic crisis. Halal cosmetics exporters must ascertain which halal certification is required by the importing country that may be subjected to a different School of Islamic Jurisprudence and practice different halal standards. Other "country" specificities must be considered regarding the definition of a halal cosmetic. For example, while alcohol is strictly prohibited as a residue in cosmetics in Saudi Arabia or Iran, in Malaysia the threshold for accepting alcohol is 0.5% and 1% in Indonesia and Dubai.

Apart from the formulation, the entire value chain must meet the *halal* preconditions, from raw materials to the production line, including research and development, packaging, logistics, etc. Halal must include *tayyib* which stresses wholesome, hygiene, and safety and covers not only the product but also the production process. In addition, the perseverance of the traceability and integrity of the halal product is vital.

3.3.4 Overview of Internet of Everything (IoE)

The initial idea of the Internet was conceptualized 80 years ago by Leonard Kleinrock in his article "Information Flow in Large Communication Nets" published in 1961 (Bay 2021). This was followed by J. C. R. Licklider, an American psychologist and computer scientist, who in 1962, commented that "*the future of computing will be in the form of a huge intergalactic computer network connecting people from all over the world*" (Britannica). Indeed today, the Internet is a vast network that links computers all over the world, to enable the sharing of information and communication anytime, from anywhere where if there is Internet accessibility. Today, this vital immense computer network furnishes us with a platform for communicating, socializing, teaching, learning, shopping, and entertainment and has become a necessity to humankind.

The COVID-19 pandemic has taught us that digital platforms offer viable options to continue to work and communicate virtually almost without the need for physical presence. In the third quarter of 2021, it was reported by the Department of Statistics Malaysia that the e-commerce income generation increased by 17.1% year-on-year to reach RM279 billion. Murali Raman (2022) stressed that like what happened to the e-commerce "dot-com to dot-gone" phenomenon, Metaverse, when assimilated with business strategy and social ideals, will endure the test of time.

3.3.5 IoT Versus IoE

IoT describes the network of physical devices (things) that are implanted with sensors, software, and other technologies over the Internet, devoid of human action, to connect and exchange data, while IoE is based on the concept of all-round connectivity, intelligence, and cognition of people, things, data, and processes, that are connected seamlessly and combined in a common interrelated system. IoE is aimed at improving individual experiences and enabling smarter decision-making (Auger 2018). IoT is a part of IoE which is all-encompassing.

As part of IoE, IoT has many advantages, including access to information anywhere in real-time, making it convenient for people to do their job without being physically present in their office. Human efforts can be minimized as devices of IoT interact and communicate (M2M) and complete designated tasks without human intervention. IoT is also time efficient and cost saving since data packets can be transferred faster over a connected network. As an example, IoT for transportation/logistics ecosystem starts with data sensing/collection, retrieved from device connection, which is then passed to data transportation/logistics, then is then analyzed to obtain data analytics to result in enhanced data value, then eventually can be used by humans to create value (i-SCOOP website).

Consequently, IoE technology will enable relevant information to be used to facilitate informed decision-making, thus ensuring that more effective action can be taken, thereby providing new capabilities and enrich experiences.

Figure 3.3, a proposed system by author, demonstrates the overall ecosystem of IoE comprising industrial IoT and consumer IoT. Digitized devices are linked to a shared network of other digital devices, people, and processes, to generate and exchange valuable data and information, therefore expediting pertinent decision-making (Di Martino 2018; Sakovich 2019; Gwangwava 2021). Decentralization and innovative decision-making are enabled when data are processed not in one center, but in many distributed nodes. The process of digital transformation leverages on cloud computing, fog computing, Artificial Intelligence (AI), machine learning (ML), IoT, Big Data (BD), Blockchain (BC), and others.

3.3.6 Interaction Between Pillars of IoE

Industry players and stakeholders must be vigilant about the ever-changing business ecosystem to increase revenue, save costs, improve efficiency, enrich experiences, and create better engagement. Application of digital transformation of the ecosystem will enable this to happen on condition that establishments are easily adaptable and adequately agile to tackle the challenges and seize opportunities that are logical and delivers benefits. IoE requires advanced information sharing capabilities beyond IoT. Real-time data, extracted from a variety of heterogeneous IoE environments, from simple sensors to complex robotic devices, and from autonomous service agents to human actors, will be analyzed instantly (Raj 2018). Miraz (2018) reported that smart devices installed with AI enabled people and things to network interchangeably in a multi-user environment demonstrating the innovative use of IoE. His study was followed by Langley (2021) who declared that processes embody the links within the network, with instantaneous data/information flowing between IoE hubs, and create the core of IoE. The results are smart and intelligent insights that are combined in real-time and working in harmony (Masoud 2019; Vandebroek 2016). IoE is beyond IoT context disruptions, as it addresses the public and organizational needs for more data and actionable information. A study conducted by Nonaka (2015) published that activities and exchanges within the IoE environment generate and develop valuable information in a transcending process during which the four pillars gain new knowledge and interfaces, which results in a continuous knowledge-creation cycle.

The interactions between the four pillars of IoE are as below:

- People: Insights given by people via websites, smart/mobile applications, or connected devices are converted into AI algorithms and are then analyzed by other technologies to understand human issues, and information is then delivered accordingly to assist stakeholders to solve problems or make decisions quickly and efficiently.
- Things: Sensors and actuators embedded in multiple physical items with product data that are then conveyed to designated destinations throughout the network.
- Data: Raw data are reviewed, condensed, organized, and evaluated. Armed with this meaningful information, network members can then control various systems and provide better informed resolutions.
- Processes: The correct information is sent to the right person at the right as the various processes are conducted based on Artificial Intelligence, machine learning, social networks, or other technologies.

3.3.7 Current IR4.0 Technology Applications in Halal Cosmetics Industry

The halal sector must adapt to the new technologies accessible in IR 4.0, as current technical developments can strengthen and improve not only a nation's economy but the global economy. History has shown that companies and industries that fail to adapt to new technologies may lead to them their demise, as demonstrated in the case of KODAK and Nokia. Exploring the potential of integrated IR4.0 platforms connecting local halal industries to global halal players will equip the industry with a competitive market edge.

The applications of emerging technology in the cosmetics industry are limitless and able to offer infinite solutions to satisfy new consumers. The following are some examples of technological advances that have occurred in recent years:

1. Artificial Intelligence (AI) is currently applied to offer personalized products via diagnosing consumers' skin at that time and some apps will show a visualization of the affected before and after the use of the recommended beauty product. AI will also be able to understand consumer preferences and buying behavior based on the collected data. A good example is the app currently used by ModiFace.
2. Online try-on apps using VR and AR technology offered to consumers, from MAC and Sephora, enables them to try eye shadows and lipstick anywhere and anytime before they make their purchase on mobiles. New Kinpo Group utilizes

AR to scan consumers' faces for any imperfection and then give their recommendation on skincare that is suitable to tackle such problems. Meanwhile, to address hair and skin problems, Olay and Clinique have introduced an innovative app that their consumers can use to facilitate their buying experience. Other brands offer real-time assistance to consumers to suggest the right combination of products and tutorials on how to apply make-up products (Thomas 2019). AR used as a real-time marketing tool has created a joyful, fun, and practical experience for customers.

3. Some companies have benefitted from 3D printing, facial recognition technology, and AI-powered image analysis. P&G launched the Opté wand, to hide, whereas Seymour Powell used a printer-like tool to print images chosen by consumers directly to their faces.

Another major brand such as Yves Saint Laurent uses AI and RFID to assist their customers to identify their chosen perfume and its contents. Sephora has launched a contactless store equipped with mirrors, robots, vending machines, and connected sensors.

The beauty and cosmetics sector has increased in performance, agility, and adaptability by having implemented technological innovation at the heart of its value chain. Another indirect benefit is it simplifies and facilitates beauty staff's daily job by providing them with more efficient work premises, products, and systems.

Meanwhile, demand for *halal* cosmetic products from e-commerce and social media platforms has steadily increased compared to that offered from conventional distribution channels. Moreover, manufacturers are influencing consumers on a huge scale through social media platforms such as YouTube, Facebook, Instagram, and now TikTok.

3.4 Discussion

As mentioned in the literature review, the *halal* ecosystem is a dynamic all-encompassing network of components that are linked in an eternal progressing connection, whereby each component has its own unique activities that are interrelated with each other, creating a constant evolving relationship (Fig. 3.2). While IoE comprises people, IoT, data, and processes connected to each other seamlessly and combined into a common interrelated system (Miraz 2015), each pillar of IoT has its own set of activities where the output (information) is shared and integrated into the ecosystem to enable decision-makers (consumers, producers, support agencies) to make better and opportune judgment.

Author propose Fig. 3.4 to illustrate the IoE ecosystem that would be viable to be implemented within the *halal* ecosystem. Sensors are planted in the devices or machines (manufacturing line, logistics, retailers, computers), and the interconnected world will enable IoT processes to take place. Data generated by this process are then stored and secured with Blockchain, shared, and transmitted to

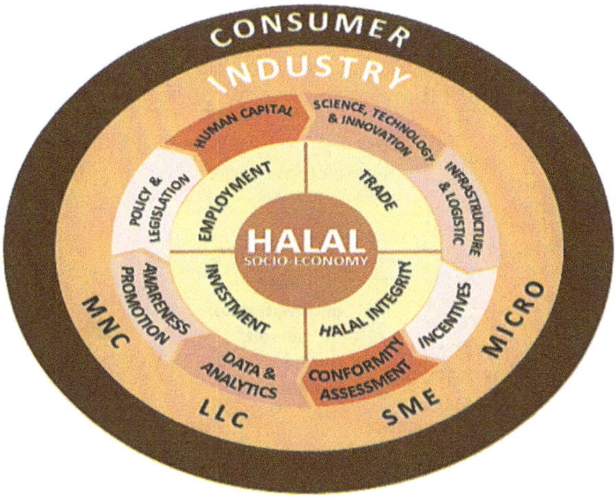

Fig. 3.2 Malaysia's halal ecosystem. *Source* Adapted from HDC website

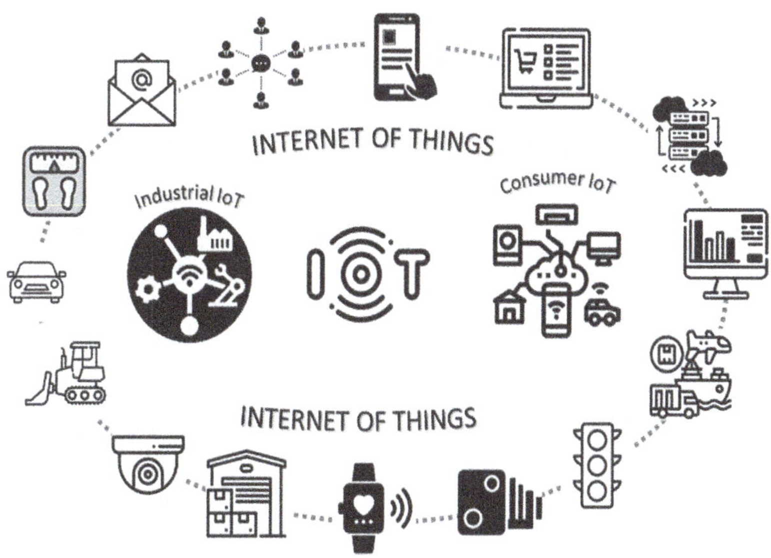

Fig. 3.3 Internet of everything (IoE), industrial IoT (IIoT), and consumer IoT (CIoT)

a station for analysis using big data analytics and AI technology. The processed information is then presented to the industry players (MNCs, LLCs, SMEs, and micro-enterprises), agencies (JAKIM, HDCs, MOF, NPRA, MITI etc.), and consumers (customers and users), in a timely manner, who will then use the processed information to make the necessary decision by accessing the Internet or app. The

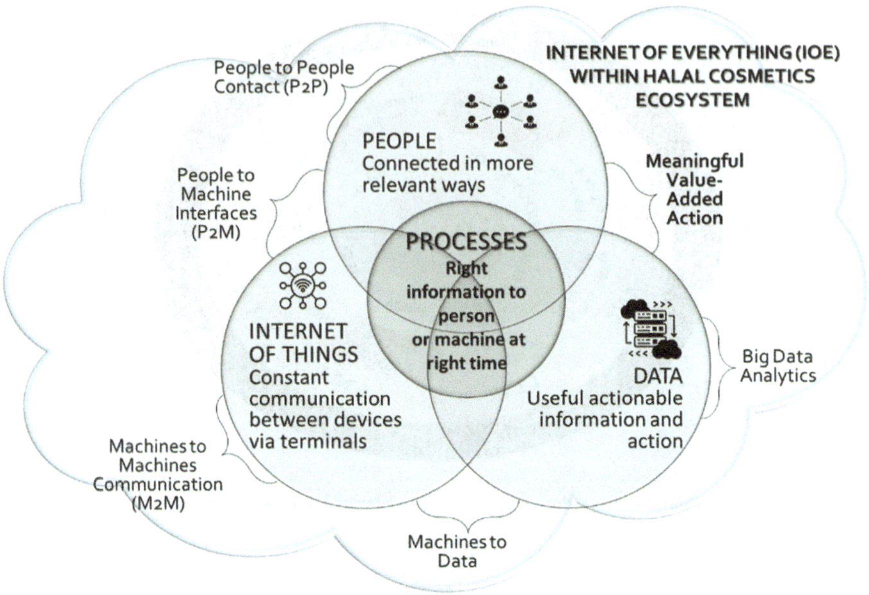

Fig. 3.4 Proposed viable IoE ecosystem model for halal cosmetics

processes that have taken place would involve people to people (P2P) as well as people to machine (P2M).

The interconnectedness of the IoE pillars, with information securely stored and made available in actual time, greatly supports stakeholders within the *halal* ecosystem. The absence of one universal *halal* certifying body, the presence of differing *halal* certifications across the globe, and discrepancies in *halal* regulations and transparency can partly be resolved with the application of IoE. This solution will empower industry players immensely as they would have almost immediate access to *halal*/Shariah-compliant resources (e.g., finance, information, expertise, raw material supply, human resource) and equip them to make accurate informed decisions to address current and future problems. For instance, stakeholders can obtain information on sources of *halal* compliant resources required, its *halal* certification verification, and increased accessibility to information on *halal* export requirements of importing countries. Getting relevant timely information about on-situ progress, in the halal supply chain, will be advantageous to stakeholders to strategize accordingly thereby avoiding disrupting the supply chain and saving time and costs. Consumers can also obtain truthful information about the *halal* product they intend to purchase as they are assured of the *halal* integrity of the said product, in that way effectively positively influencing consumers' behavior. This added assurance will bode well for *halal* companies and assist them in increasing customer loyalty and the company's brand value.

3.5 Conclusion and Implications

Today, the *halal* industry is acknowledged globally as one of the most competitive and fast-moving industries with significant untapped growth potential. Internationally, there is a growing recognition that *halal* is associated with SDGs 8, 9, and 12 and represents a new paradigm for addressing global issues such as food safety, environmental consciousness, economic sustainability, and animal welfare.

However, the *halal* business, which relies significantly on an uninterrupted *halal* supply chain, is plagued by issues with the preservation of *halal* integrity across the supply chain and logistics management, as well as the absence of a standardized *halal* standard and global halal certification organization. The stated obstacles have continued to impede the international expansion of local halal industries.

Internet of Everything (IoE) application to the *halal* ecosystem may help resolve the problems highlighted previously. Similarly, to IoE the *halal* ecosystem also consists of things such as devices/equipment/machines; people, including various stakeholders; processes comprising logistics, production, transactions, communications, marketing; and data, for example, *halal* standards, import/export requirements, suppliers, ingredients, etc. The pandemic and other global crises have not only accelerated digital transformation, but also demonstrated that all stakeholders are willing to adopt it, as it has become a requirement for continuous trade and transaction and provides a method for communication, interconnection, and accessibility. Many businesses had even resorted to investing more in digitally altering their business operations and activities to not only ensure a continuous supply of products to users but also to maintain their industry relevance. In fact, the food and beverage, cosmetics, and pharmaceutical industries have already used IoE to some extent. The same can thus be used for the *halal* industry if it wants to continue to expand and avoid major disruptions.

The capacity and competence of implementing IoE into the *halal* ecosystem will be beneficial and future-proof the *halal* industry, while many efforts must be made by the many stakeholders to ensure the industry's viability, sustainability, and growth.

Regardless of the future, we must remember that IoE is merely a tool to facilitate our undertakings, just like any other technology just like the Internet. Though there remain numerous possibilities, options, and clear value propositions, its use for business and in society in general should consider several aspects. Issues such as security, privacy, cost, and access to proper infrastructure must also be addressed to avoid stolen data and incorrect information and even identity thefts from occurring.

Acknowledgement This research for this chapter was funded by the Ministry of Higher Education (MOHE) of Malaysia under the Fundamental Research Grant Scheme (FRGS/1/2021/SS02/USIM/02/2).

References

Ab Talib MS, Rubin L, Zhengyi VK (2013) Qualitative research on critical issues in halal logistics. J Emerg Econ Islamic Res 1(2):131–150
Aliénor (2020) Halal cosmetics: technical and technological challenges. Accessed from https://www.alcimed.com/en/alcim-articles/market-access-challenges-of-halal-cosmetics/
Auger A, Exposito E, Lochin E (2018) Towards the internet of everything: deployment scenarios for a QoO-aware integration platform. In: Proceedings of the IEEE 4th world forum internet things, Singapore, 5–8 Feb 2018, pp 499–504
Azam MSE, Abdullah MA (2020) Global halal industry: realities and opportunities. Int J Islamic Bus Ethics 5(1):47. https://doi.org/10.30659/ijibe.5.1.47-59. Accessed from http://jurnal.unissula.ac.id/index.php/ijibe
Bay M (2021) Leonard Kleinrock internet pioneer. Manage Bus Rev. Accessed from https://mbrjournal.com/2021/01/26/leonard-kleinrock-internet-pioneer/
BIS Central Bankers' Speeches (2021) Abdul Rasheed Ghaffour: opening remarks—world halal business conference 2021. Accessed from https://www.bis.org/review/r210910c.pdf
Britannica "J. C. R. Licklider". Accessed from https://www.britannica.com/biography/J-C-R-Licklider
Di Martino B, Li K-C, Yang LT, Esposito A (2018) Internet of everything: algorithms, methodologies, technologies, and perspectives. Springer, Singapore. ISBN 978-981-10-5860-8
Elasrag H (2016) Halal industry: key challenges and opportunities. Accessed from https://mpra.ub.uni-muenchen.de/69631/1/MPRA_paper_69631.pdf
Gerstell E, Marchessou S, Schmidt J, Spagnuolo E (2020) "How COVID-19 is changing the world of beauty" McKinsey and Company. Accessed from https://www.mckinsey.com/industries/consumer-packaged-goods/our-insights/how-covid-19-is-changing-the-world-of-beauty#
GIER (2019/2020) The Global Islamic Economy Report
GIER (2020/2021) The Global Islamic Economy Report
GIER (2022) The Global Islamic Economy Report
Gwangwava N, Mubvirwi TB (2021) Design and simulation of IoT systems using the cisco packet tracer. Adv Internet Things 11:59–76
i-SCOOP. Digital transformation and technologies through the human prism. Accessed from https://www.i-scoop.eu/digital-transformation/human-prism
Islam T, Chandrasekaran U (2013) Halal marketing: Growing the pie. Int J Manage Res Rev 3(12):3938
Langley DJ, van Doorn J, Ng I, Stieglitz S, Lazovik A, Boonstra A (2021) The internet of everything: smart things and their impact on business models. J Bus Res 122:853–863. ISSN 0148-2963.https://doi.org/10.1016/j.jbusres.2019.12.035
Market Research Report (2017) Halal market size and forecast, by product (food and beverages, travel, media and recreation, finance, fashion, pharmaceutical, cosmetics), and trend analysis, 2014 to 2024. Hexa Res. Accessed from https://www.hexaresearch.com/research-report/halal-market
Masood A, Zaidi A (2021) Empowerment of SME's sustainability in halal cosmetics' ecosystem by diagnosing growth constraints. J Islamic Mark. https://doi.org/10.1108/JIMA-12-2020-0371
Masoud M, Jaradat Y, Manasrah A, Jannoud I (2019) Sensors of smart devices in the internet of everything (IoE) era: big opportunities and massive doubts. J Sens: 1–26. https://doi.org/10.1155/2019/6514520
Miraz MH, Excell PS, Picking R (2015) A review on internet of things (IOT) and internet nano of things (IoNT) in proceedings of internet technologies and applications. Wrexham, pp 219–224. Accessed from https://doi.org/10.3390/fi10080068
Miraz MH, Ali M, Peter S (2018) Excell, and richard picking. Internet of nano-things, things and everything: future growth trends. Future Internet 10(8):68. https://doi.org/10.3390/fi10080068

Nonaka I, Toyama R (2015) The knowledge-creating theory revisited: knowledge creation as a synthesizing process. In: The essentials of knowledge management. Palgrave Macmillan, London, UK, pp 95–110

Özlem Sandıkcı (2011) Researching islamic marketing: past and future perspectives. J Islamic Market 2(3):246–258. https://doi.org/10.1108/17590831111164778. (https://www.researchgate.net/publication/254192679_Researching_Islamic_marketing_past_and_future_perspectives/citations#fullTextFileContent)

Pacific A (2010) Global halal industry: an overview. Global Islamic Finance Report 2013, 140–159. Accessed from http://www.gifr.net/gifr2013/ch_13.PDF

Raj A, Prakash S (2018) Internet of everything: a survey based on architecture, issues and challenges. In: Proceedings of the 5th IEEE Uttar Pradesh section international conference on electrical, electronics and computer engineering, pp 1–6

Raman M (2022) Rise of the metaverse. The Star. Accessed from https://www.thestar.com.my/news/education/2022/06/12/rise-of-the-metaverse

Sakovich N (2019) What is the internet of everything (IoE)? SaM Solutions. Accessed from https://www.sam-solutions.com/blog/what-is-internet-of-everything-ioe/

Salaamgateway (2020) Halal cosmetics 2020: more traction in Asia led by demand from Malaysia, Indonesia, and exports from South Korea. Accessed from https://www.salaamgateway.com/story/halal-cosmetics-2020-more-traction-in-asia-led-by-demand-from-malaysia-indonesia-and-exports-from-so

The Global Islamic Economy Report 2019/2020 (GIER 2019/2020)

The Global Islamic Economy Report 2020/2021 (GIER 2020/2021)

The Straits Times (2022) Security trumps obesity in Britain's first food strategy plan. Accessed from https://www.straitstimes.com/world/europe/security-trumps-obesity-in-britains-first-food-strategy-plan

Thomas D (2019) Five tech trends shaping the beauty industry. BBC News. Accessed from https://www.bbc.com/news/business-48369970

United Nations Sustainable Development Goals. Accessed from https://www.un.org/sustainabledevelopment/economic-growth/

Vandebroek SV (2016) Three pillars enabling the internet of everything: smart everyday objects information-centric networks, and automated real-time insights. In: Proceedings of IEEE international solid-state circuits conference (ISSCC). IEEE, pp 14–20

Digital Gold Investment Platform in Shariah Perspective: A Case Study of Quantum Metal

4

Mohammad Aiman bin Mohammad Zakir, Mohamad Zaharuddin bin Zakaria, Ahmad Zaki Salleh, Abdul Manan Ismail, Mushaddad Hasbullah and Mohd Nasir Abdul Majid

Abstract

Gold investment has been popular among the investors for its good hedge against inflation. This is due to the high value of the commodity that the gold stores. There is a question, however, whether digital gold investment is permissible or not? The Accounting and Auditing Organization for Islamic Financial (AAOIFI) shariah Standard in Gold Investment and Gold Investment Parameter issued by 96th Muzakarah of The National Fatwa Committee (MFK) in 2011 have been referred to ensure the halal compliant of the digital gold investment. Quantum Metal is a new digital platform that offers three premier products which are the Gold Savings Account (GSA), Gold Currency Account (GCA), and Gold Advance Enhancement (GAE). This paper highlights the halal principles underlying these products in addition to the approval of the International shariah Research Academy for Islamic Finance (ISRA) consultancies and how it could make the Quantum Metal (ISRA 2021), as a shariah-compliant gold digital investment platform, to improve the economic growth during the economic crisis. This research is conducted using primary data sources, digital gold investment contracts, journals, and interviews. It is found that the Quantum Metal Digital Gold investment platform is compliant to the halal standard of gold investment.

Keywords

Halal · Digital gold investment · Shariah compliant · Quantum metal

M. A. b. M. Zakir · M. Z. b. Zakaria (✉) · A. Z. Salleh · A. M. Ismail · M. Hasbullah · M. N. A. Majid
Faculty of Syariah and Law, Universiti Sains Islam Malaysia, Nilai, Negeri Sembilan, Malaysia
e-mail: zaharuddin@usim.edu.my

A. Rafiki (ed.), *Digitalization in Halal Management*, Contributions to Environmental Sciences & Innovative Business Technology,
https://doi.org/10.1007/978-981-99-5146-8_4

4.1 Introduction

In August 2022, department of statistics Malaysia announced that the inflation rate has increased to 4.7% where gauge's food and non-alcoholic beverage components were the main contributor towards increasing inflation in Malaysia by 7.2%. Hence, the rise of the inflation rate has contributed to the increase in production input and fuel prices. The consumers do not realise that this may be a threat to the economic growth in Malaysia. By time, the whole family may have difficulty in affording a week's meal and people will restrict their use of money because of the rises in general price levels. The scenario has resulted in the reduction of the purchasing power of money. At one point, our money values will shrink and give a big impact on the investment's part such as in property investment. They noted that the policymakers and central banks are in interest to expand the money supply by ratifying the inflationary pressures, rather than high unemployment rate or jeopardise consumption and investment behaviour (Pinga and Nelson 2001). Under this view, the causal relationship between money supply and aggregate prices is expected to run from aggregate prices to the money supply.

Therefore, the characteristics of the gold itself, that is sacred, valuable and did not change throughout the period despite the changes in economic conditions, have been seen as a great potential for recovery of economic growth after a few episodes of recession. Investors have been looking for an alternative to prevent taking high risk with low return on investment. This is because they realised that the strength of currency is depending on the economic situation of their government.

Additionally, gold's low to negative correlation with bonds and stocks makes it an excellent and excellent portfolio diversifier. In fact, many financial advisors recommend keeping 4–12% of an individual's total wealth in gold. Gold is considered a stable means of hedging against inflation, a store of intrinsic value, and preserving wealth in addition to generating high returns. This is an opportunity for the recovery of the economic crisis and sustains the economic growth despite the changing in currency. However, it becomes an issue whether the standard of the gold investment is following the shariah principles in Islamic Economic and Muamalat or not. Hence, the 96th Muzakarah of The National Fatwa Committee (MFK) on 13th–15th of October 2011 has agreed on the Gold Investment Parameter that content of a clear guidance that must be followed by the Muslim investors on gold investment. The parameters have perfectly covered the basic conditions of transactions in gold trading (JAKIM 2011).

In this contemporary era, there are so many things that are convenient for the purpose of living and one of them is digital gold investment where there is a platform in the form of banks or monetary institutions that offer a gold investment account or "paper gold". In this gold investment account, the buying and selling of the gold still can happen, but the physical gold will be kept in a storage vault. That being so, the benefits of investing in gold still can be obtained with the facilities of keeping the gold safe and secure (Agha et al. 2015).

4.2 Literature Review

Gold as Investment Instrument and Haven Asset

Hasyim (2017) has described gold investment as a way of buying physical gold or registered in any financial gold-based product with the purpose of gaining profit. According to (The World Gold Council 2018), India and China have been listed as the biggest markets in demanding the gold in global markets. However, demands have been weak among the central banks and investors due to its mix in demand after it fell to its lowest as the current conventional market.

According to (Mas'ad et al. 2018), Islamic banking institutions offer two options: either the investor wants to keep the gold by themselves, or let the gold be kept in the bank. An account book will be given by the bank institution to every investor that registered for the investment account. The account book will assert the amount and the values of their gold as record for the investors to keep as a proof and for any transaction purpose later.

Since gold has high property values and the prices will always rise by time, this can benefit an asset to the financial both individual and community. Allah (s.w.t) has recognised and emphasised the important of gold in human life since the Prophet Muhammad's era as mentioned in (Al-Quran. Ali-Imran: 91) as follows:

> إِنَّ ٱلَّذِينَ كَفَرُواْ وَمَاتُواْ وَهُمْ كُفَّارٌ فَلَن يُقْبَلَ مِنْ أَحَدِهِم مِّلْءُ ٱلْأَرْضِ ذَهَبًا وَلَوِ ٱفْتَدَىٰ بِهِۦٓ أُوْلَٰٓئِكَ لَهُمْ عَذَابٌ أَلِيمٌ وَمَا لَهُم مِّن نَّٰصِرِينَ.
>
> which means: "Indeed, those who disbelieve and die while they are disbeliever- never would the (whole) capacity of the earth in gold be accepted from one of them if he would (seek to) ransom himself with it. For those there will be a painful punishment, and they will have no helper".

In this verse, Allah takes the gold as the comparison as the valuable thing that non-Muslim (Al-Kafirun) cannot compensate for even though they give all the gold they have, but still Allah and the earth would not accept their regret. Allah did not mention other commodities as the comparison like fruit, farm, or land because they will never be as conservative estimates as the gold (Aunurrochim and Ezzuddeen 2018).

The strength of the gold in its price motivates and drives people to grab the gold's wealth as a safe haven despite the current economic uncertainty. McCown and Zimmerman (2006) have suggested that gold has the criteria of zero beta assets and bearing zero market risk. Others have found evidence of the exchange-rate hedging potential of gold by analysing the role of the gold as a hedging tool against the dollar (Capie et al. 2005). This has been proven in one of the research papers that gold has the potential to be the safe haven asset from deprivation in financial markets (Lucey and Tully 2006).

Gold Investment Practise in Malaysia

Normally, there are two ways in the application of gold investment which are gold physical investments and gold savings' account where they invest in gold price that is being floated in the market, while (Aunurrochim and Ezzuddeen 2018) explained that there are two ways in applying gold investment. First was conventional gold investment where gold can be bought at a certain value from the gold seller who intended to sell it (Louis Ng 2009). Early in times, gold will physically be brought back home, but the risks are very high where it is exposed to any possibilities that are unpredictable such as theft, broke, and natural disasters. However, in this era of the new norm, a few trusted platforms have been introduced as the third parties between the buyers and the sellers to ensure the safety of the gold. They exist as a bank institution or an insurance company to keep the gold in top security. This was the second practice in Malaysia where they invest their money in gold by a monetary institution.

Gold investment through a monetary institution or known as Gold Investment Account whereby the investors do not have to keep the gold by themselves, but the bank will keep the gold for the investors. It is convenient since the risk of losing gold is low. Other than it being a secured and hassle-free investment, it holds high liquidity to the account investor as they can sell their gold to the bank back during trading hours. The function is still the same as they hold a good hedge against inflation. However, different companies and different banking institutions offered different kinds of gold investment schemes in order to attract the public to invest in theirs.

In Malaysia, there are bank institutions and insurance companies that offer gold investment as they see this as a potential to attract the investor and the public to enrol and grow their company other than can help in economic growth in Malaysia. However, only a few truly follow the shariah compliance and free from any limits imposed by Allah. Currently, there are some institutions and companies offer gold saving account to the public which is in compliant with the shariah principles such as Kuwait Finance House (KFH), Public bank, Bank Muamalat, Affin Islamic Bank, HelloGold, Goldina, Public Gold and Quantum Metal Sdn. Bhd. as another way to invest in gold other than physical gold.

A good example of practising shariah compliance in gold investment is Kuwait Finance House (KFH) where they invest in Malaysia to offer to the majority of the Muslims in this country a shariah-based banking product that is free from any prohibited element in Islam (Aunurrochim and Ezzuddeen 2018). This is the type of gold investment through monetary institutions where KFH is an international bank that uses the concept of *Bay' Al-Sarf* and *Qard al-Hassan.* They are offered as a third party to keep the investors' gold under *Qard al-Hassan contract*, and the investor or the owner of the gold is permitted to gain profit by selling their gold back to the bank or to another buyer. Any investor who invests gold in KFH will have KFHMB Gold Account-i to record all the transactions and trading of the gold and prevent any doubtful or *gharar* (Kuwait Finance House Malaysia Berhad 2017).

Following the instruction from the Bank Negara Malaysia Resolution, "Hand to Hand" method is followed to fulfil the condition of the gold as "*ribawi*" item. The transaction should be "*lani*" (on the spot) without delaying the gold exchange and free from the element of *Riba' al-Nasiah* where the gold can be withdrawn from the bank and the time will be arranged by the bank in two to three days for the gold to transport from the saving vault. Other than that, KFH was also free from elements of *Bay' al-Ma'dum* where they offered the physical gold as low as 1 g and the customer was aware of the existence of the gold kept in the savings vault.

Another known and familiar bank institution that offered gold investment is Maybank. Nowadays, Maybank has been approved as a bank institution that offers gold investment following the shariah compliance where they follow the shariah standard and free from any elements that are prohibited in muamalat in Islam. Whoever is interested in gold investment through Maybank, they will have a Maybank Islamic Gold Account-1 (MIGA-i). They can buy and sell the gold as low as 1 g and the 999.9 fine gold commodity will be stored at Produits Artistiques Metaux Precieux (PAMP S.A.) and can withdraw the physical gold at selected Maybank branches.

Public gold also has been approved following the shariah standard in gold investment by the shariah consultancy from Amanie Advisors. Public Gold collaborated with Ar-Rahnu to carry out buy and sell gold activities where they used the concept of *Qard al-Hassan* by offering the customer to make a loan by selling their golds but can withdraw when they can pay back the loan without a time period given. Public Gold offered only 99.99% gold commodity that they own by themselves, and the customer will get either a passbook or monthly statement depending on the product they invested into as the proof of the transaction and free from any element of *gharar.*

This is the best scenario that complies with the shariah standard and does not have any element of *gharar* and *riba'* that can be prejudiced towards the buyer. The scenario given allowed the investor to choose whether they want to keep the gold by themselves or let the gold be kept by the bank under the contract of *Qard Hassan* (Aunurrochim and Ezzuddeen 2018). Moreover, they can withdraw their gold from 1 g of gold and proven the acceptance of the gold as *al-Qabd* which is parallel to the Gold Investment Parameter no. 13 from the Muzakarah of The National Fatwa Committee (MFK), 2011, which are:

The delivery of purchased goods (gold) must take place by actual delivery or a reliable method that can replace the actual delivery. A reliable delivery of purchased goods will have the same effect as actual delivery, namely:

a. Transferring dhaman (warranty responsibility) from the seller to the buyer.
b. The buyer's ability to get his purchase at any time without any obstacles.

4.3 Methodology

This study is based on the qualitative methodology whereby the data were derived from the library sources, namely books, journals, thesis, and gold investment contracts extracted from the QM gold investment platform. Data were also gathered from semi-structured interviews with the experts and scholars who did involve directly in the shariah review of the platform. All data were analysed to ensure that the QM digital gold investment is incompliant to the Sharing rulings and principles.

4.4 Results and Discussion

4.4.1 Shariah Compliance of Gold Investment

Nordin et al. (2018) found that the sample of population in Terengganu has a moderate level in having knowledge and understanding about shariah-compliant gold investment. They also explained the shariah parameters on gold investment that have been issued by the Fatwa Committee of Malaysian National Council for Islamic Religious Affairs to be able to use gold at its fullest and to recover from the economy crisis, together with the launching of the AAOIFI shariah standard on gold in collaboration with the World Gold Council on 2016 to provide a clear guidance, especially in the usage of gold in financial and investment transaction.

The basic rule and principle of transaction in Islam are it must be in line with the shariah criteria where all the types of transaction of buying and selling must not contain any element from *riba*, *gharar* and cruelty and these transactions or trading must begin with *aqad* with *ijab* and *qabul*. This must also be applied in the gold investment to prevent the exploitation of the poor and increasing poverty in riches.

Muamalat in Islam must involve a contract between a seller and a buyer. The aqad or sighah in a contract required ijab as offering and qabul as acceptance. These are very important in any contract or muamalat because it will advocate whether the transaction is acceptable or not alongside the shariah standard. The words that can be portrayed as Ijab is "I sell", while for qabul is "I buy", so that they have a clear agreement and they may use it in various ways such as physically, message, or call. As long as it means buying and selling, where ijab must be followed by the qabul in words without any postponing (Al-Sharbayni 1997). While the aspect from Law, contract is any type of agreement or promises that are made of the willingness of each party who are eligible to enter into a contract, for a legitimate consideration, and with a legitimate purpose.

According to Imam Hanbali, these are the conditions that need to be followed in ijab and qabul.

1. The buyer and the seller must present in the same place.
2. The *ijab* and *qabul* must not have things that can separate between them where it can cause rejection in aqad.
3. The *ijab* and *qabul* must not involve things that are not related to the *aqad.*

Following the conditions, the seller and the buyer that make the aqad need to be someone who is Muslim, mature, and sensible. While from imam Syafi'i, the person needs to be someone who is adult, not being forced, Muslim and a buyer should not be enemies. Other than that, the item needs to be clean according to shariah, beneficial, owned by the seller, capable in giving the item at the place that they agreed and the items need to be acknowledged and exist.

In the exchange of prices for prices in the form of currency or *Bai' al-sharf* (بيع الصرف), Islam has introduced certain rules to strictly limit the use of money as a medium of exchange (Mahmood n.d). The goal is to prevent the "pricing of money" and use it as a commodity. As reflected in a hadith reported by Abu Sa'id al-Khudri, where the Prophet (saw) had said: "*Gold for gold, silver for silver, wheat for wheat, barley for barley, dates for dates and salt for salt, should be exchanged like for like, equal for equal, and hand-to-hand. If the types of exchanged commodities are different then sell them as you wish, if they are exchanged on the basis of a hand-to-hand transaction*".

From Maqasid shariah perspective, Al-Ghazali explained that it is a blasphemy if we did not appreciate gold and silver as a metal that have a high value and have the potential as a measurement of currency in muamalat. However, in Islamic jurisprudence, there is no hindrance from the shariah to buy gold with a low price and sell it with a high price as long as the parameters of the shariah standard in gold investment is followed. Hussain et al. (2012) Gold as a ribawi item and can be traded with other types of currency such as money as allowed in the contract *bay' al-sarf.* Hence, gold investment is permissible as long as the conditions and the parameters of the *bay' al-sarf* are fulfilled as follow:

1. There is *taqabud* or parallel taking possession between the gold and the other currencies in selling and buying. Therefore, there are two types of transactions which are through *qabad hakiki* (actual possession) or *qabad hukmi* (constructive possession).
2. "*Aqad*" or contract must happen on the spot basis and free from delay in exchange of goods and price.
3. "*Aqad*" must be free from the cooling-off period which is the cancellation period.
4. A different type of currency that wants to be traded needs to have the same value.

Muamalat in Islam emphasises the prohibition of *riba'* or interest which is mostly found conventional loan. This is because it has elements of injustice and prejudice towards the poor. Hence, the concept of justice, mercy, and charity in muamalat is known as *Qard al-Hassan*. Qard al-Hassan means a good debt where it emphasises doing charity towards people in need, especially in financial matters. The basis in Debt or "*Qard*" in Islam is a contract that contains values of helping each other, willingness (*tabarru*'), and mercy towards the people in need. To obtain the objective of doing charity in Qard, Islam obligates in every debt to return the money with the same values without any interest. If there is an element of interest, it is included in the category of *Riba al-Nasiah;* hence, it is forbidden in Islam (Zakaria et al. 2017).

An analysis from the International Shari'ah Research Academy for Islamic Finance (ISRA 2021), there are different types of gold investment that can be offered to the investors. One of them is physical gold, gold account, and digital gold. Types of physical gold that can be obtained are in coins or bars and jewellery. Gold coins are usually in pure 24 karat and available in sizes such as 1, 1/2, 1/4–1/20 oz, while in bars, they have the same purity and are available in grams such as 5, 10, and 100 g. However, for jewellery, they made the gold with a specific quality and weight in different sizes and shapes. Next is a gold account that is offered by some banks nowadays where the customers do not have to keep their gold physically but by a record from an account that has been provided and make profits with the rise in gold prices in the market. Lastly, digital gold offers gold that can be traded for investment but not physically. They use electronic platforms as a medium to monitor the gold prices in the markets and the investors have to follow the instruction by the brokers such as forex, but they cannot demand the gold physically. Conversely, every open position will end with a close position where they have to accept any profit and loss in money currency.

Transactions of buying and selling gold through the internet or online also have been controversial due to its delay in gold exchange. The issue is that the seller must give the gold on the spot and the buyer should make the full payment in the same time period. However, the gold is a *ribawi* item that may contain elements of riba. From the *Al-Kafi li al-Fatawi, Al-Kafi* chapter 706 stated that the transaction is still permissible with the condition, and there is no statement of delaying the gold for a period of time. In addition, *Majma' al-Fiqh al-Islami* Council for *Rabitah al-Alam al-Islami* for the 11th time that convened in Makkah on 19–26 February 1989H discussed the issue of *taqabudh hakiki* and *hukmi* that transaction documentation in the form of receipts and account records is acceptable as one form of *taqabudh;* thus, it is valid and accepted in Islam. Therefore, when the gold has been deposited into the buyer's account and recorded in the system and receipt, the buyer can withdraw the gold at any time with the quantity based on the agreement or contract, then the delivery is considered complete. So, it is permissible to buy and sell gold online if you meet these conditions.

These will also prevent any elements of *gharar* where a transaction concerning the uncertainty of the existence of the goods is against in Islamic finance. This type of transaction is known as *gharar* which is very prohibited in Islam. In *al-Fiqh al-Islami wa Adillatuhu* conclude that *gharar* contains the element of betting whereby the existence of the item is uncertain. The items may not exist at all or there is less information about the existence or inability to deliver the items. From the AAOIFI shariah standard, *gharar* defined as a situation that has an element of uncertainty in the process of the transactions occurring. In other words, *gharar* refers to a decision state that may or may not happen (Awang 2016).

Indeed, *gharar* is very strict in Islamic finance, especially in gold investment that uses a gold account rather than buying physical gold. This is because the gold will not be given physically but will be kept by the banks. Thus, the elements of *gharar* may happen in the operation of gold investment accounts in every bank or monetary institution if the transaction does not follow the shariah rulings.

4.4.2 Quantum Metal

Introduction to Quantum Metal

Quantum Metal Sdn Bhd imports and distributes gold throughout Malaysia and ASEAN. Quantum Metal was registered on the 30th of July, 2012, in Malaysia and is one of the largest gold bullion distributors in Malaysia, offering 99.99% premium gold products to government bodies, banks, and financial institutions on the Malaysian market. The Quantum Metal is not a money scam or involved in money laundering because it reports to the Anti-Money Laundering and Anti-Terrorism Financing Act (AMLA) 2001, which means that all business and money coming in and going out will be reported to and audited by Bank Negara Malaysia (BNM). Most companies involved in get-rich-quick schemes or money laundering do not report to AMLA. In addition, all of Quantum Metal's gold products are certified by the London Bullion Market Association (LBMA). Moreover, the gold supplier is The Perth Mint Australia, which is wholly owned by the Western Australian government and whose purity and weight are 100% guaranteed. This also means that Quantum Metal only sells gold products of the highest quality. The Perth Mint Australia is accredited as a refiner, weight master, and assayer by the LBMA, New York Commodity Exchange, Shanghai Gold Exchange, Tokyo Commodity Exchange, and Dubai Multi Commodities Centre, which are the five global precious metal exchanges.

In Quantum Metal, the gold can be purchased both physically and digitally. By withdrawing physical gold, it is proven that the customer demonstrates legal ownership towards the gold. Hence, the gold as *Al-Muthman* is fulfilled when they will have an allocated gold account if they store the physical gold in secure, insured vaults, such as those provided by Lloyd's of London Insurance. The gold remains in the vault and cannot be used for anything else. In contrast, the risk associated with unallocated gold, also known as paper gold, is generally higher. Unallocated gold accounts are not required to store physical gold. During the time that a company or bank holds your gold on your behalf, it may be lent out.

By maintaining an account with Quantum Metal, your gold will be securely stored in the Brink's gold vault. Since 1859, Brink's has safeguarded valuables. With 150 years of experience, Brink's provides services of the highest quality and sets the industry standard. Quantum Metal has been working closely with

Brink's as the official bullion vault storage for the Gold Storage System in Brink's Singapore. Brinks serves as a third-party vault to safeguard the ownership of gold stored through the Gold Storage Account system. In addition, Lloyd's of London Insurance is the leading insurance company. At Brinks, the gold is completely protected from all potential dangers.

Quantum Metal accounts are not bank accounts, but they function similarly. When a customer stores their currency at a bank, Quantum Metal stores their currency as gold. The gold currency is only redeemable for other gold products within Quantum Metal. Additionally, the value can be withdrawn in your local currency to your bank account at your convenience.

The customer may withdraw physical gold from their Gold Storage Account (GSA) at any time, with the exception of GAE5X/GAE10X, which must be withdrawn upon full payment settlement to QM. This is possible by selecting "Physical Withdrawal" in GSA and the gold can be redeemed physically at the Ar-Rahnu Tekun with an appointment set by the Quantum Metal.

All other purchased and buyback transactions on Quantum Metal's gold investment platform are conducted securely via the FPX payment gateway. Quantum Metal will immediately credit an equivalent amount of gold to the customer's gold account upon receiving confirmation of payment. Quantum Metal also offers a buyback guarantee on all gold products sold. In addition, all gold transaction records for Quantum Metal are independently audited by ISRA Consulting, which is wholly owned by ISRA, a subsidiary of Bank Negara Malaysia, and registered with the Securities Commission of Malaysia.

Shariah-compliant transactions presented the greatest difficulty for Muslim investors. Gold, including the primary digital products Gold Saving Account (GSA), Gold Advance Account (GAA), and Gold Leverage Account (GLA), has been deemed shariah-compliant by Quantum Metal's shariah Advisor, ISRA Consulting, which is registered with the Securities Commission of Malaysia. ISRA, a wholly owned subsidiary of Bank Negara Malaysia, has established ISRA Consulting. All operations and processes adhere to the gold shariah Standard. Together with the Accounting and Auditing Organisation for Islamic Financial Institutions (AAOIFI), the World Gold Council, and Bursa Suq Al-Sila, the shariah Standard for gold was developed. Dr. Tuan Badrul Hisyam bin Tuan, the ISRA Consulting shariah Advisor, has also discussed Quantum Metal products in mainstream media such as TV Al Hijrah on 1 February 2021. Quantum Metal meets two crucial requirements. It is backed by physical Gold, and the buying and selling of physical gold are completed within the standard's timeframe (Quantum Metal n.d).

Products in Quantum Metal

Quantum Metal offered three premier products which are Gold Storage Account (GSA), Gold Currency Account (GCA), and Gold Advanced Enhancement (GAE).

Gold Storage Account

A Gold Storage Account (GSA) is a saving product. It allows customers to convert their currency to gold. A customer can already own a small portion of gold for as little as USD2.50 (RM 10) equivalent to the gram of gold. Customers may store as much gold as desired without any limits. No contract for storage is created by Quantum Metal, and no storage fee is charged. The spread charged is 7%, and customers can withdraw gold in physical form or in cash at any time. The spread is not a profit for Quantum Metal due to the fact that all activities of selling and buying gold, including buying and selling in the gold savings account, incur a spread. Customers will gain profit from the increase in gold price globally.

Furthermore, this saving product is similar to the shariah saving product which is called *(Wadi'ah)*. It means that this Quantum Metal storage account is shariah compliance. This can be proved by no limitation for the storage period and no storage fee charged. Hence, there will be no prejudices towards the customer, also no riba element during the transaction and storage process occurs.

Gold Convert Account

A Gold Convert Account (GCA) enables the preservation of wealth value. GCA is analogous to an Ar-Rahnu Quantum Metal digital account. However, GCA is not the same as Ar Rahnu. Because the Ar-Rahnu has a time limit, after 18 months it is required to change the letters. This will interfere with cash flow. While GCA is a gold overdraft with no period and time limit, the contract will be automatically renewed. Quantum Metal also provides flexi gold *tawarruq* cash conversion for GCA products. In shariah terms, it is also known as *tawarruq murabahah* financing.

Furthermore, GCA is designed to offer customers the opportunity to convert up to 85% of their gold into cash when they request GCA. In the meantime, 100% of the value of gold is retained in the holding inventory. This result represents that a customer has received a flexi payment; hence, flexi gold *tawarruq* cash conversion is applied. For instance, a customer holding 10 g of gold worth USD10 can instantly withdraw USD8.50 while maintaining a total gold holding of USD10. A customer is only required to pay a 3.5% annual management fee based on the cash amount. It is calculated daily at 0.00959% of cash received. This 3.5% fee is the lowest management fee or profit rate charged on the gold investment market. In subsequent years, the customer has the option to defer the management fee until account termination, at which point the contract will renew automatically. In addition, customers can redeem gold if they wish to return their grammy taken to the GSA. Additionally, it can be sold in the future for a profit if the price of gold rises.

Gold Asset Enhancement

GAE is a form of investment. Investments in GAE are quantified by unit count. GAE is divided into two types which are GAE5X (multiple of five) and GAE10X (multiple of 10). The purpose of this product is to increase the customer's assets or wealth by twofold. This product is exclusive to Quantum Metal, one of the best products, and yields a huge profit. If the profit from the global gold price is 1%, GAE5X and GAE10X customers will receive five to ten times that amount. Plus, Quantum Metal offers flexi gold *tawarruq* asset enhancement for both GAE products and it is also known as *tawarruq murabahah* in shariah terms.

By utilising GAE5X, a customer has the advantage of holding USD125 equivalent grams of gold for only USD25. In order to hold USD125 equivalent grams of gold, a customer only needs to provide a paid-up capital of USD25. Meanwhile, the USD100 is made easily through flexi gold *tawarruq* asset enhancement or the *tawarruq murabahah* financing which is provided by Quantum Metal. This indicates that the flexi payment is equal to 80% of the purchase price. A customer can see on the website that the Quantum Metal GAE5x price includes a 4% spread on top of the London Metal Exchange (LME) rate and the 3.5% of the annual management fee must be served at daily rest at the end of the customer contract. Another benefit is that contracts will be automatically renewed, and customers will be able to cash out through a bank account or GSA account at any time.

On top of that, with GAE10x, a customer can hold USD2500 equivalent grams of gold for only USD250. In order to hold USD2500 equivalent grams of gold, a customer only needs to provide a paid-up capital of USD250, while the remaining USD2250 can be acquired through flexi payment or *tawarruq murabahah* financing, which is offered by Quantum Metal. This demonstrates that the flexi payment or *tawarruq murabahah* financing is equal to 90% of the purchase price. A customer can see on the website that Quantum Metal's GAE selling price includes a 3% spread in addition to the London metal exchange rate. The first year of a 3.5% annual management fee is due upon the execution of a customer contract, while subsequent management fees may be deferred and deducted upon the customer's contract termination at daily rest. Same with the GAE5x, the contracts will be automatically renewed and the customers will be allowed to cash out the money from their bank account or GSA account at any time.

Concept of Tawarruq in Quantum Metal

Tawarruq technically entails a series of sale contracts in which a buyer (*al-mustawriq*) purchases an asset from a seller for future payment before selling the asset at a low price to a third party. This is done without planning to use the asset that was bought, instead with the ultimate goal of obtaining cash or liquidity at that specific moment.

The *tawarruq* is applied when a member makes an agreement with Quantum Metal (QM) for Flexi *Tawarruq* Cash Conversion or *Tawarruq* Asset Enhancement. When the contract is drawn down, the member will buy from QM a specified commodity bought from commodity broker A or supplier A at the QM's purchase price through Bursa Suq-Al-Sila (BSAS). QM then sells the specified commodity to the member at the QM's purchase price plus with a profit which is the management fee, which will be paid. Then, the member will appoint the QM as an agent to sell the commodity to commodity broker B or supplier B at a price equal to QM's purchase price on a spot basis.

In a nutshell, the transaction of buying and selling a commodity such as gold with deferred payment through *murabahah* contract is a purchase of asset with the profit margin from a buyer to the seller. The commodity will be sold to the third party to gain the cash. The gained cash that the buyer get from selling the commodity is the Quantum Metal's endowment for us to multiply the profit. However, the endowment that we get should be payback to the QM once the buyer gains the profits.

The shariah Advisory council of Malaysia's central bank alleged *tawarruq* as permissible (Bank Negara Malaysia 2014). They see *tawarruq* as a trade contract, and its permissibility is supported from the views of the schools of thought such as Hanafi, Hanbali, and Syafii. In 2006, Accounting and Auditing Organisation for Islamic Financial Institution—AAOIFI's shariah Board cleared that the adaptation of *tawarruq* principle in contemporary Islamic banking is subjected to strict controls and restrictions (Mohamad and Ab Rahman 2014).

Quantum Metal and Get-Rich-Quick Scheme

By registering as a member of Quantum Metal, an individual becomes an investor in Quantum Metal. With the facilities and products provided, investors and members of Quantum Metal can earn more income. By becoming a business partner with Quantum Metal, a member is able to maintain wealth and increase income every month. Quantum Metal allows them to achieve financial independence and career advancement in the end.

However, Quantum Metal is not a company for money-laundering platforms and get-rich-quick schemes in which one must find new people or profitable downlines in order to earn money without working. The Quantum Metal business programme is an affiliate programme, similar to shopee and lazada, which also employ affiliate programmes, including the Mae by Maybank application's referral campaign concept and the Luno application's rewards and promotions concept. This indicates that when a consultant approaches a customer and promotes Quantum Metal, the consultant will be compensated if the customer registers and makes a purchase as a result of the promotion. Moreover, being a consultant for Quantum Metal is a career, meaning that a consultant works for a Quantum Metal company. In the promotion of Quantum Metal products, consultants serve as agents. When that occurs, they will receive a commission or salary proportion determined by the company. In addition, even if a person does not become a consultant or business partner, they can still profit from investing in or storing gold. Through GSA, GCA, and GAE products, one can achieve cost savings, cash conversion, or asset enhancement. Clearly, the profit is not derived from membership fees.

Into the bargain, anyone wishing to become a business partner as a dealer of Quantum Metal is required to pass a test with 80 questions as a qualification requirement. One is ineligible to become a Quantum Metal dealer if they do not pass the examination. In addition, Quantum Metal was founded in 2012, and this business programme has been applied to corporate groups such as banks, with no negative reports or records from Bank Negara Malaysia or the Securities Commission. ISRA Consulting has in fact granted Quantum Metal shariah compliance certification. This demonstrates that Quantum Metal is not a get-rich-quick scheme or a platform for money laundering.

The levels of business partner or consultant in Quantum Metal are as follows: Introducer (I), Dealer (D), Dealer Developer (DD), Merchant (M), Merchant Developer (MD), Senior Merchant (SM), Senior Merchant Developer (SMD). Different commission percentages apply to each consultant tier. For instance, commissions for GAEX10 range from 2% as an introducer to 20% as a senior merchant developer for every customer purchase. The percentage difference demonstrates the consultant's efforts. For instance, to become an introducer, a person must invest GAEX5 up to 3 units or GAEX10 up to 1 unit. The introducer will then receive a 2% commission on all purchases and investments made by themselves or customers who register using the introducer's referral code. The senior merchant developer then earns 20% based on the results of their work, which has helped and recruited a large number of consultants to advance to a higher level using their referral code. However, the high percentage of commission does not indicate that Quantum Metal is a money-laundering organisation or that it contradicts Islamic teachings due to the existence of exaggeration (*ghuluw*) on profits, as the percentage is determined by the company and not by business partners or consultants.

4.4.3 The Ethical Issue

Even though, in general, Quantum Metal has been proven to follow the shariah compliance in gold investment, there are some reseller or agent that work under Quantum Metal that did not fully understand the guideline of the shariah standard and the framework of maqasid syariah in muamalat and investments. As a result, there is misconduct of business practises among them and portraying the Quantum Metal almost the same as the multilevel marketing (JAKIM 2013). In some cases, they are promising a guaranteed profit with high return and no risk. They also use spamming methods towards some targeted customers to convince them to join the Quantum Metal. This is seen as dishonesty and forcing a person to register and join the investment without taking concern of their financial status; thus, financial bubbles are getting controversial in investment and currency markets.

In Imam al-Syatibi's opinion, the ultimate goal of Islamic law is to achieve the intended goals for the implementation of a law. Thus, even though these goals are the same as the ultimate goal of the Islamic law, it is still forbidden in Islam if they do not follow the shariah rulings. The frameworks of the *Maqasid shariah* in muamalat and investments can be divided into many meanings from different scholars' opinions, but generally it is to protect the assets and the peace between Muslims and community and to give benefits to the country in economic growth (Asmadi 2016). Allah says in (Al-Quran. Al-Rum: 41)

> ظَهَرَ ٱلْفَسَادُ فِى ٱلْبَرِّ وَٱلْبَحْرِ بِمَا كَسَبَتْ أَيْدِى ٱلنَّاسِ لِيُذِيقَهُم بَعْضَ ٱلَّذِى عَمِلُوا۟ لَعَلَّهُمْ يَرْجِعُونَ.
> which means: Corruption has appeared throughout the land and sea by [reason of] what the hands of people have earned so He may let them taste part of [the consequence of] what they have done that perhaps they will return [to righteousness].

However, Quantum Metal is very vigilant particularly in ethical issues related to the dignity of the quantum metal. They are mentioning about misconduct in business practice in reseller terms and conditions that whoever disobeyed the conditions either their accounts will be suspended or blocked. In addition, the merchandise or experienced reseller or agent in Quantum Metal is providing platforms such as talks and webinars for people that are willing to know better about gold investments and Quantum Metals. Dr Asmadi Yusof is one of the speakers explaining Quantum Metal in the Angel Group. Every week he and his team will explain about Quantum Metal. Among his most interesting slots are the ones titled Gold and World Economy. They will explain thoroughly the shariah concepts in every product in contrast to forcing people to invest in them by spamming and deception. Some of them are also forming an agency that is legally registered to help and guide them with a proper approach for those who are interested.

4.4.4 Impact of Digital Gold Investment

Looking back at the benefits of the gold investment and digital gold platform, they may give so many impacts towards sustainability of economic growth both individually and community. Investing in gold has been recommended for many years since the 1990s and one of them is Tuan Azizi Ali, the Malaysian Financial Expert said that gold saving will be guaranteed by its value for a long-term wealth, and he advised the people to keep at least 10% from their savings to invest in gold. Individually, gold can have financial freedom and aid the investors to generate their income and saving for the purpose as an emergency fund. Data recorded that there was an increase in bankruptcy cases from 2015 until 2019 with total cases of 27,199 due to personal loan (Juisin 2021). Hence, gold saving may prevent a person from taking loans and reduce their burden. Moreover, gold as a diversification investment will help balance and neutralise other negative effects of other high-risk investments such as stock markets and mutual funds. The proof is when the gold price is increasing despite the high inflation with increase in oil prices globally due to Covid-19 crisis. Therefore, gold investment performs well in volatile markets and helps in growing assets.

Apart from that, digital gold platforms such as KFH, Public Gold-i, and Quantum Metal have been introduced to facilitate the people that have intentions to buy gold by providing a storage vault to keep their money safe and secure. The impact of that will lessen the chance for robbery from keeping the physical gold at home. Other than that, it holds high liquidity from the buying and selling at any time hence, while the value is still preserved, the investors can gain profit and generate income for long-term wealth. Not only that, it opens the chances for the people that seek for a job as part of the management. While it sustains economic growth, it also raises employment opportunities and helps the jobseekers improve their living standards.

Specifically, by introducing the benefits of investing in Quantum Metal, it not only obtained the impact from the gold investment, but it helps more people to grow and expand their money in gold currency. Unlike other banks such as Public Gold and Maybank, Quantum Metal offered another two products which are Gold Currency Account (GCA) and Gold Advanced Enhancement (GAE) that clearly gives double profits from normal gold investment accounts. While the Public Gold is only offer the gold investment account where the investors solely gain profit from the rises of the gold prices globally, the GCA and GAE provide facilities for the investors to maintain their livings while expanding and growing their savings in gold currency from the *tawarruq murabahah* concept. This can help more people to keep them as emergency funds and lessen the dependency on the private loans and debts. With the help of Quantum Metal, many people will start investing in gold and this may be the assets for future gain and economic growth. While the gold investment acts as a hedge against inflation, if many people will invest in gold, it may be helpful for them to sustain their livings during an economic crisis.

4.5 Conclusion

To conclude the findings and discussions, Quantum Metal is an initiative to develop another platform for the people to invest their money in gold more conveniently and efficiently for the sake of economic growth. It acts as a digital gold platform that helps the investors keep their gold in a storage vault and makes it easy for the investors to buy and sell at any time without worrying about the existence of the gold. In order to prevent the prejudice towards the customers, shariah compliance in gold investment has to be followed. Suitably, Quantum Metal has been pronounced by the ISRA Consultancy for its products, GSA, GCA, and GAE that follow the shariah standard. Therefore, they succeeded in gaining people's trust, especially among Muslims. The advantages that they gain from the gold investment as the haven asset and hedge against inflation give big impacts to both individually and the community. While the investors can learn to save their money, they also can expand their assets; hence, the chance of bankruptcy and loans are reduced. At the same time, it helps in economic growth knowingly and there will be an economic crisis in 2023 which is the effect of the Covid-19 crisis. Conclusively, Quantum Metal can be among the best choices for the people to invest their money to gold currency.

References

Awang AH (2016) Elemen Gharar Dalam Akaun Pelaburan Emas Menurut Hadith. Kolej Universiti Islam Antarabangsa Selangor, Bangi
Agha S, Saafi AR, Qayoom OA (2015) Gold investment from Islamic perspective: the case of Malaysia. Int J Econ Fin 7(5)
Al-Quran. Ali Imran: 91, Al-Rum: 41
Bank Negara Malaysia (2014) The Shariah Advisory Council of Bank Negara Malaysia (the SAC) 151st Meeting. Retrived from https://www.bnm.gov.my/-/sac-151st-meeting
Hashim S (2017) Analisis Fundamental Harga Emas. One Publisher, Kuala Lumpur
Hussain L et al (2012) Analisis Syariah Terhadap Produk-Produk Pelaburan Emas di Malaysia. ISRA research paper, no 36
Ibrahim MH (2012) Financial market risk and gold investment in an emerging market: the case of Malaysia. Int J Islam Middle East Financ Manag 5(1):25–34
ISRA (2021) Statement by ISRA consulting on quantum metal products. Retrieved from https://israconsulting.my/statement-by-isra-consulting-ic-on-quantum-metal-qm-gold-products/
JAKIM (2011) Parameter Pelaburan Emas. Keputusan Muzakarah Jawatankuasa Fatwa Majlis Kebangsaan Bagi Hal Ehwal Ugama Islam Malaysia Kali Ke-96, 13–15 Oct 2011
JAKIM (2013) Garis Panduan Perniagaan Jualan Langsung Secara Pemasaran Berbilang Tingkat (Multi-Level Marketing) Menurut Syariah. Keputusan Muzakarah Jawatankuasa Fatwa Majlis Kebangsaan Bagi Hal Ehwal Ugama Islam Malaysia Kali Ke-95, 16–18 June 2011
Juisin HA (2021) Shariah gold investment in Malaysia: prospects and challenges. Labuan Bull. Int. Bus. Fin. UMS Sabah 19(2):95–108
Konsep Shariah dan Produk-Produknya (n.d.) Retrieved from Bank Rakyat https://www.bankrakyat.com.my/c/perniagaan/konsep_shariah_b/konsep_shariah-7
Kuwait Finance House Malaysia (2017) Kuwait Finance House Malaysia annual report. Retrieved from https://www.kfh.com.my/malaysia/reports/malaysia/annual-reports/KFH-Malaysia-Annual-Report-Year-2017/document_en/KFH%20Malaysia%20Annual%20Report%20Year%202017.pdf.pdf
Lucey B, Tully E (2006) Seasonality, risk and return in daily COMEX gold and silver 1980–2002. Appl Fin Econ 16:519–533
Mahmood AM (n.d) Konsep Akad dan Jenisnya Dalam Muamalat Islam. JAKIM. Retrieved from https://e-muamalat.islam.gov.my/en/bahan-ilmiah/artikel/156-konsep-akad-dan-jenisnya-dalam-muamalat-islam
Mas'ad MA, Ezzuddeen M, Fattah WA (2018) Gold investment practices in malaysia: a Shariah review. J Fatwa Manage Res
Mohamed Naim A (2016) Memahami Kerangka Maqasid Al-Syariah Dalam Kewangan dan Pelaburan. Pasaran Modal Islam Malaysia 2016. Institute Kefahaman Islam Malaysia
Ng L (2009) Public Gold: Kelebihan Pelaburan Emas Fizikal. Prosiding Seminar Antarabangsa Penjanaan Ekonomi Melalui Transaksi Wang Dinar. Bangi
Nordin N, Aziz MI, Embong R, Daud N, Aziz SA (2018) Shariah compliant gold investment: an understanding among academicians in Terengganu, Malaysia. Int J Acad Res Bus Soc Sci 8(10):226–235
Pinga VEB, Nelson GC (2001) Money, prices and causality: monetarist versus structuralist explanations using pooled country evidence. Appl Econ 33(10):1271–1281
Quantum Metal (n.d) Pelaburan Emas Patuh Syariah Dari Quantum Metal. Retrieved from https://shariahgold.net/.Quantum, Metal Malaysia. Quantum Metal Sd. Bhd. Retrieved from https://www.quantummetal.com.my/
Rahman ZHA (2006) Kos Pengurusan dalam Implementasi 'Qard Hasan'. Metro Ahad, 10 Feb
Sahalani QM (2022) Cara Jadi Consultant Berjaya di Quantum Metal [video]. Youtube. https://youtu.be/kidCV887eC0

Tuan Haji Zaharuddin M (n.d) Perniagaan multi lavel marketing (MLM) dari Perspektif Fiqh Islami. JAKIM. Retrieved from https://e-muamalat.islam.gov.my/ms/bahan-ilmiah/kertas-ilmiah/414-perniagaan-multi-lavel-marketing-mlm-dari-perspektif-fiqh-islami-2

World Gold Council (2018) The world gold council 2017 annual review. World Gold Council Publisher, London

Yusof A (2022) Konsep Tawarruq dalam Quantum Metal [Video]. Youtube. https://youtu.be/tQPpd13ciFE

Zakaria MZ, Salleh AZ, Hasbullah M, Hamad H, Safyan YH (2017) Al-Madkhal Ila Fiqh Al-Muamalat, USIM PRESS, Negeri Sembilan, Malaysia

5 Overview of Halal-Integrated Platform (HIP) Adaption as a Halal Digital Economy Hub for Small–Medium Enterprises (SMEs) in Malaysia

Zahariah Sahudin, Mohamad Idham Md Razak, Lennora Putit, and Mohammed Hariri Bakri

Abstract

The world halal industry and the domestic market are expected to reach up to RM20.92 trillion and RM473.57 billion, respectively, in 2030. The local halal industry has contributed as much as 7% toward the Gross Domestic Product (GDP) of the country in 2020. It indicates that Malaysia has the potential to be one of the world's halal market hubs that have contributed to local economic growth. As stated in the 12th Malaysia Plan (MP12), the major focus of local economic synergy is to strengthen the local halal industry by creating more opportunities and positive competition among key industrial players to ensure the sustainability of economic growth. The Halal Development Corporation (HDC) has launched the Halal-Integrated Platform (HIP) to introduce Malaysia as a halal digital economy hub. This initiative allows the country to maximize the benefits of multiple aspects of the Fourth Industrial Revolution (IR 4.0). This study will put attention to the overview of HIP adaption as a halal digital economy hub for Small–Medium Enterprises (SMEs) specifically in Malaysia. The existence of HIP enables local halal industry players to connect with global halal players effectively and sustainably as well as helps companies to grow their business capabilities, backed by halal certification.

Keywords

Halal-integrated platform • Small–medium enterprise • Industrial revolution

Z. Sahudin (✉) · M. I. M. Razak · L. Putit
Universiti Teknologi MARA, Shah Alam, Malaysia
e-mail: zahariah128@uitm.edu.my

M. H. Bakri
Universiti Teknikal Malaysia, Melaka, Malaysia

A. Rafiki (ed.), *Digitalization in Halal Management*, Contributions to Environmental Sciences & Innovative Business Technology,
https://doi.org/10.1007/978-981-99-5146-8_5

5.1 Introduction

5.1.1 Background of the Research

Small and Medium Enterprise (SMEs) are the backbone of the Malaysian economy, accounting for about 97% of total businesses in Malaysia. SMEs also have been making waves and build their reputations in recent years as a result of technology improvement and the digital revolution. This is due to the fact that the efficacy of logistics and transportation of goods is a crucial aspect in the rising trend of SMEs in Malaysia. It is proven when FedEx Express, one of the world's top logistics and transportation companies, agreed to form a joint venture with the Malaysian Small and Medium Enterprise Association. With this agreement, Malaysian SMEs will be able to conveniently expand and liaise directly with the clients across Asia Pacific region for example, China, Japan, Taiwan, Korea, and Indonesia (SME Malaysia 2022).

Research done by SME Malaysia (2022) confirms that more than 50% of the respondents are looking forward to expand their current businesses and the remaining balance is interested in sustaining their current businesses due to the pandemic Covid-19. From the result, it can be concluded that business owners nowadays realize that it is a need and necessity to learn the digital business technology platforms because they must struggle with competitors. This is because with the help of this technological advancement, business owners will be empowered with precious information.

Not only that, they also will get to know their customer behavior and preferences toward their product. In addition, the transformation into a digital business technology platform will help SMEs to unify the needs across the enterprise from the stakeholders to the customers, partners, and employees. However, it is not an easy task to have a completely digital economy in the country. Furthermore, there are a lot of factors affecting digital business technology platforms and the digital economy. Among the factors are lack of new technology enhancement and innovation, new regulations by the authority body, international threats as well as new customer requirements (Loh et al. 2021).

Other than that, it has become necessary for all business owners to utilize digital platforms to compete with each other. The facilities offered by digital technology will improve the company's business performance and growth toward digital transformation. Those companies which unable to adopt this new digital transformation will be left behind and incompetent to compete with their competitors in today's market. For instance, digitalization and transformation are a unification of all functionalities in a company, including information systems, analytical tools, and customer engagement. As a result, it is easier for the company to know their customer and competitor better and planned well for its future performance.

Despite the advantage that the latest technological advancement brought, SME owners will be more likely to avoid this technological advancement because of the complexity and lack of familiarity with the tools and not to mention it is expensive (Daniela et al. 2020). As mentioned before, digital technologies are the pioneer in a company's digital transformation and if the companies are lacking in digitalization, they cannot close the gap between them and big companies (Eller et al. 2020). Consequently, all the processes of marketing, operations, logistics, and information systems will heavily be affected due to the fierce market competition. Besides, the involvement of Muslim scholars in the halal trade, in which they have set the halal standardization based on the Quran and Hadith, will also be tackled by one of the digital business platforms (Florence et al. 2016). Overall, this paper will emphasize on *halal* digital economy for Small–Medium Enterprises in Malaysia.

5.2 Literature Review

5.2.1 Overview of the Halal Industry in Malaysia

In general, 60 percent of Malaysian are Muslims, and they are required to eat and consume *halal* food in accordance with the guideline set by the *Quran* and *Hadith.* Both sources mentioned by our Prophet Muhammad (PBUH) clearly stated that Muslims are only allowed to eat *halal* food, not *haram.* In conjunction with it, creating a global *halal* market to satisfy the needs of Muslim users and *halal* consumers is a compulsory task as *Khalifah* on the earth. The development of the *halal* market is growing tremendously in Malaysia as well as in other Muslim countries around the world. There are a lot of initiatives that have been done to make it successful such as the *halal* logo being familiarized to ensure that Muslims can easily find *halal* foods to be consumed, establishing *halal* restaurants and hotels, and many more. In addition, Malaysia also has been verified as an acceptable *halal* logo based on the preparation and the ingredients of such foods (Sahar et al. 2019).

In concurrence with the growing popularity of Muslims around the world, the request or demand for *halal* products and foods is getting greater and the competition is getting tougher. Consistent with a research accomplished by Sahar et al. (2019), the Muslim population globally by 2030 is expected to spread up to 2.2 billion, and by 2050 it will be 2.6 billion. This volume will represent 30% of the world's total population. Hence, it is a sign that *halal* industrial players in Asia, the Middle Eastern, and North Africa have got to empower themselves with all the digital advancements to furnish the growing Muslim market. The success of becoming the global *halal* hub makes Malaysia to be the leading *halal* food benchmark in the world. Additionally, Malaysia has been recognized as the best example of benchmarking of *halal* food in the world as the *halal* standard is being practiced across Malaysia to evade confusion and it is being used globally by *halal* food industries. It is in accordance with the Codex Alimentarius Commission which embraced the general guideline for the use of the term *halal* in Geneva in 1997.

5.2.2 The Growth of Small-Medium Enterprises (SMEs) and Digitalization in Malaysia

As mentioned earlier in the introduction part, SMEs in Malaysia are one of the highest contributors to the Malaysian economy, which is about 38.4% of the country's gross domestic product (Adan and Hussain 2021). The involvement of SMEs in services, transportation, and manufacturing sectors, greatly helps them in pushing such numbers into Malaysia country's gross domestic product (Mohamad et al. 2021). In addition, credit is also given to the Malaysian government which implemented a model that goes by the name of the New Economic Model which focuses on SMEs in Malaysia. For all its pomp and glory, SMEs in Malaysia are facing countless issues and challenges throughout their first journey since 2001 based on a recent study by Salleh et al. (2020). It is clearly stated that the main problem for SMEs is unable to remain competitive when the wave of the market changes. In addition, it is difficult for SMEs to sustain in tough times such as the pandemic which resulted in closing the business for good.

5.3 Methodology

There are various approaches and methods that have been employed by the previous researcher in their studies regarding digital platform. Some of them use case studies and some use analytical modeling in their research. Both methods are being used and relatively frequent. However, there are only a few studies that employed the grounded theory, simulations, as well as action design research. There are only few studies which have been done by using this method (Asadullah et al. 2018). Webster and Watson's (2002) recommended that literature review in the Information System (IS) discipline must be done. Therefore, this paper adheres to their recommendations to accomplish the goals of digital platform.

5.4 Analysis of Results/Discussion

Digital platform offers a combination of business and technologies whereby the business can be operated in an innovative and transformative ways. The business owner now can reach the customer globally as well as can deliver the product efficiently with the implementation of technology applications such as digital platforms, digital marketing, and financial technology. Those platforms play a vital role in bringing the businesses and customers together. It also helps and facilitates the buying and selling of goods and services processes.

The digital platforms are being used widely and it became important lately because it has implications toward the socio-economics of the country. It makes the Information System (IS) which becomes relevant to our daily life nowadays. This provides an overview of the existing body of knowledge and serves as a foundation for future research on digital platforms, supporting the demand of IS (de Reuver et al. 2017; Asadullah et al. 2018). The findings demonstrated that during the past four years, interest in digital platform research has grown tremendously and significantly. The topic is being discussed and debatable until now. It also being explored as the number of publications has been rising gradually, and this trend is anticipated to remain in the future.

The goal of this paper is to look at the literature on management and IS for digital platforms, particularly *halal* digital platforms. Nonetheless, future advancements in the sector will encounter several difficulties. This is because lack of agreement on a precise definition and conceptualization of Halal-Integrated Platform (HIP) in the IS literature represents one of these difficulties. Despite that, this study also has some constraints. The economics literature has addressed the subject of digital platforms in detail, but the previous research did not address it in the way this paper is presented. This paper also has referred to a significant findings and theoretical advancements from the prior findings.

5.4.1 Halal-Integrated Platform

The relationship and synergy between all sectors and *halal* industry players in Malaysia are imperative. This is to ensure that all the products and services produced by all sectors, especially SMEs, will meet the requirements of the potential customers. Hence, it will lead to better socio-economics as well as the growth of the country. Other than that, the introduction of the Halal-Integrated Platform (HIP) by the Malaysian Halal Development Corporation is another technique for upgrading the *halal* market ecosystem. It is hoped that with this digital platform, businesses can reach out to potential customers globally. Moreover, business owners could also utilize this digital platform to promote their products to increase their business transactions. Besides, it also facilitates the entrepreneurs of the small and medium industries to get support in terms of copyright or intellectual property protection, financial technologies as well as market access (MDEC 2021). Consequently, this digital platform enables SMEs in Malaysia to know their potential customer's preferences, and therefore, they will deliver better halal products and services that can be offered locally and globally (The Halal Time 2022).

On the other hand, the Malaysian government also has played a very important role to establish Malaysia Digital Hub. The purpose of this establishment is to facilitate the entrepreneurs of the small and medium industries to get support and facilities such as fintech and digital technologies. Subsequently, SMEs need to enter into the digital economy, especially into this Halal-Integrated Platform (HIP), because it will connect the local business companies with ASEAN through the global digital ecosystem as well as brings a better opportunity for the SMEs. The government also providing grants to all SMEs, to encourage them in R&D innovation and quality improvement, new product development and skills training.

Moreover, the SMEs will obtain a verified *halal* certification after the government authorities approved their application through the Malaysian Halal

Development Corporation platform. But they must be familiar with all the *halal* guidelines, obtaining suppliers, and finding a local marketplace. Other than that, the SMEs must establish the connection between SMEs in Malaysia and Muslim countries. Additionally, the launching of the Halal-Integrated Platform is a very good move by the Malaysian Halal Development Corporation as it provides opportunities for SMEs as well as helps them to accommodate the halal market in a very efficient manner (The Halal Time 2022).

Furthermore, the purpose of the Halal-Integrated Platform is to promote e-commerce services that have been made available for every single-registered SME. Likewise, it will also empower and facilitate SMEs by obtaining their *halal* certification, and eventually, they will become part of the leading troupers in the *halal* food industry. Consequently, with all the advantages provided by the Halal-Integrated Platform (HIP), it could help SMEs to sustain their business as well as encourage them to become the big player that will outfit the *halal* market for all Muslims around the globe. Effendi (2023) in his study stated the need to enhance human resources readiness in SMEs to support the halal industry, emphasizing the role of government, academia, and business actors in optimizing the halal product guarantee system. On the other hand, Machfud et al. (2011) proposes the use of Halal-Integrated Platform (HIP) to promote and improve the condition of local Halal SMEs, making their products more accessible to customers and contributing to national development. These findings collectively suggest that the Halal Integrated Platform (HIP) can provide advantages for SMEs to sustain their business and thrive in the global halal market.

5.5 Conclusions and Implications

The *halal* product and services as well as sector takes the lead in the global businesses nowadays. More and more people are aware of the *halal* product and services. The volume of businesses and international trade transactions is expanding along with the growth of Muslim populations around the world. This is because of the transparency and the utmost importantly is it comply with the Islamic teaching. Muslims are required to adhere to the guidelines set out by Islam to carry out the businesses. *Allah SWT* says in *Surah Al Baqarah*, verse 172: "*O believers! Eat from the good things We have provided for you. And give thanks to Allah if you truly worship Him alone*".

This verse clearly indicates that *Allah* prohibit human to eat and consume other food except the *halal* product. Hence, the *halal* industry has a promising future in the social and economic spheres. These *halal* products can be traded across the world and the businesses is becoming interconnected globally. The consumers can make the transactions efficiently and effectively by using the new technology.

Consequently, by introducing and promoting the HIP, it will build a community consists of industry players, training providers, researchers, post and undergraduate students, federal and state agencies, and others authority bodies which involve in the development of the halal industry to strengthen its leadership in the global *halal* industry.

This analysis demonstrates that despite the rise in number of publications on digital platforms, there is still considerable ambiguity in the definition of HIP within the IS and management literature, particularly in term of the emphasis on the technological core versus the broader ecosystem. Additionally, it is discovered that case study methods predominate in the studies of digital platforms. This paper explores the main HIP research strands and classify some of the variables that are examined in each of them. Finally, this paper analyzes prior categorization schemes and highlights the major characteristics that distinguish different kinds of *halal* digital platforms.

5.6 Future Study

For future study, the Unified Theory of Acceptance and Use of Technology (UTAUT) is highly recommended. This model is usually applied to find the degree of acceptance among individuals who are introduced to the latest technology. Moreover, according to Alaa (2020), the UTAUT model focuses on five independent variables which are in the form of expectancy such as performance expectancy (PE). It can be defined as the individual who believes that systems and technological advancement can simplify the existing work. Other than performance expectancy (PE), the next one is effort expectancy (EE). It is when the individual expects the ease level regarding the system usage. Next in terms of behavioral intention (BI). Behavioral intention is the intention of the customer to use self-service technology. The fourth independent variable is social influence (SI). It is when individuals perceive others toward their behavior in believing that they need to have a new system. Finally, facilitating conditions (FC). Facilitating conditions is an individual beliefs that there are technical organizations that will assist during the usage of the system (Venkatesh and Morris 2003).

In a conclusion, UTAUT is one of the most powerful tools which can be used to determine the degree of acceptance within an individual. Not only that, but this theory is also able to examine individuals' intentions and acceptance of new technology. Thus, the objective of this paper is achieved by answering the question of how the UTAUT model can become one of the most vigorous models compared to other technology acceptance theories and models. And the second question is how UTAUT's determinants can give impact technology adoption and their relation to the software quality requirements.

Acknowledgement The author no. 2, Dr. Mohamad Idham Md Razak is currently a Postdoctoral Fellow funded by the Ministry of Higher Education Malaysia and Universiti Teknologi MARA. He is currently attached to the Faculty of Technology and Technopreneurship Universiti Teknikal Malaysia Melaka.

References

Adan M, Hussain SI (2021) The relationship between economic growth, small and medium enterprises and the number of employees in Malaysia using vine copula approach. Sains Malaysiana 50(8):2455–2468. https://doi.org/10.17576/jsm-2021-5008-26

Asadullah A, Faik I, Kankanhalli A (2018) Digital platforms: a review and future directions

Alaa M (2020) The unified theory of acceptance and use of technology: a new approach in technology acceptance. Int J Sociotechnol Knowl Dev 12:79–98. https://doi.org/10.4018/IJSKD.2020070105

de Reuver M, Sørensen C, Basole RC (2017) The digital platform: a research agenda. J Inf Technol: 1–12

Daniela I, Adewale OS, Lucian-Florin O (2020) Using a digital platform to support market research for SMEs. https://doi.org/10.2478/9788395815072-005

Effendi J (2023) Human resources readiness of small medium enterprises (SMEs) for Halal Product in Indonesia. Global review of islamic economics and business, 10(2):83–95

Eller R, Alford P, Kallmünzer A, Peters M (2020) Antecedents, consequences, and challenges of small and medium-sized enterprise digitalization. J Bus Res 112:119–127. https://doi.org/10.1016/j.jbusres.2020.03.004

Florence B-B, Johan F, John L (2016) Halal matters, Islam, politics, and markets in global. Perspective. https://doi.org/10.4324/9781315746128

Loh YX, Ahmad NAA, Seah CS, Yo JJ, Law YH, Tan SY, Chung HL, Liew YL, Chong C (2021) The factors and challenges affecting digital economy in Malaysia. Conf Manage Bus Innov Educ Soc Sci 1(1)

Machfud AK, Khatib J, Haji-Ahmed AA, Ahmad Dahlan AR (2011) Halal product price indicator portal: promoting malaysia's local halal SMEs. Software engineering and computer systems. Springer berlin heidelberg 254–262

MDEC (2021) A hub for start-ups and internet. Retrieved from https://mdec.my/what-we-offer/digital-hubs/

Mohamad A, Mustapa AN, Razak HA (2021) An overview of Malaysian small and medium enterprises: contributions, issues, and challenges. In: Sergi BS, Jaaffar AR (eds) Modeling economic growth in contemporary Malaysia (entrepreneurship and global economic growth). Emerald Publishing Limited, Bingley, pp 31–42. https://doi.org/10.1108/978-1-80043-806-420211004

Salleh F, Yatin M, Radzi M, Kamis S, Zakaria S (2020) Malaysian's new digital initiative to boost e-commerce–where we are. Int J Acad Res Bus Soc Sci 10(11):1138–1154

Sahar S, Fauzi M, Ab Rahman H, Hassin W (2019) Halal economic: challenges and emerging opportunities in Malaysia. https://doi.org/10.5281/Zenodo.5196917

SME Malaysia (2022) https://smemalaysia.org/smes-in-malaysia-looking-to-tap-growth-in-asia-pacific-survey-reveals/

The Halal Time (2022) https://www.halaltimes.com/malaysia-launches-platform-to-connect-halal-smes-with-the-world/

Venkatesh V, Morris MG, Davis GB, Davis FD (2003) User acceptance of information technology: toward a unified view. JSTOR 67(3):95–98.https://doi.org/10.1016/j.inoche.2016.03.015

Webster J, Watson RT (2002) Analyzing the past to prepare for the future: writing a literature review. MIS Q 26:13–23

6 Sustainable Development Goals Concept Overview in the Digital Service Halal Certification in Indonesia

Zulfia Hanum Alfi Syahr, Dian Hafit Syaifullah, Buyamin, and Novian Uticha Sally

Abstract

Business actors receiving a halal certification can contribute directly to realizing Sustainable Development Goals (SDGs) dimensions: economic, social, and environmental. The halal certification application has been developed using digital technology to facilitate access and speed up the service process. The halal policy for food products indicates the sustainability of beverage, pharmaceutical, cosmetic, and service providers. However, the problem arises of how products that have passed the halal certificate can be reviewed for sustainability in terms of economic, social, and environmental aspects in their development in the era of digital product services. The goal is that consumers and business actors increasingly believe in the urgency of halal certification to support the sustainability of products and services produced from an economic, social, and environmental perspective. Thus, it will impact business actors who are increasingly motivated to register their products and services to have halal certificates, especially with the application of digital technology to make it easier for users. Therefore, in this study, we will examine the concept of Sustainable Development Goals (SDGs) on products that have passed the halal certification. This concept will review how to analyze the social, economic, and environmental impacts on the life cycle of halal products. The method used is qualitative, with a literature review in the form of a narrative review of the latest globally indexed scientific articles. They were then analyzed with a classification approach based on objectives, research questions, results evaluation, and knowledge contribution. The result is that the concept of SDGs in halal certification is critical to

Z. H. A. Syahr (✉) · D. H. Syaifullah · Buyamin
Research Center for Public Policy, National Research and Innovation Agency, Jakarta, Indonesia
e-mail: zulf008@brin.go.id

N. U. Sally
Faculty of Da'wah and Islamic Communication, Kudus State Islamic Institute, Kudus, Indonesia

A. Rafiki (ed.), *Digitalization in Halal Management*, Contributions to Environmental Sciences & Innovative Business Technology,
https://doi.org/10.1007/978-981-99-5146-8_6

study with the conceptual framework of its analysis. This paper will explore more deeply because it aligns with the global SDGs program.

Keywords

Digital era • Environment • Halal certification • Social-economic • Sustainable

6.1 Introduction

Halal certificates and symbols guarantee what they consume or use according to Islamic law and encourage manufacturers to meet halal standards. The halal label means that products are under Islamic shari'a, safe, and valuable for use by the public. Halal product services cannot be separated from the shariah label, where all forms of services related to halal products must be adjusted to shari'a values. These sharia values are reflected in the halal products. Therefore, to make it easier for the public to recognize whether a product is included in the halal product category, the government issued a policy to provide a halal logo on every product with a halal certificate.

The Indonesian Ulema Council has also issued several criteria related to the requirements of halal products, management, product materials, and facilities (Jumiono et al. 2020). In addition, a product is said to meet the standards as a halal product; of course, it must go through 4 (four) stages. The first must meet the basic requirements of a halal product (Wibawa et al. 2021). Second, it must meet the criteria as a halal product, namely a qualitative or quantitative benchmark that describes the provisions that must be heeded or carried out to make a decision related to halalness. Third, it must go through the halal testing method, which measures characteristics or auditing, the results of which are compared with halal standards to be used as a conformity assessment. Fourth, in conducting its production business, it refers to halal production guidelines.

These four stages are stages in the halal certification process. They are currently an essential attribute for producers in providing information to consumers that their products are not classified as prohibited according to Islam. In addition, looking at most of Indonesia's Muslim population can be seen as an economic opportunity. The government utilizes this opportunity to encourage business actors to be able to apply halal principles to each of their products. The government shows this effort by issuing regulations about the halal certification policy. This halal certification is a tangible form to practice public awareness in buying and consuming halal products as an essential thing for a Muslim (Kurniaputri 2020).

The research (Rafiki and Abdul Wahab 2016) explained that there are several challenges faced by companies when trying to apply for halal certification, including the following: (1) institutional policies; (2) changes in costs and organization; (3) information where the owner or manager needs to update information related to halal standardization, exchange, and control. Initially, the discussion of halal issues was only limited to the law in Islam to separate between things that are allowed and things that are prohibited in Islam. However, the development of technology and information impacted halal products that transformed into a *lifestyle* followed by people who are not only Muslims. Halal under the development and demands of needs is not just a statement but needs to be proven by the existence of halal labeling through certification. It requires proactive action with employees to manage the flow of information regarding the material, both forward and backward.

An authorized institution will issue halal certification after going through a long series of processes to ensure a product meets halal elements in terms of materials, processes, processing, storage, and other processes related to the product. The institution has created and set several standards that must be met and completed in product submissions to obtain halal labels. Halal certification in Indonesia is carried out by the Indonesian Ulema Council and the Ministry of Religious Affairs, with standards based on Islamic law.

6.2 Method

The method used is descriptive qualitative, with a literature review in the form of a narrative review of the latest globally indexed scientific articles. Then perform an analysis with a classification approach based on objectives, research questions, evaluation of results, and the impact on knowledge.

6.3 Discussion

6.3.1 Halal Certification in SDGs Corridor in Indonesia and Globally

Halal consumption is a must for all Muslims. The guidance for halal consumption is mentioned in the Holy Al-Qur'an, which becomes Muslim's first and supreme law source. The command from Allah to consume only halal products is stated in the QS. Al Baqarah: 168 and 172. Other than that, the guidance to consume halal is also stated in other verses such as Al Maidah verse 88, Al Anfaal verse 169, and An-Nahl verse 114. The repetition of this guide shows the importance of it being done. It shows that halal certification products must adhere to Islamic guidance. From an Islamic perspective, many requirements must be met to achieve halal certification. Those requirements must be met in obtaining raw materials, developing the materials, cooking processes, distribution, and marketing up to the products that come to the hand of the consumer. Some of the requirements for halal consumption are free from pork, lard, or any substance from pig or dog; an animal that is slaughtered without Islamic requirements contains blood or alcohol.

The fact that halal products must undergo rigorous controls and checks through all processes of the supply chain makes halal products more appealing and trustworthy. There is a trend that even halal products are preferable by the non-Muslim community. The previous study by (Chong et al. 2022) shows that customers are now more educated to choose a high-quality and safer product. It shows how halal products may have a more significant market in the future.

Aside from halal, Muslim product consumption also has to be *tayyib. Tayyib* means good or, in other words, technically giving reassurance of cleanliness, safety, and quality of the products. It usually comes together with the words halal, the term halal. The term halal relates to Sustainable Development Goals (SDGs). Both terms aim to benefit humans and the environment in the long run. The fact that halal products have a set of ethics in the production process of its products ensures that it is in line with the SDGs.

In Islam, Muslims are seen as the leader (Khalifah) of this world. The consequence is that Muslims must take care not only of themselves but also others and the environment. Muslims and others worldwide also realize the importance of caring for nature and wise consumption. It implies making MDGs in 2000, focusing on the eight developmental millennium goals. Unfortunately, there were some critics of the MDGs. Some critics highlight limitations to its framework, including constraints in its development, formulation, and content (Larionova 2020). MDG 8 failed to re-energize the spirit of a democratic and equitable international order, a credible multilateral trading system, or a solid and stable international financial architecture. So, the key target for developing a predictable, rule-based, open, and non-discriminatory trading and economic system still needs to achieve. Albeit failed to meet the goal perfectly; MDG has already paved the way for a better program, namely SDGs.

SDGs consist of 17 points of sustainable goals and 169 targets to be achieved up to 2030. Halal industry can contribute directly to realizing SDGs in economics, society, and the environment. In the halal management process carried out by the Ministry of IBRA in the publication of the Masterplan of Indonesian Islamic Economics 2019–2024, through the halal concept by design based on the Islamic Shari'a regarding the production of goods and services processing, namely halal and *tayyib*, can become a new paradigm in the world's industrial sector. International halal industry opportunities from the international community recorded in a report of The Global Islamic Economy year 2018 that the world's Muslim community consumed about US$2.1 trillion in 2017 (Spear and El-Rahim 2020). Indonesia contributed 10% of halal global economic revenue from that amount in 2017. This amount is even more significant than Indonesia's total export, which only constitutes 3.8%.

The publication Masterplan of Indonesian Islamic Economics, 2019–2024, recorded four things in order to build a digital economy strategy lawful (Yakin and Christians 2021); (1) Halal marketplace and system sharia financing; (2) build the incubator facilities that can facilitate the growth of start-up companies can strengthen the halal value chain national and have global coverage; (3) collect the information system for traceability of halal products. In the field of the social, the halal industry can help to boost the integration of different systems in the supply chain (Sumarliah et al. 2021). Society will also receive training in the halal industry, which will build their skill. While in the environmental element, the halal industry has long had a good effect on nature. The emergence of the halal industry also opens many opportunities for the halal tourism industry. The concept of halal tourism is to protect nature while providing a halal alternative for tourists (Chebab et al. 2021). In Indonesia, ten locations are labeled as the halal tourism industry.

6.3.2 Indonesian and Global People's Perceptions and Preferences Toward Halal Product

The concept of Halal guides the life of a Muslim so that his daily activities follow their standard of living, including consumer goods and services and the purchasing process. Malaysian fast food restaurant manufacturers, a Muslim-majority country, are well aware of this importance. They do various consistent advertising content to form a Halal image of their products but with different nuances among the various brands (Yarar 2020). The same response also occurred to Malaysian retailers of consumer goods. For retailers, halal certification reflects the safety and quality of the products. Considering Malaysia's Muslim-majority population, halal certification is a consumer goods retailer's most crucial guarantee (Ibrahim et al. 2019). In Turkish chefs/cooks, another Muslim-majority country, the awareness of choosing halal foodstuffs is the source of the ingredients for the dish varies. The level of chef education influences the high awareness of choosing halal ingredients for cooks/chefs. The higher the chef's education will be, the higher the awareness of selecting halal foodstuffs and the lower the level of chef or cooking education, resulting in lower awareness of selecting halal foodstuffs (Seçim 2022).

Kuwaiti consumers' preference for halal products relies only a little on the halal certification attached to those products. They also consider the origin of the country issuing the product's halal certification, whether it is an Islamic state or an Islamic-majority country. Products certified halal by the Islamic Gulf/Middle East countries such as Kuwait, or Saudi Arabia has a higher level of trust than those certified halal by Muslim-majority countries in Asia such as Malaysia and Indonesia. The preference for trust in the halal certification of a Kuwaiti consumer product is in order contributed by: (1) trust value from the country of origin of the issuer of halal certification, (2) relation between consumers and halal country, and (3) the brand from a halal country as the original product (Rios et al. 2014).

In addition to the materials that make up halal products, halal transportation services are another component in producing halal products. A halal product and service involve complicated regulations and execution that require a significant investment. Transportation costs for a halal product will be transferred to the end user, consequently increasing the price of the final product. The attitude and control of behavior perceived by Malaysian halal consumers positively correlate with the willingness to pay it. Meanwhile, subjective norms and religiosity are inversely proportional because both do not significantly affect the willingness to pay more for Halal transportation. It is shown that knowledge can influence attitudes and willingness to pay for halal transportation (Hafaz et al. 2019).

6.3.3 Halal Preferences in Muslim Minority Countries

In secular Muslim minority countries, consumers' preferences for a slaughtered animal product are spread on several attributes, including halal label, price, method of slaughter, and country of origin. Austrian consumers consider the halal label the essential attribute of halal meat. This condition clearly shows the importance of labeling halal products correctly and trustworthily (Meixner et al. 2018). The preferences of UK halal consumers have different preferences toward certified Halal meat. For halal-certified meat, most UK consumers (70%) prioritize how animals are slaughtered. Namely, choosing animals that are slaughtered without prior pre-stunning/slaughtering so that they consume poultry meat more often than lamb and beef (Fuseini and Knowles 2020).

For example, the demand for halal products in China is quite incredible. However, counterfeiting of halal products often occurs, and the product's originality is doubtful due to the unknown origin of the product. As a Muslim minority country, China has had various local halal certifications in various provinces and particular autonomy regions, along with halal Industrial Estates aimed at meeting the domestic market and 20% for export purposes to Malaysia, the United Arab Emirates, and Saudi Arabia (Dubé et al. 2016). Even though personal care products are halal, their availability in China is quite limited. Muslim consumers who adhere to their beliefs and have experience buying halal products are willing to pay more. Chinese Muslims prefer low-priced halal products to high-priced ones and imported products to locally made ones (Hong et al. 2020). The country of origin of halal cosmetic products is one of the essential selection criteria for (consumer confidence in products) Chinese halal consumption (Hong and Kamaruddin 2020).

In India's Muslim minority community, religious commitment and halal certification influence the purchasing behavior of consumers of halal cosmetics, which drives their repurchase intentions. There was no effect of religious value on purchasing behavior and repurchase intent on halal cosmetics. In addition, the actual purchase behavior positively affected customers' repurchase intentions (Shahid et al. 2022).

6.3.4 Halal Preference in Indonesia

For Indonesia, halal certification and labeling requirements are necessary because, in addition to its large Muslim population, it is also one of the standards recently set by export destination countries (Krisharyanto et al. 2019). The formation of Indonesia's halal value is a contestation between the state and society over the sharia of goods in Indonesia. The contestation concerns not only lifestyle issues but also commodification. Shari'a that started as cultural shari'a has more continuation and legitimation than 'structural' shari'a imposed by the state (Hasyim 2022). The consequence of this is that the halal standard of an item is no longer centered on food snacking fiqh but also on non-religious problems, including lack of meat production and protection in the poultry sector, pre-slaughter and severance approval authority, prohibition of mechanical slaughter. In this regard, the standard halal criteria have a religious dimension and as a tool for economic development and facilitation of international trade (Suryawan et al. 2022). The aims are to give protection by guaranteeing food safety issues and protecting the national economy from the adverse consequences of global markets (Suryawan et al. 2022).

6.3.5 Consumer Preferences

Indonesian Muslim consumers view perceived value as a predictor of consumer attitudes and intentions. They are essential in consuming halal food. In this case, it is necessary to emphasize the "value" of food products with a halal logo that reflects the product's quality, safety, and comfort (Asnawi et al. 2020). The halal label has a limited influence on Indonesian consumers' perception of a product compared to the product's country of origin. Indonesian consumers with high faith-based purchasing behavior do not trust food produced in countries other than Islamic customs, even though it looks like it is labeled halal (Maison et al. 2018). Most Indonesians consider the food produced in Indonesia is halal because of the positive perception that all domestic products produced by Indonesians who are Muslims are automatically halal, even though the food and drinks in circulation are not all labeled halal (Nirwandar 2018). These facts from the consumer side cause the slow awareness of Indonesian halal product certification by Indonesian producers. Information technology assistance is needed to strengthen the halal image of a certified product so that the product can be traced halal by online search (Sayogo 2018).

Halal certification is proven to improve restaurant business performance due to significant results (Yama et al. 2019). However, awareness of halal restaurants has recently emerged not only due to the obligation of Muslims to obey Islamic rules but also influenced by attitudes, norms, and perceived behavioral control (Aisyah et al. 2019). Muslim and non-Muslim Indonesian consumers also have a good perception of food with a halal logo (Farhan and Sutikno 2022). They are more concerned with halal logos and product quality than other variables, namely halal awareness, food safety, and halal perception (Arifin et al. 2021).

6.3.6 Halal Product Certification Digitalization Service Standards

So far, halal certificates issued by institutions in a country are generally only valid in the country where the halal certificate is issued. It is because each country has different halal certification bodies. It is undoubtedly a problem with the trade-in of halal products between countries worldwide. For many years, the Muslim community has demanded and yearned for a comprehensive and consistent halal certification that applies internationally. Responding to these demands through the Diplomatic Council (DC), the Global Halal Forum has established the Halal Certification Institute (HCI), which is authorized to issue halal certification of products submitted as halal products. It is supervised by the ambassadors of Muslim countries worldwide to ensure that the certification process is consistent.

This halal certificate issued by HCI is one of the most rigorous halal certificates in the world. The food manufacturers must be proven to comply with more than 100 criteria for applying for halal certification to obtain a halal certificate from HCI. HCI will send a team of world halal experts to the product's factory submitted for halal to check its compliance with the established halal rules. This halal certificate from HCI also addresses various problems worldwide, including the horrendous meat scandal that has made the Muslim community question the purity of the meat. Through halal certificates that are valid internationally, of course, the world community, both Muslims and non-Muslims, can judge the goodness of halal products that the Muslim community has consumed. To the point that this halal product is stated as one of the purest food products ever. Then the Diplomatic Council (DC) is an institution consisting of experts with the primary objective of contributing the power of diplomacy to realize international understanding and peace. DC members are from the diplomatic and consular corps and are figures in the economic, cultural, sports, political, and religious fields. Besides Indonesia, many other countries form institutions or organizations to run halal labeling. Halal labeling institutions already have predetermined provisions and standards based on an understanding of Islamic law that can vary.

Malaysia takes a chance to be one of the countries in the world with a government-supported halal food certification plan and has aspirations to become a global hub. Albeit the Muslim population in this country is relatively lower than in Indonesia, this ambition shows the visionary plan of the government. It is shown that the State Islamic Department and State Islamic Council only issue certificates for domestic markets for food and daily-used items for domestic and international markets. Halal cuisine must adhere to the standard created by that council. Another standard component of a Halal Assurance Management System applies to retailing, warehousing, and the transportation of commodities. Malaysian Halal Directory has recognized food and beverage manufacturers, food establishments where the food and beverage business is performed, consumer goods, slaughterhouses, and hotel and resort food and beverage operations. A site audit and monitoring are part of the certification procedure, and when it is successful, a formal halal logo can be shown in public (Henderson 2016).

Unlike Malaysia, which is well known as an Islamic country, Singapore is secular. It means the fundamental constitution of that country is not constructed on or derived from any religious faith (Lončar 2020). This policy gives the believer of any religious freedom to practice their belief freely without any interference from the government. Muslim adults (15 and older) made up 459,800, or 14.04% of the population. However, the halal sector is taken into account. The halal business in Singapore has developed five times in the last ten years. To ensure that its Muslim population could practice their faith, the Islamic Religious Council of Singapore gave more than 4000 halal certificates in 2016 (Ridwan 2020). This council has dealt with halal certification since 1978. It promotes international initiatives and maintains a Halal Quality Management System that encompasses the entire food supply chain, from sourcing to production, storage, logistics, and sales and marketing. Food preparation spaces, poultry abattoirs, goods, storage facilities, complete plants, and endorsements (imported, exported, or re-exported items issued with halal certificates) are the seven certification categories for eating establishments (Henderson 2016). This data shows how rigorous halal certification in Singapore is done, despite its relatively small Muslim population. The geographical position of Singapore, which is surrounded by majority Muslim populated countries, has its benefit. It widely opens the chance to develop halal-based products in Singapore to distribute worldwide.

In its history, since the Islamic Ulema Council of Singapore started using the halal label on food goods in 1978, the halal industry in Singapore has even started to expand. Additionally, the Islamic Ulema Council of Singapore's halal certification is a halal certification that is fully acknowledged and used by numerous multinational corporations throughout the world. Several multinational corporations acknowledge Singapore's halal product certification from governments, the Gulf Cooperation Council, and numerous Southeast Asian nations, including Indonesia, Brunei Darussalam, Malaysia, and Singapore (Yusuf et al. 2022). The hard work of its government to promote its halal products is effective since many Muslim countries accept halal certification products from Singapore. There are six stages of halal certification in Singapore by MUIS. The first stage is the question stage, in which the applicants are asked to answer some questions and declare the purpose of seeking certification for their products. The next stage is registering online through the e-halal website, and from now the process will be done by accessing the License One site. The applicants need to upload the required documents. The primary information of each product must be provided and mentioned in the MLS for this scheme. The product's description must list the producer, supplier, and halal certification body of the raw materials used to make it. It relates to the following requirements, which stipulate that all ingredients.

6.4 Conclusion

By seeing this vast opportunity for national and international Muslim market share, it can economically open up sustainable income opportunities for business actors. In addition, the benefit for consumers as part of a digital society is that they will get safe products to consume and use sustainably, as well as guaranteed credibility in Islamic values socially. Another benefit is that product certification can support environmental sustainability because halal product processing always pays attention to ecological sustainability aspects. This study implies that the SDGs and the halal industry have the same formation philosophy, especially in the economic field, namely efforts to bring human welfare to meet the needs of life by utilizing and maintaining the potential of natural resources, as well as humans.

References

Aisyah M, Suzanawaty L, Said M (2019) The demand for halal certified restaurants in Indonesia. Int J Supply Chain Manage 8(5). https://doi.org/10.3390/su12030820

Arifin A, Ferdinand Wyman S, Hendijani RB (2021) An investigation of the aspects affecting Non-Muslim customers' purchase intention of halal food products in Jakarta, Indonesia. Future Food J Food Agric Soc 9(2):1–13.https://doi.org/10.17170/kobra-202102163257

Asnawi N, Haris A, Shamakov N, Fanani MA (2020) Consumer perceived value and halal food products consumption in the Indonesia and Russia. Int J Adv Sci Technol 29(5):57–81

Chebab D, Hassan MK, Rabbani MR (2021) Integrating Islamic finance and halal industry: current landscape and future forward. Int J Islamic Market Brand 6(1):60. https://doi.org/10.1504/ijimb.2021.10040922

Chong SC, Yeow CC, Low CW, Mah PW, Tung DT (2022)Non-Muslim Malaysians' purchase intention towards halal products. J Islamic Market.https://doi.org/10.1108/JIMA-10-2020-0326

Dubé FN, HaiJuan Y, Lijun H (2016) Halal certification system as a resource for firm internationalisation: comparison of China And Malaysia. Int J Asia-Pacific Stud 21(1):73–88

Farhan F, Sutikno B (2022) The acceptance of halal food products among Non-Muslim consumers in Indonesia. J Int Food Agribus Mark

Fuseini A, Knowles TG (2020) The ethics of halal meat consumption: preferences of consumers in England according to the method of slaughter. Veterinary Record 186(19). https://doi.org/10.1136/vr.105287

Hafaz NA, Jeevan J, Salleh NHM, Lee TTH, Ruslan SMM (2019) Willingness to pay for halal transportation cost: the moderating effect of knowledge on the theory of planned behavior. Int J Eng Adv Technol 8(6 Special Issue 3):479–488. https://doi.org/10.35940/ijeat.F1089.0986S319

Hasyim S (2022) The politics of 'halal': from cultural to structural shariatisation in Indonesia. Austr J Asian Law 22(1):81–97

Henderson JC (2016) Halal food, certification and halal tourism: insights from Malaysia and Singapore. Tourism Manage Perspect

Hong M, Kamaruddin R (2020) The effects of country of origin (COO) on halal consumption: evidence from China. Pertanika J Soc Sci Human 28(1):483–501

Ibrahim MA, Michael Hall C, Ballantine PW (2019) Food certification: the relationships between organic and halal certification in Malaysian food retailing. In: The Routledge handbook of halal hospitality and Islamic tourism, pp 264–277

Jumiono A, Barang EKSH, Jumiono A, Rahmawati SI (2020) KRITERIA SERTIFIKASI HALAL BARANG GUNAAN DI INDONESIA. JURNAL ILMIAH PANGAN HALAL 2(1):10–16

Krisharyanto E, Retnowati E, Hastuti NT (2019) Regulation and provisions for supervision of halal products in Indonesia. J Legal Ethical Regul Issues 22(1)

Kurniaputri MR (2020) Brand Equity Dan Labelisasi Halal Dalam Pengaruhnya Terhadap Minat Beli Produk Lifebuoy. Jurnal Ilmiah Ekonomi Islam 6(3):450. https://doi.org/10.29040/jiei.v6i3.1200

Larionova M (2020)The challenges of attaining the millennium development goals (MDGs). Int Org Res J.https://doi.org/10.17323/1996-7845-2020-01-07

Lončar M (2020) Secular states, religious politics: India, turkey and the future of secularism'. Polit Religion J. https://doi.org/10.54561/prj1402507l

Maison D, Marchlewska M, Syarifah D, Zein RA, Purba HP (2018) Explicit versus implicit 'halal' information: influence of the halal label and the country-of-origin information on product perceptions in Indonesia. Front Psychol 9. https://doi.org/10.3389/fpsyg.2018.00382

Meixner O, Friedl R, Hartl B (2018) Preferences for attributes of halal meat: empirical evidence from the Muslim community in Vienna, Austria. Int J Food Syst Dyn 9(3):265–278. https://doi.org/10.18461/ijfsd.v9i3.935

Nirwandar S (2018) Halal lifestyle industry, high growth and attractive investment in Indonesia. Malaysian J Consum Family Econ 22(Special Issue 2):144–152

Rafiki A, Wahab KA (2016) The human capital and the obtainment of halal certification. J Islamic Market 7(2):134–147. https://doi.org/10.1108/JIMA-03-2014-0020

Ridwan A (2020) Authorization of halal certification in Indonesia, Malaysia and Singapore. Int J Psychosoc Rehabil 24(08):7992–8011

Rios RE, Riquelme HE, Abdelaziz Y (2014) Do halal certification country of origin and brand name familiarity matter? Asia Pac J Mark Logist 26(5):665–686. https://doi.org/10.1108/APJML-03-2014-0046

Sayogo DS (2018) Online traceability for halal product information: perceptions of Muslim consumers in Indonesia. J Islamic Market 9(1):99–116. https://doi.org/10.1108/JIMA-07-2016-0057

Seçim Y (2022) The views and knowledge levels of the personnel working in tourism in Turkey on halal food. J Culinary Sci Technol.https://doi.org/10.1080/15428052.2022.2038333

Shahid S, Parray MA, Thomas G, Farooqi R, Islam JU (2022) Determinants of Muslim consumers' halal cosmetics repurchase intention: an emerging market's perspective. J Islamic Market.https://doi.org/10.1108/JIMA-08-2021-0265

Spear S, Abd El-Rahim IHA (2020) The halal food handbook

Sumarliah E, Tieke Li B, Wang AM, Sackey I (2021) The impact of customer halal supply chain knowledge on customer halal fashion purchase intention. Inf Resour Manag J 34(3):79–100. https://doi.org/10.4018/IRMJ.2021070105

Suryawan AS, Hisano S, Jongerden J (2022) Negotiating halal: the role of non-religious concerns in shaping halal standards in Indonesia. J Rural Stud 92:482–491. https://doi.org/10.1016/j.jrurstud.2019.09.013

Wibawa BM, Pranindyasari C, Bhawika GW, Mardhotillah RR (2021) Discovering the importance of halal tourism for Indonesian Muslim travelers: perceptions and behaviors when traveling to a Non-Muslim destination. J Islamic Market.https://doi.org/10.1108/JIMA-07-2020-0210

Yusuf M, Rangkuty DM, Krisna R, Raditio ME (2022) Economic analysis of halal certification by the halal product assurance administration agency. World J Adv Res Rev.https://doi.org/10.30574/wjarr.2022.13.1.0077

Yakin AU, Christians L-L (2021) Rethinking halal: genealogy, current trends, and new interpretations

Yama A, Jermsittiparsert K, Saregar A (2019) Impact of halal food certification on restaurant business in Indonesia: the mediating role of ethical compliance. Int J Innov Creat Change 9(7):306–322

Yarar AE (2020) Halal advertising on Youtube: a case study of fast food brands in Malaysia. Food Res 4:157–163. https://doi.org/10.26656/fr.2017.4(S1).S34

Innovation in Halal Supply Chain Management (HSCM) 7

Budi Harsanto and Egi Arvian Firmansyah

Abstract

Amid the rapidly growing literature on halal supply chain management (HSCM), previous researchers highlighted the scarcity of knowledge on innovation in the context of HSCM. Besides halal regulation, HSCM is concerned with the value-driven approach and ethical business practices. This paper aims to understand the innovations in the context of HSCM. The method used is a systematic literature review on the academic database of Scopus. Content analysis is used with a structured and objective process of data extraction. A coding list is developed based on the typology of technical/technological–administrative/organizational and theoretical concepts of supply chain innovation type. The findings show that the study on innovation in HSCM is interdisciplinary, with technological innovation, including digitalization, being less researched than organizational innovation. It is also found that many studies focused on internal HSCM rather than externals, both upstream and downstream, indicating the current challenges faced by the firms. This study contributes to the HSCM literature by being the first to link innovation with HSCM and be a baseline for future development.

Keywords

Business ethics • Halal supply chain management (HSCM) • Innovation • Systematic review

B. Harsanto (✉) · E. A. Firmansyah
Universitas Padjadjaran, Bandung, Indonesia
e-mail: budi.harsanto@unpad.ac.id

A. Rafiki (ed.), *Digitalization in Halal Management*, Contributions to Environmental Sciences & Innovative Business Technology,
https://doi.org/10.1007/978-981-99-5146-8_7

7.1 Introduction

Attention to halal supply chain management (HSCM) has been rising, especially in the last decade (Haleem et al. 2020; Indarti et al. 2020). HSCM is a specific study in SCM that focuses on design and execution along the supply chain—from source to consumer, from farm to fork which complies with halal regulations or is acceptable in the Islamic principle (Khan et al. 2018a, b; Tieman 2011). The HSCM concept has evolved along with the development of a business concept known as a value-driven approach that appreciates the values of consumers and ethical business practices (Khan et al. 2018a, b; Kotler 2010). The development of HSCM is also in line with the complexity of the supply chain, which makes ensuring the provision of halal products highly challenging (Lam and Alhashmi 2008). That makes practice and research related to activities to ensure the halalness of products along the supply chain critical (Tieman et al. 2012).

The development of research in HSCM and related area to date tends to focus on the definition of HSCM (e.g., Khan et al. 2018a, b), the model (e.g., Omar and Jaafar 2011; Tieman et al. 2012), consumer perceptions and adoption (e.g., Harsanto and Jatnika 2017; Ngah et al. 2014; Tieman et al. 2013), risks (Fujiwara 2017; Khan et al. 2020), or barriers for HSCM (e.g., Ab Talib 2021; Khan et al. 2019a, b, c). The dominance of these fundamental topics in HSCM is understandable because studies on this subject only appeared in the past 10 years, and the field is still developing. Despite this development, it is still rare for the study to focus on the innovation aspects of HSCM (Secinaro and Calandra 2020). The innovation aspect in HSCM is essential because, as in conventional SCM, the firm's competitive advantage is most likely higher when it is underpinned by innovation in its SCM (Arlbjørn et al. 2011).

Innovations in the supply chain can be in the form of introducing new products or processes or entering new markets (Arlbjørn et al. 2011). Innovation goes further than invention because innovation includes not only the invention itself but also the phase of commercialization or implementation of ideas (Freeman 1974). For example, appreciation for innovation in the supply chain was given by the council of supply chain management professionals (CSCMP), granting the supply chain innovation award in the last 15 years (CSCMP 2021). Arlbjørn et al. (2011) revealed that there are three elements in supply chain innovation, including supply chain business process (such as demand management or customer relationship management), supply chain technology (such as application of radio frequency identification/RFID or enterprise resource planning/ERP), and supply chain network structure (such as collaboration or distribution channel management). The current implementation of supply chain technology is closely related to digitalization. A study by Kilay et al. (2022) reveals a positive and significant effect of e-payment and e-commerce service variables on the supply chain performance of MSMEs. In addition, non-profit entities, such as disaster relief organizations, benefit from digitalization in their humanitarian logistics because it will create early warnings that will save many lives, and the victims will get proper and timely assistance (Iqbal and Ahmad 2022). Furthermore, supply chain digitalization is more critical than ever due to the current worldwide disruptions brought on by the COVID-19 pandemic and the rush of businesses migrating toward the digital economy (Aamer et al. 2022).

With this background, this paper focuses on the innovation aspect of HSCM. This study analyzes academic publications in scientific journals indexed in an academic database using a systematic literature review guided by the Preferred Reporting Items for Systematic Reviews and Meta-Analyses (PRISMA) flow diagram (Moher et al. 2009). In particular, this paper has two aims. First, it explores various aspects of innovation in HSCM publications, studied using content analysis. Second, the paper explores the interconnection between concepts related to innovation in HSCM. This paper contributes to the literature on HSCM by being an early study focusing on innovation in HSCM.

This paper is structured as follows. The next section describes the methods used in this study, including how the data were collected and analyzed. The third section contains the results of the analysis, which are divided into two specific subsections as the aims of this paper. A detailed discussion of the research findings is presented in the fourth section. Finally, the conclusion of this study, including limitations and direction for future research, is provided in the last section.

7.2 Materials and Methods

The method used in this study is a systematic literature review. The systematic review is a methodology for synthesizing literature and knowing the development of a field transparently and reliably (Tranfield et al. 2003). The systematic review was initially used in the health domain to obtain factual information based on evidence from previous research on specific interventions or actions (Clarke and Chalmers 2018). In its development, this method was used in various other fields of science to synthesize knowledge using structured and transparent techniques (Petticrew 2001). A protocol is needed in conducting systematic reviews so that the synthesis results are reliable and reproducible (Denyer and Tranfield 2009; Harsanto et al. 2020a). This paper follows the protocol proposed by Tranfield et al. (2003), which consists of three main stages, namely: (a) planning the review, (b) conducting a review, and (c) reporting and dissemination.

The need for this review was identified at the planning stage, as described in the Sect. 7.1. Knowledge was scarce regarding innovation on HSCM in the scholarly literature. In the second review stage, key terms for the search and the academic database where the search was located were identified. The initial key terms used in the search process are: innovate* and "halal" and "supply chain management" in the article title, abstract, and keyword. This process, however, returns no results because of the scarcity of studies on this specific topic. Therefore, the key terms are modified into "halal" and "supply chain management", then the results are refined using the features in the database and filtered by manual reading titles, abstracts, and full papers to extract implicit discussion about innovation on HSCM (Table 7.1).

Table 7.1 Coding scheme for content analysis

Code	Sub-code	Definition
Innovation typology (INTY)	HSCM technical innovation (TE)	Implementation of ideas related to the technical aspects of the firm's HSCM or commercialization into products or services
	HSCM organizational innovation (OR)	Implementation of the idea in terms of management practices for HSCM
Supply chain innovation type (SCIN)	HSCM business process innovation (BP)	Improvement in the design of HSCM to produce the specified output
	HSCM technology innovation (TC)	Application of specific technology in HSCM
	HSCM network structure innovation (NW)	Improvement in the vertical and/or horizontal HSCM structure
Supply chain (SC) direction (SCDR)	Upstream (UP)	Processes that focus on the firm-supplier interface
	Internal (IN)	Processes that focus on the firm internally
	Downstream (DW)	Processes that focus on the firm-customer interface
Industry focus (INFC)	Food and beverages (FB)	Industry focus on the food and beverage business
	Other than food and beverages (NF)	Industry focus on the non-food and beverage business

The initial search results returned 88 articles which were then refined based on article type (journal article only; conference paper, book chapter, and others are excluded), language (English only, other languages are excluded), and title/abstract screening resulting in 40 articles to be read manually. This process resulted in 30 articles being included in the synthesis. This number of articles is reasonable for this specific field (e.g., Gupta et al. 2019). The search process was carried out on the academic database of Scopus. Scopus was selected because it is currently the academic database with the most extensive coverage and maintained quality (Gupta et al. 2019). A detailed filtering process using the adapted PRISMA flow diagram (Moher et al. 2009) is shown in Fig. 7.1.

7.3 Results and Discussion

The articles reviewed come from a diverse range of scholarly journals. Some journals specialized in SCM and logistics (such as Supply Chain Management and Transportation Research Part E: Logistics and Transportation Review), food

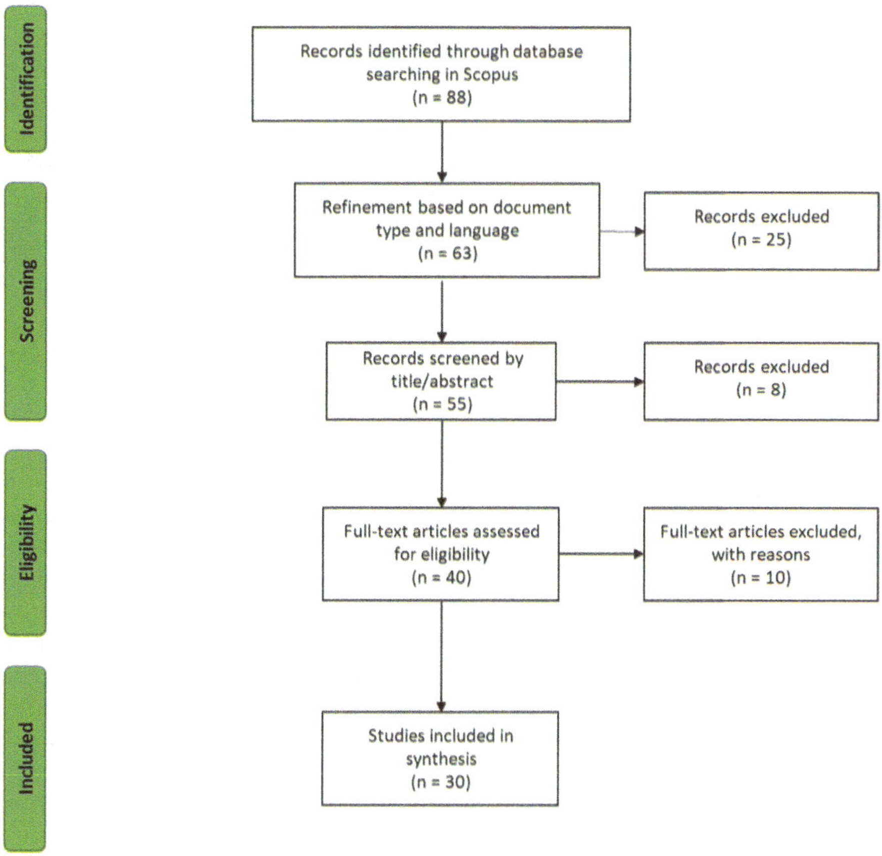

Fig. 7.1 Flow diagram adapted from PRISMA framework (Moher et al. 2009)

and beverages (such as British Food Journal), Islamic studies (such as Journal of Islamic Marketing), entrepreneurship (such as Journal of Small Business and Entrepreneurship) or other fields (such as Sustainability and Journal of Modeling in Management). Journals with the highest impact factor are Supply Chain Management (4.725), Transportation Research Part E: Logistics and Transportation Review (4.690), and Industrial Management and Data Systems (3.329) (Clarivate Analytics 2021). Not many articles are published in high-impact factor journals, signaling that studies related to innovations at HSCM have an excellent prospect for publication in outstanding journals.

Based on the best quartile data, most articles were published in Q3 (fifteen articles), followed by Q1 (eight articles) (Scimago 2021). The five journals originating from Q1 are Supply Chain Management, Transportation Research Part E: Logistics and Transportation Review, Industrial Management and Data Systems, British Food Journal, and Journal of Big Data. The journal with the most reviewed articles is the Journal of Islamic Marketing (nine articles) from Emerald Group Publishing. It is understandable because this journal has readers interested in Islamic business, specifically marketing relevant to HSCM, especially (although not limited) to the downstream supply chain. The detail on the year of publication is shown in Fig. 7.2.

The articles included in the review are all from the last 10 years (2011–2021). The early years in which articles were published were from 2011 and 2012 (three articles) written by Tieman (2011), Manzouri (2012), and Tieman et al. (2012). The year with the most publications was 2020, with seven articles. From Fig. 7.2, it can be seen that there is an increasing trend from year to year, especially since 2017. Furthermore, based on the research context, most papers used the Malaysian context (63.64%), followed by the Indian context (9.09%). The remainder is divided into seven different contexts in Asia, Europe, and Australia (see Table 7.2).

A summary of the content analysis is given in Table 7.3. As explained in the Sect. 7.2, the content-coding is analyzed based on innovation typology, SC innovation type, SC direction, and industry focus. Innovation on HSCM discussed in the reviewed articles varies, e.g., blockchain for HSCM, HSCM network optimization, or halal trade cluster. The majority of the innovation typology discussed is organizational innovation. For the SC innovation type, the types discussed more in the literature are the SC business process and the SC network structure with a balanced proportion. In terms of SC direction, the internal firm and the firm interface with upstream and downstream are much discussed. Finally, as expected, the industry focus mostly comes from the food and beverage industry.

This study shows that the innovation in HSCM is discussed in many journals indicating its interdisciplinary and intersectoral nature (Indarti et al. 2020; Rahman and Zailani 2017). Not only in the HSCM field journals, but this topic is also discussed in journals related to Islamic studies, food research, and entrepreneurship.

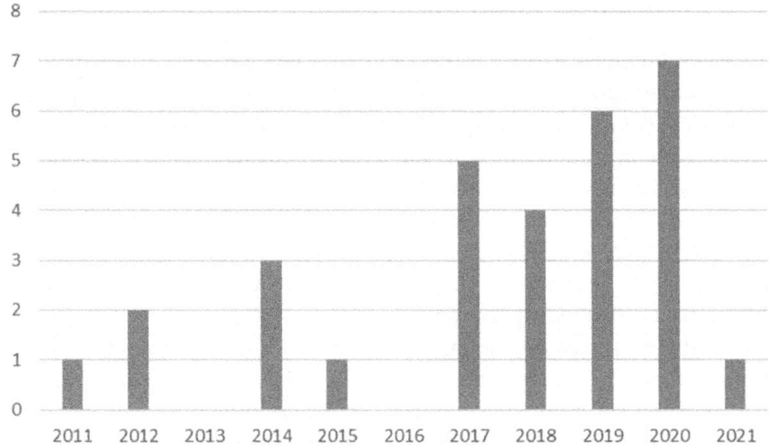

Fig. 7.2 Publication year

Table 7.2 Context of reviewed articles

Context	Frequency	Percentage (%)
Malaysia	21	63.64
India	3	9.09
Australia	1	3.03
Brunei Darussalam	1	3.03
China	1	3.03
Indonesia	1	3.03
The Netherlands	1	3.03
UAE	1	3.03
UK	1	3.03
NA	2	6.06

Note The cumulative value may not be equal to *N* since each article may have studied more than one context; the percentage is calculated based on *N*

Although not many, several reviewed articles were published in journals with high impact factors indicating that HSCM, in general, is an emerging field (Haleem et al. 2020), and, in particular, innovation in the HSCM field has the potential to grow. Publications relating to the reviewed topics were published in the last 10 years. Malaysia is the most researched context, and the authors from Malaysia contributed the most papers. A more detailed discussion regarding the four main themes in the content analysis, including innovation typology, supply chain innovation type, supply chain direction, and industry focus, is presented in the following paragraphs.

From the perspective of socio-technical system theory, there are two common typologies of innovation, namely technical/technological innovation and administrative/organizational innovation (Azar and Ciabuschi 2017; Birkinshaw et al. 2008; Damanpour and Evan 1984). Technical or technological innovation for HSCM found in the articles reviewed includes blockchain technology (Hew et al. 2020; Surjandari et al. 2021), the Internet of Things (Ahmad Tarmizi et al. 2020), radio frequency identification (RFID), and other technologies such as near-field communication (NFC) or DNA barcoding (Khan et al. 2018a, b; Mohammed et al. 2017). The use of these technologies is primarily for traceability systems to ensure halalness and transaction efficiency (Surjandari et al. 2021).

Blockchain technology, for example, is used for traceability purposes by integrating information exchange and reliable documentation in the entire process of HSCM (Hew et al. 2020). Blockchain, a result of digitalization, has affected multiple aspects of life, including the HSCM practices in a company. Besides ensuring halalness and transaction security, blockchain can increase the effectiveness of data transfer speeds and transaction efficiency (Hew et al. 2020; Surjandari et al. 2021). Ahmad Tarmizi et al. (2020) revealed that in the era of industry 4.0, traceability could also be supported by the Internet of Things (IoT) through various means, such as tracking and tracing along the production line, management of

Table 7.3 Content analysis summary

Authors and year	Innovation on HSCM	INTY			SCIN			SCDR		INFC	
		TE	OR	BP	TC	NW	UP	IN	DW	FB	NF
Surjandari et al. (2021)	Blockchain	✓	–	–	✓	✓	✓	✓	✓	✓	–
Hew et al. (2020)	Blockchain	✓	–	–	✓	✓	✓	✓	✓	✓	–
Haleem et al. (2020)	Fair-trade and sustainability practices	–	✓	✓	–	–	–	–	✓	–	–
Mohamed et al. (2020)	HSCM integrity assurance	–	✓	✓	–	✓	✓	✓	✓	✓	–
Zainuddin et al. (2020)	Halal traceability system	–	✓	✓	–	–	✓	✓	✓	✓	–
Muhamed et al. (2020)	Halal certified SMEs	–	✓	✓	–	–	–	✓	✓	✓	✓
Ag Majid et al. (2020)	Aiming non-Muslim consumers	–	✓	✓	–	–	–	–	✓	✓	✓
Ahmad Tarmizi et al. (2020)	Internet of Things	✓	–	–	✓	–	–	✓	–	✓	–
Ahmad et al. (2019)	Collaboration with competitors	–	✓	–	–	✓	✓	–	–	✓	–
Zainuddin et al. (2019)	Halal packaging and labeling	–	✓	✓	–	–	–	✓	–	✓	–
Kwag and Ko (2019)	HSCM network optimization	–	✓	–	–	✓	✓	–	–	–	–
Randeree (2019)	Global halal ecosystem; Halal branding	–	✓	–	–	✓	✓	✓	✓	✓	–
Khan et al. (2019a, b, c)	Halal logo	–	✓	✓	–	–	–	–	✓	✓	✓
Khan et al. (2019b, c)	Reward system for supplier commitment	–	✓	–	–	✓	✓	–	–	✓	–
Yaacob et al. (2018)	A halal committee in a logistics firm	–	✓	✓	–	–	–	✓	–	–	✓
Khan et al. (2018a, b)	Technology for a traceability system	✓	✓	–	✓	✓	✓	✓	✓	✓	–
Fujiwara and Ismail (2018)	Risk mitigation strategy	–	✓	✓	–	–	✓	✓	–	✓	–
Zulfakar et al. (2018)	Halal integrity in the non-Muslim-majority country	–	✓	✓	–	–	✓	–	–	✓	–

(continued)

Table 7.3 (continued)

Authors and year	Innovation on HSCM	INTY			SCIN			SCDR		INFC	
		TE	OR	BP	TC	NW	UP	IN	DW	FB	NF
Fujiwara (2017)	No-animal based; local sourcing; local manufacturing	–	✓	✓	–	–	✓	✓	–	✓	–
Ali et al. (2017)	Supplier integration	–	✓	–	–	✓	✓	✓	–	✓	–
Rahman and Zailani (2017)	Intersectoral collaboration, e.g., medical-tourism	–	✓	–	–	✓	✓	✓	✓	–	✓
Tan et al. (2017)	Halal assurance system; supplier-customer integration	–	✓	✓	–	–	✓	✓	✓	✓	–
Mohammed et al. (2017)	RFID-enabled HMSC network	✓	–	–	✓	✓	✓	✓	✓	✓	–
Tieman (2015)	Halal trade cluster	–	✓	–	–	✓	–	✓	✓	✓	–
Tieman and van Nistelrooy (2014)	Halal logistic system	–	✓	–	–	✓	✓	✓	–	✓	–
Manzouri et al. (2014)	Waste elimination in HSCM	–	✓	✓	–	–	–	✓	–	✓	–
Ab Talib and Hamid (2014)	Supplying talents in HSCM	✓	✓	–	✓	✓	–	✓	✓	✓	–
Manzouri (2012)	Lean HSCM	–	✓	✓	–	–	–	✓	–	✓	–
Tieman et al. (2012)	Innovative logistics cargo solutions	–	✓	✓	–	✓	✓	✓	✓	✓	–
Tieman (2011)	Compliance in transportation, storage, and handling	–	✓	✓	–	✓	✓	✓	✓	✓	–
Sum		6	26	16	6	16	19	23	17	26	5

See Table 7.1 for detail of the coding code

product information, or authentication of product ingredients, which in turn could strengthen HSCM integrity. RFID, which has long been used in manufacturing, can also be useful for traceability and decision-making optimization in designing HSCM networks (Khan et al. 2018a, b; Mohammed et al. 2017).

For organizational innovation, the topics include the HSCM integrity assurance system (Mohamed et al. 2020), HSCM packaging, labeling, and branding (Khan et al. 2019b, c; Saudi et al. 2019), risk mitigation system (Fujiwara and Ismail

2018), halal trade cluster (Tieman 2015), supplier–consumer integration (Ahmad et al. 2019; Ali et al. 2017), targeting new markets (Ag Majid et al. 2020), waste elimination in HSCM (Manzouri et al. 2014), or fair trade and sustainability practices (Haleem et al. 2020). Discussing sustainability practices has made this topic intersect with sustainability innovation (Harsanto et al. 2018; Harsanto and Permana 2020; Permana and Harsanto 2020). Organizational innovation is essential in developing HSCM because, theoretically, the firm's successful technological innovation depends on the changes or improvements that occur in its organization (Azar and Ciabuschi 2017). For example, a firm policy regarding HSCM integrity assurance can be in the form of a halal traceability system for improving HSCM performance (Zainuddin et al. 2020).

Innovation in HSCM based on supply chain innovation type can be divided into three: HSCM business process innovation, HSCM technology innovation, and HSCM structure network innovation. HSCM business process innovation is related to improvements in HSCM design to produce specific outputs. This design is primarily related to designing business processes, such as policies, to ensure halalness in the logistics system. Therefore, it enables order fulfillment and demand management to run smoothly (Tieman and van Nistelrooy 2014). Another example of business process innovation, according to Arlbjørn et al. (2011), is supplier relationship management. In the context of HSCM, supplier relationship management in HSCM could be realized, e.g., in the form of a reward system for suppliers who are committed to maintaining the halalness of the product (Khan et al. 2019b, c).

HSCM technology innovation relates to technical innovation in the typology of technological-organizational innovation. HSCM technology innovation is a specific application of technology in HSCM. In the supply chain, Arlbjørn et al. (2011) explained that the specific application of technology can be in the form of using a global positioning system (GPS) application, barcoding, RFID, electronic data interchange (EDI), enterprise resource planning (ERP), warehouse management system (WMS), manufacturing execution system (MES), the Internet, e-auction or other technologies to support a reliable supply chain. From the reviewed articles, not all of these technologies emerged as part of the innovation on HSCM. The widely mentioned technology is RFID and barcoding to help traceability. In addition, the more sophisticated one is blockchain technology (Hew et al. 2020; Mohammed et al. 2017; Yaacob et al. 2018). It shows that there is still an area open for HSCM technology application by reflecting on its application in the conventional SCM.

HSCM network structure innovation deals with improvements to the vertical and/or horizontal HSCM structure. Arlbjørn et al. (2011) revealed that network structure innovation could be in the form of, e.g., in, and outsourcing, partnership, collaboration, or improvement in distribution channels. Regarding sourcing, Fujiwara (2017) mentioned the idea of local sourcing and local manufacturing to mitigate risks in HSCM. Partnership and collaboration are widely discussed in the reviewed articles; for example, customer and supplier integration contributed significantly to HSCM integrity (Ali et al. 2017). Another interesting idea is a network between halal authorities across the world to be able to provide a more robust and more acceptable global halal standard and collaboration between stakeholders in supplying talent for HSCM (Ab Talib and Hamid 2014). Concerning the distribution channel, one of the reviewed articles proposed an interesting algorithm regarding the network optimization design for the halal food logistic network (Kwag and Ko 2019).

Concerning the macro process of the supply chain, there are three directions in the supply chain: the upstream (supplier), the internal firm itself (internal supply chain management), and the downstream (customer) (Chopra et al. 2013). Considering that these three elements are highly interrelated, it is not surprising that many articles discuss those topics (Rahman and Zailani 2017; Randeree 2019; Tan et al. 2017; Tieman et al. 2012). Individually, the elements of the supply chain that have been lot studied are those that focus on internal SCM, e.g., the halal committee in the logistic firm (Yaacob et al. 2018), waste elimination for HSCM efficiency (Manzouri et al. 2014), or lean HSCM practices to improve inventory management, quality management, and customer satisfaction (Manzouri 2012).

The high level of attention to internal SCM is understandable because many internal aspects of HSCM need to be considered before dealing with external aspects, such as the internal HSCM assurance system (Mohamed et al. 2020), mitigation risk (Fujiwara and Ismail 2018), or internal efficiency strategy (Manzouri et al. 2014). Meanwhile, the attention toward upstream and downstream is relatively balanced. Many authors have discussed the importance of collaborating with suppliers because of the awareness that it is almost impossible to provide halal guarantees for products without suppliers' guarantees (Ali et al. 2017; Randeree 2019; Tieman et al. 2012). Toward the downstream, the aspect that has received significant attention is how consumers can be confident in the halalness of the products and how good HSCM can attract a broader potential market, such as non-Muslim consumers (e.g., Ag Majid et al. 2020; Randeree 2019).

As expected, almost all articles focus on the food and beverage industry. That is perfectly reasonable for at least two reasons. First, halal is synonymous with the Muslim diet, particularly in food sourcing (Kashim et al. 2015; Randeree 2019). Providing food and beverages that are guaranteed halal at the point of consumers is the HSCM's primary concern. Lack of information or transparency in HSCM makes it difficult for consumers to ensure that unwanted cross-contamination occurs in production, transportation, or storage (Bonne and Verbeke 2008; Tieman 2011). Second, the development of the food and beverage industry is getting bigger in value and broader in scope. The global halal food industry is estimated at 1.2 trillion USD, with a growth rate of around 8.5% (Armanios and Ergene 2018).

Apart from food and beverage, other sectors mentioned in the reviewed articles are logistics, cosmetics, pharmaceuticals, medical, and tourism. Rahman and Zailani (2017) have conducted a study that focuses on the medical tourism industry with data collected from hospitals, medical agencies, and hotels, with the main finding that the effectiveness of supply chain practices significantly influences HSCM. One study on 284 halal-certified SMEs across industries, including pharmaceutical and logistics, shows that the credibility of supply chain partners played a major role in determining firm performance (Muhamed et al. 2020). Firm performance is a common measure in firm-level research (Harsanto et al. 2020b; Harsanto and Roelfsema 2015; Widianto and Harsanto 2017). Several articles also highlighted the logistical aspects of inventory handling, including cargo, to ensure the simplicity and effectiveness of HSCM (e.g., Tieman 2011; Tieman et al. 2012; Tieman and van Nistelrooy 2014).

7.4 Conclusion

Amid increased attention to HSCM, the study of innovation on HSCM is among the prospective to be developed further. This study found that innovation in HSCM is interdisciplinary, with the discussion of organizational innovation still more dominant than technological innovation. Regarding the supply chain innovation type, many articles discuss the ideas of the HSCM business process and the HSCM network structure to help all stakeholders obtain halal products with guaranteed quality and quantity. Internal SCM has also received significant attention, indicating that the current focus is more on internal aspects than external ones (upstream and downstream HSCM).

As an initial attempt to develop a field regarding innovation on HSCM, this study indeed has limitations. This study systematically reviews the literature to discover ideas about innovation in HSCM. Literature is obtained from the academic database of Scopus. Even though it is considered the largest academic database with maintained quality, the search limitation on the Scopus academic database makes it possible that other articles relevant to this topic and of good quality are not indexed in that database. Another limitation is that this specific area has not yet been established, so not all of the articles studied were explicitly mentioned in the title, abstract, and keywords that the research is about innovation on HSCM. This challenge is overcome by synthesizing data through manual reading of the full articles.

This study could be a baseline for some future research directions. As this study uses systematic review, empirical research with a specific focus on innovation on HSCM is highly encouraged. This empirical study could use a quantitative, qualitative, or mixed-method approach focusing on the entire supply chain or a particular part of the supply chain (upstream, internal SCM, and/or downstream). This study found that research on technological innovation is still rare, so research on technological innovation beyond RFID and blockchain technology would be an important area for future investigation. This study also found that so far, Malaysia still dominates studies related to innovation HSCM both in terms of productivity and collaboration, so it is encouraged that scholars from other countries such as Indonesia or other regions such as the Middle East and Africa could contribute more to the development of this field.

References

Aamer A, Sahara CR, Al-Awlaqi MA (2022) Digitalization of the supply chain: transformation factors. J Sci Technol Policy Manag. https://doi.org/10.1108/JSTPM-01-2021-0001

Ab Talib MS, Hamid ABA (2014) Halal logistics in Malaysia: a SWOT analysis. J Islam Mark 5(3):322–343. https://doi.org/10.1108/JIMA-03-2013-0018

Ab Talib MS (2021) Identifying halal logistics constraints in Brunei Darussalam. J Islam Mark 12(6):1145–1158. https://doi.org/10.1108/JIMA-09-2019-0189

Ag Majid DKZ, Abdul Hanan S, Hassan H (2020) A mediator of consumers' willingness to pay for halal logistics. Br Food J. https://doi.org/10.1108/BFJ-01-2020-0047

Ahmad H, Sabar R, Udin ZM, Abd Latif MF, Zainuddin N (2019) Quality management of internal supply chain in Halal food manufacturer. Int J Supply Chain Manag 8(4):1047–1052

Ahmad Tarmizi H, Kamarulzaman NH, Abd Rahman A, Atan R (2020) Adoption of internet of things among Malaysian halal agro-food smes and its challenges. Food Res 4(1996):256–265. https://doi.org/10.26656/fr.2017.4(S1).S26

Ali MH, Zhan Y, Alam SS, Tse YK, Tan KH (2017) Food supply chain integrity: the need to go beyond certification. Ind Manag Data Syst 117(8):1589–1611. https://doi.org/10.1108/IMDS-09-2016-0357

Clarivate Analytics (2021) InCites. J Citation Rep

Arlbjørn JS, de Haas H, Munksgaard KB (2011) Exploring supply chain innovation. Logist Res 3(1):3–18. https://doi.org/10.1007/s12159-010-0044-3

Armanios F, Ergene BA (2018) Halal food: a history. Oxford University Press

Azar G, Ciabuschi F (2017) Organizational innovation, technological innovation, and export performance: the effects of innovation radicalness and extensiveness. Int Bus Rev 26(2):324–336. https://doi.org/10.1016/j.ibusrev.2016.09.002

Birkinshaw J, Hamel G, Mol MJ (2008) Management innovation. In: Academy of management review, vol 33, issue 4, pp 825–845. https://doi.org/10.5465/AMR.2008.34421969

Bonne K, Verbeke W (2008) Religious values informing halal meat production and the control and delivery of halal credence quality. Agric Hum Values 25(1):35–47

Chopra S, Meindl P, Kalra DV (2013) Supply chain management: strategy, planning, and operation, vol 232. Pearson Boston, MA.

Clarke M, Chalmers I (2018) Reflections on the history of systematic reviews. BMJ Evid Based Med 23(4):121–122. https://doi.org/10.1136/bmjebm-2018-110968

CSCMP (2021) Supply chain innovation award

Damanpour F, Evan WM (1984) Organizational innovation and performance: the problem of "organizational lag." Adm Sci Q 29(3):392–409

Denyer D, Tranfield D (2009) Producing a systematic review. In: The SAGE handbook of organizational research methods, pp 671–689

Freeman C (1974) The economics of industrial innovation. Penguin Books

Fujiwara T (2017) Supplier management in halal food supply chain: a preliminary case study. Pertanika J Social Sci Hum 25(December):25–38

Fujiwara T, Ismail RM (2018) Supplier management system in Halal food supply chain: a case study approach. Int J Supply Chain Manag 7(3):216–220

Gupta S, Altay N, Luo Z (2019) Big data in humanitarian supply chain management: a review and further research directions. Ann Oper Res 283(1–2):1153–1173. https://doi.org/10.1007/s10479-017-2671-4

Haleem A, Khan MI, Khan S (2020) Conceptualising a framework linking halal supply chain management with sustainability: an India centric study. J Islam Mark. https://doi.org/10.1108/JIMA-07-2019-0149

Harsanto B, Permana CT (2020) Sustainability-oriented innovation (SOI) in the cultural village : an actor-network perspective in the case of Laweyan Batik Village. J Cult Herit Manag Sust Dev. https://doi.org/10.1108/JCHMSD-08-2019-0102

Harsanto B, Roelfsema H (2015) Asian leadership styles, entrepreneurial firm orientation and business performance. Int J Entrep Small Bus 26(4):490. https://doi.org/10.1504/ijesb.2015.072759
Harsanto B, Michaelides R, Drummond H (2018) Sustainability-oriented Innovation (SOI) in emerging economies: a preliminary investigation from Indonesia. IEEE Int Conf Indus Eng Eng Manag 2018:1553–1557. https://doi.org/10.1109/IEEM.2018.8607473
Harsanto B, Kumar N, Zhan Y, Michaelides R (2020a) Responsible research and innovation (RRI) in emerging economies: a preliminary review. In: IEEE international conference on engineering, technology and innovation (ICE/ITMC)
Harsanto B, Kumar N, Zhan Y, Michaelides R (2020b) Firms' ICT and innovation in Jakarta Metropolitan area. Int Conf Technol Entrepr Virtual 2020:1–4
Harsanto B, Jatnika D (2017) Analysis of intention toward halal products: an empirical study of young consumers. J Econ Bus Account Ventura 20(2):203–212. https://doi.org/10.14414/jebav.v20i2.760
Hew JJ, Wong LW, Tan GWH, Ooi KB, Lin B (2020) The blockchain-based Halal traceability systems: a hype or reality? Supply Chain Manag 25(6):863–879. https://doi.org/10.1108/SCM-01-2020-0044
Indarti N, Lukito-Budi AS, Islam AM (2020) A systematic review of halal supply chain research: to where shall we go? J Islam Mark. https://doi.org/10.1108/JIMA-05-2020-0161
Iqbal T, Ahmad S (2022) Transparency in humanitarian logistics and supply chain: the moderating role of digitalisation. J Hum Logist Supply Chain Manag 12(3):425–448. https://doi.org/10.1108/JHLSCM-04-2021-0029
Kashim MIAM, Majid LA, Adnan AHM, Husni ABM, Nasohah Z, Samsudin MA, Yahaya MZ (2015) Principles regarding the use of Haram (Forbidden) sources in food processing: a critical Islamic analysis. Asian Soc Sci 11(22):17–25. https://doi.org/10.5539/ass.v11n22p17
Khan MI, Haleem A, Khan S (2018a) Defining Halal supply chain management. Supply Chain Forum 19(2):122–131. https://doi.org/10.1080/16258312.2018.1476776
Khan S, Haleem A, Khan MI, Abidi MH, Al-Ahmari A (2018b) Implementing traceability systems in specific supply chain management (SCM) through critical success factors (CSFs). Sustainability 10(1). https://doi.org/10.3390/su10010204
Khan MI, Khan S, Haleem A (2019a) Analysing barriers towards management of Halal supply chain: a BWM approach. J Islam Mark. https://doi.org/10.1108/JIMA-09-2018-0178
Khan S, Khan MI, Haleem A (2019b) Evaluation of barriers in the adoption of halal certification: a fuzzy DEMATEL approach. J Model Manag 14(1):153–174. https://doi.org/10.1108/JM2-03-2018-0031
Khan S, Khan MI, Haleem A, Jami AR (2019c) Prioritising the risks in Halal food supply chain: an MCDM approach. J Islam Mark. https://doi.org/10.1108/JIMA-10-2018-0206
Khan S, Haleem A, Khan MI (2020) Risk management in Halal supply chain: an integrated fuzzy Delphi and DEMATEL approach. J Model Manag. https://doi.org/10.1108/JM2-09-2019-0228
Kilay AL, Simamora BH, Putra DP (2022) The influence of e-payment and e-commerce services on supply chain performance: implications of open innovation and solutions for the digitalization of micro, small, and medium enterprises (MSMEs) in Indonesia. J Open Innov Technol Mark Compl 8(3). https://doi.org/10.3390/joitmc8030119
Kotler P (2010) Marketing 3.0: values-driven marketing. MarkPlus Inc
Kwag SI, Ko YD (2019) Optimal design for the Halal food logistics network. Transp Res Part E Logist Transp Rev 128(May):212–228. https://doi.org/10.1016/j.tre.2019.06.005
Lam Y, Alhashmi SM (2008) Simulation of halal food supply chain with certification system: a multi-agent system approach. In: Pacific Rim international conference on multi-agents, pp 259–266
Manzouri M (2012) How lean supply chain implementation affect halal food companies. Adv Nat Appl Sci 6(8):1485–1489
Manzouri M, Ab-Rahman MN, Zain CRCM, Jamsari EA (2014) Increasing production and eliminating waste through lean tools and techniques for Halal food companies. Sustainability 6(12):9179–9204. https://doi.org/10.3390/su6129179

Mohamed YH, Abdul Rahim AR, Ma'aram A (2020) The effect of halal supply chain management on halal integrity assurance for the food industry in Malaysia. J Islam Mark. https://doi.org/10.1108/JIMA-12-2018-0240

Mohammed A, Wang Q, Li X (2017) A cost-effective decision-making algorithm for an RFID-enabled HMSC network design: a multi-objective approach. Ind Manag Data Syst 117(9):1782–1799. https://doi.org/10.1108/IMDS-02-2016-0074

Moher D, Liberati A, Tetzlaff J, Altman DG, Altman D, Antes G, Atkins D, Barbour V, Barrowman N, Berlin JA, Clark J, Clarke M, Cook D, D'Amico R, Deeks JJ, Devereaux PJ, Dickersin K, Egger M, Ernst E, Tugwell P (2009) Preferred reporting items for systematic reviews and meta-analyses: the PRISMA statement. PLoS Med 6(7). https://doi.org/10.1371/journal.pmed.1000097

Muhamed AA, Salim N, Ab Rahman MN, Hamzah FM, Ali MH (2020) Effects of supply chain orientation on firm performance: insights from a Malaysian case study of halal-certified small and medium-sized enterprises. J Small Bus Entrepr 1–17. https://doi.org/10.1080/08276331.2020.1848328

Ngah AH, Zainuddin Y, Thurasamy R (2014) Adoption of halal supply chain among Malaysian halal manufacturers: an exploratory study. Procedia Soc Behav Sci 129:388–395

Omar EN, Jaafar HS (2011) Halal supply chain in the food industry—a conceptual model. In: 2011 IEEE symposium on business, engineering and industrial applications (ISBEIA), pp 384–389

Permana CT, Harsanto B (2020) Sustainable city planning concepts and practices in emerging economies: a systematic review. J Indonesia Sustain Dev Plann I(1):67–82

Petticrew M (2001) Systematic reviews from astronomy to zoology: myths and misconceptions. BMJ 322(7278):98–101

Rahman MK, Zailani S (2017) The effectiveness and outcomes of the Muslim-friendly medical tourism supply chain. J Islam Mark 8(4):732–752. https://doi.org/10.1108/JIMA-11-2015-0082

Randeree K (2019) Challenges in halal food ecosystems: the case of the United Arab Emirates. Br Food J 121(5):1154–1167. https://doi.org/10.1108/BFJ-08-2018-0515

Saudi MHM, Sinaga O, Gusni G, Zainudin Z (2019) The effect of green innovation in influencing sustainable performance: Moderating role of managerial environmental concern. Int J Supply Chain Manag 8(1):303–310

Scimago (2021) Scimago. J Country Rank

Secinaro S, Calandra D (2020) Halal food: structured literature review and research agenda. Br Food J 123(1):225–243. https://doi.org/10.1108/BFJ-03-2020-0234

Surjandari I, Yusuf H, Laoh E, Maulida R (2021). Designing a permissioned blockchain network for the halal industry using hyperledger fabric with multiple channels and the raft consensus mechanism. J Big Data 8(1). https://doi.org/10.1186/s40537-020-00405-7

Tan KH, Ali MH, Makhbul ZM, Ismail A (2017) The impact of external integration on halal food integrity. Supply Chain Manag 22(2):186–199. https://doi.org/10.1108/SCM-05-2016-0171

Tieman M (2011) The application of Halal in supply chain management: In-depth interviews. J Islam Mark 2(2):186–195. https://doi.org/10.1108/17590831111139893

Tieman M (2015) Halal clusters. J Islam Mark 6(1):2–21. https://doi.org/10.1108/JIMA-05-2014-0034

Tieman M, van Nistelrooy M (2014) Perception of Malaysian food manufacturers toward halal logistics. J Int Food Agribus Mark 26(3):218–233. https://doi.org/10.1080/08974438.2013.833572

Tieman M, van der Vorst JGAJ, Ghazali MC (2012) Principles in halal supply chain management. J Islam Mark 3(3):217–243. https://doi.org/10.1108/17590831211259727

Tieman M, Ghazali MC, Van Der Vorst JGAJ (2013) Consumer perception on halal meat logistics. Br Food J 115(8):1112–1129

Tranfield D, Denyer D, Smart P (2003) Towards a methodology for developing evidence-informed management knowledge by means of systematic review. Br J Manag 14:207–222. https://doi.org/10.1111/1467-8551.00375

Widianto S, Harsanto B (2017) The impact of transformational leadership and organizational culture on firm performance in Indonesia SMEs. In: Muenjohn N, McMurray A (eds) The Palgrave handbook of leadership in transforming Asia. Palgrave Macmillan, UK, pp 503–517. https://doi.org/10.1057/978-1-137-57940-9_27

Yaacob TZ, Rahman FA, Jaafar HS (2018) Risk categories in halal food transportation: a preliminary findings. Int J Supply Chain Manag 7(6):453–461

Zainuddin N, Saifudin AM, Deraman N, Mahidin N (2019) Effect of Halal certification and labelling process on Halal supply chain performance. Int J Supply Chain Manag 8(4):1075–1081

Zainuddin N, Saifudin AM, Deraman N, Osman AA (2020) The effect of halal traceability system on halal supply chain performance. Int J Supply Chain Manag 9(1):490–498

Zulfakar MH, Chan C, Jie F (2018) Institutional forces on Australian halal meat supply chain (AHMSC) operations. J Islam Mark 9(1):80–98. https://doi.org/10.1108/JIMA-01-2016-0005

Halal and Digitalization: A Bibliometric and Content Analysis

8

Anita Priantina and Devinna Pramitha

Abstract

Research on halal has been conducted in various disciplines and on various topics. Bibliometric analysis on halal-related studies has also been conducted several times. As information technology and computer-based applications develop rapidly, digitalization in the halal industry has become an emerging concern. This study aims to analyze the articles indexed by Google Scholar on the topics of halal and digitalization. This index is used to allow for wider coverage and to include articles in Bahasa Indonesia. This is important since halal certification has now become mandatory in the country. This has attracted researchers to conduct more studies on halal-related research. The keywords used are limited to halal and digital. Bibliometric and content analysis is used to structure the discussion and answer the research question. The result for articles in Bahasa Indonesia and articles in English are separated to see whether there is any difference. The result shows that the research topics in Bahasa Indonesia are dominated by Muslim-friendly tourism and digital marketing for almost all halal industry sectors. Papers in English have more varied topics including cryptocurrency but are dominated by cases in Indonesia and Malaysia. Therefore, research in Bahasa Indonesia related to halal and digitalization may cover uncovered topics such as halal bitcoin and cryptocurrency. Research in English may adopt cases from the two most mentioned cases, but with implementation in other countries.

Keywords

Halal • Digitalization • Bibliometric • Google Scholar

A. Priantina (✉) · D. Pramitha
Institut Agama Islam Tazkia, Bogor, Indonesia
e-mail: anitapriantina@tazkia.ac.id

A. Rafiki (ed.), *Digitalization in Halal Management*, Contributions to Environmental Sciences & Innovative Business Technology,
https://doi.org/10.1007/978-981-99-5146-8_8

8.1 Introduction

Halal is a topic of various inter and intra-related disciplines including natural sciences and social sciences. The number of research related to halal worldwide has increased significantly, especially for cases in Indonesia. Since Indonesia announced the halal certificate to be mandatory in 2014, interest to study on halal has attracted researchers more than before.

There are at least twelve academic journals based in Indonesia dedicated to halal research. Four of them are indexed by Science and Technology Index (SINTA) under the Ministry of Education. They are the Indonesian Journal of Halal Research (IJHAR), Journal of Digital Marketing and Halal Industry, Journal Agroindustri Halal, and Indonesia Journal of Halal. These four journals and another eight journals are listed under *Garba Rujukan* Digital (Garuda: Digital Reference Platform) under the Ministry of Education. The eight names are Ar Rehla: Journal of Islamic Tourism, Halal Food, Islamic Traveling, and Creative Economy, Food Technology and Halal Science Journal, Halal Research Journal, IJMA: International Journal Mathla'ul Anwar of Halal Issues, Journal of Halal Product and Research (JHPR), Journal of Halal Science and Research, Journal of Industrial Engineering and Halal Industries, and *Jurnal Bisnis Ekonomi Halal*.

The worldwide increasing use of computers and the Internet, combined with the impact of the COVID-19 pandemic have disrupted manual and traditional economic and business activities. Improvements and innovations are needed to survive and to win the competition. Among others, the solution adopted globally is the further acceleration of digitalization. The use of information technology has spread to the entire value chain, from consumers to producers. The role of digitalization has enabled global Muslim spending on the halal food industry to reach a market value of USD 1.27 trillion, Islamic finance to USD 3.6 trillion, Muslim-friendly tourism to USD 58 billion, modest fashion into USD 295 billion, halal pharmaceutical into USD 100 billion, halal cosmetics into USD 70 billion, and Muslim-friendly media into USD 216 billion (Dinar Standard 2022).

As information technology and computer-based applications develop rapidly, the digitalization of halal has now become an emerging concern. The discussion on the use of digitation has expanded from halal food, halal tourism, halal cosmetics, halal healthcare, halal media, and halal finance. To examine the research that has been conducted and how far the initiative to digitize halal goes, it is important to review previous research on this topic.

In the next part of this study, the definition of digital is explained to clarify the scope of the study. The importance of the study stems from the discussion of previous studies on halal that used bibliometric analysis, such as the one conducted in this paper. Next, the methodology of data collection, data filtering, and data analysis are explained in detail. The results and interpretation of the data analysis, followed by a discussion and recommendations for further studies, conclude the study.

8.2 Literature Review

8.2.1 Halal and Digital

Definition of Digital

According to Thesaurus Dictionary, "digital" means "pertaining to, noting, or making use of computers and computerized technologies, including the internet". Oxford Learners Dictionary defines "digital" as "connected with the use of computer technology, especially the internet". Digitalization or issues related to halal and digitalization will therefore highlight how computers or the Internet as part of information technology is used on halal-related disciplines.

8.2.2 Halal and Bibliometric Studies

Bibliometric studies related to halal have covered various topics. Harzing Publish or Perish shows that there are more than twenty studies published, by Scopus and Google Scholar-indexed publications. Some researchers study halal issues in general such as Hashim et al. (2022) and Haleem et al. (2020), and some other researchers study a specific part of the halal industry.

Most halal-related bibliometric studies discuss halal tourism. Among others are Bayuny (2022), Mas'ud and Hamim (2022), Abdullah (2021), Hidayat et al. (2021), Rusydiana et al. (2021), Suban et al. (2021), and Yagmur et al. (2020). The second most studied bibliometric analysis is halal food. Some researchers such as Agrawal and Tripathi (2021), Rejeb et al. (2021), Handayani et al. (2022), Priantina and Sapian (2021), Antonio et al. (2020), Omar et al. (2020), and Wahyuni et al. (2019) link the study from supply chain perspective. Next on the list is studies related to halal cosmetics. Suryani (2022), Zakaria et al. (2022), and Azmi et al. (2021) are among the studies published. Izza (2022) used bibliometric analysis for halal fashion research.

Research that may be closely related to halal and digitalization are like Masudin et al. (2022) discussing on halal traceability system and Yanti et al. (2022) on the halal blockchain. So far, research specifically mentioning halal and digitalization in the title has not been found.

8.3 Data and Method

8.3.1 Data Collection and Filtering

Google Scholar database is chosen for wider coverage and to include papers in Bahasa Indonesia. The list of keywords or combination of keywords chosen for this study is as follows:

1. "halal" AND "digital" are to be found in the title. This combination of keywords results in 85 papers.
2. "halal" is to be found on the title and "digital" is to be found on the body of the text. This combination of keywords results in 997 papers.
3. "halal" AND "online" are to be found in the title. This combination of keywords results in 119 papers.

Note: AND is written in capital letters as a filter tool for searching code on the used application.

After searching, the next step is filtering. The second combination of the keywords results in a variety of content. It turns out that the word "digital" may be found in the body of the text or even in the reference. Papers in languages other than English and Bahasa Indonesia are also excluded from this study. Redundancy is also checked. The total number used for the analysis is 279 Bahasa Indonesia articles and 481 English articles.

8.3.2 Method

Bibliometric and content analysis are used to structure the discussion and address the research objective. This method was chosen because it is used to identify information about the selected articles. The bibliometric data used for the discussion include the author(s)'s name, title and abstract, name and year of publication, and the number of Google Scholar citations. This study refers to and modifies the steps of bibliometric analysis as introduced by Fahimnia et al. (2015). The discussion is then started with descriptive statistics on the number of articles per year, prominent articles based on the number of citations, and prominent publications based on the number of articles.

For the analysis of the network of keywords in abstracts using Vosviewer software, this research adopts Cancino et al. (2017). The Vosviewer visualization of the abstracts is elaborated to show the keywords clustering, and the overlay visualization is also applied to show the trend of the keywords used in recent years. To enrich the discussion and confirm the visualization of the keyword network, a word map analysis using Atlas.ti is added to capture the full texts of all selected articles.

In summary, the stages of the research method for this study (Fig. 8.1):

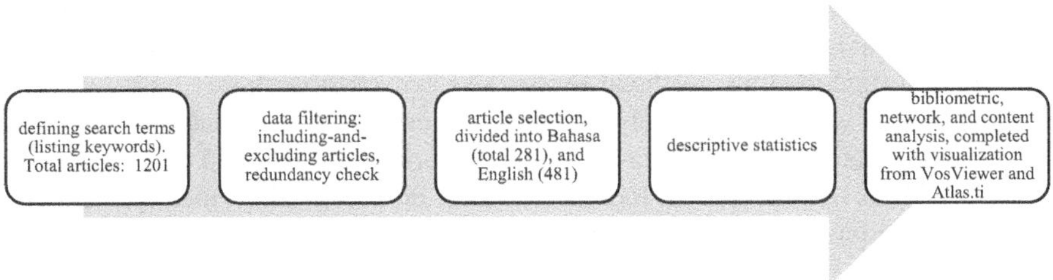

Fig. 8.1 Stages of research

8.4 Result and Discussion

8.4.1 Descriptive Statistics

The earliest article in Bahasa Indonesia was published in 2012 while for an English article, it was in 2003. Both line charts show an increasing trend with the peak in 2021, with each 85 articles in Bahasa Indonesia and 118 articles in English (Table 8.1 and Fig. 8.2).

For the Bahasa Indonesia article, the highest Google Scholar citations are Nugraha (2018), Mahardika (2020), Hermawan (2019), Zahrah and Fawaid (2019), and Kamila (2021). Three of the papers discussed on digitalization as a solution for halal tourism, especially for marketing purposes. Each of the rest discussed digitalization for halal food and the halal industry in general (Table 8.2).

English papers in the context of digitalization have more citations, probably because they have a wider coverage of readers. The coverage of the top-cited articles is also more varied. Butt and Aftab (2013) discussed on online Islamic banking service. Bohari et al. (2013) and Yousaf and Xiucheng discussed on digitalization for the halal food industry. Akyol and Klininc and Yousaf and Xiucheng study on digital marketing and halal tourism. Nurjuliana et al. (2011) is the only science paper from the top-cited papers that discussed the halal authentication method (Table 8.3).

There are three publications which have at least three Bahasa articles on halal and digital. They are *Jurnal Ilmiah Ekonomi Islam* (JIEI), *Sewagati*, and Halal Research Journal. All of them are academic journals and are home-based in Indonesia. JIEI has five articles related to this issue with varied topics. Azizuddin and Ainulyaqin (2022) and Ma'rifah and Normasyuhi discussed on halal tourism, Komariah et al. (2022) and Nugroho and Rahman (2022) discussed halal food, and Lubis and Izzah (2022) discussed on the usage of Internet related to the halal industry in general.

Sewagati is an academic journal which publishes empowerment project reports. Therefore, articles published here are related to how to use computers or Internet to empower halal-related economic activity, such as how to apply for halal certification (as discussed by Savitri and Putra 2022; Gunawan et al. 2021) and how to educate on halal industry (Gunawan et al. 2022).

Adinegoro (2022), Qoni'ah (2022), and Ashari (2021) published their article in Halal Research Journal. The topics are related to digital marketing for the halal industry and digitalization for halal logistics systems (Table 8.4).

Publications with a significant number of papers related to halal and digitalization are dominated by Indonesia-based publications. JDMHI and the proceeding of AICIEB 2021 are based in Indonesia. Halal cryptocurrency management is a book or compilation of book chapters, as the name says it covers halal finance.

Published in JDMHI, Peristiwo (2021) discussed on halal tourism, Purwanto et al. (2021) discussed on halal assurance, and Fataron and Rohmah (2020) discussed on halal purchase. Proceeding of AICIEB also has quite similar coverage. Apriliani et al. (2021) discussed on online halal food purchases, Trimulato (2021) and Nurohman and Qurniawati (2021) discussed on halal finance, Trishananto et al. (2021) discussed on halal culinary tourism, and Siswoyo (2021) discussed on using digitalization to strengthen the halal industry in general.

Table 8.1 Bahasa papers with highest citation

Cites	Authors	Title	Year	Source	Indexed by
26	Nugraha	*Analisis potensi promosi pariwisata halal melalui e-marketing di Kepulauan Riau* (Analysis of the potential for halal tourism promotion through e-marketing in the Riau Islands province)	2018	*Jurnal Penelitian dan Karya Ilmiah Lembaga Penelitian Universitas Trisakti* (Journal of Research and Scientific Work of Trisakti University Research Institute)	Sinta 5
24	Mahardika	*Strategi Pemasaran Wisata Halal* (Halal tourism marketing strategy)	2020	*Mutawasith: Jurnal Hukum Islam* (Journal of Islamic Law)	Google Scholar
23	Hermawan	*Strategi Kementerian Pariwisata Indonesia Dalam Meningkatkan Branding Wisata Halal* (Strategy of the Indonesia Ministry of Tourism in improving halal tourism branding)	2019	*Jurnal Ilmu Manajemen dan Akutansi* (Journal of Management and Accounting Sciences)	Sinta 6

(continued)

Table 8.1 (continued)

Cites	Authors	Title	Year	Source	Indexed by
23	Zahrah and Fawaid	*Halal Food di Era Revolusi Industri 4.0: Prospek dan Tantangan* (Halal food in the era of the Industrial Revolution 4.0: Prospects and challenges)	2019	Hayula: Indonesian Journal of Multidisciplinary Islamic Studies (Indonesian Journal of Multidisciplinary Islamic Studies)	Sinta 2
22	Kamila	*Peran Industri Halal Dalam Mengdongkrak Pertumbuhan Ekonomi Indonesia Di Era* New Normal (The Role of Halal Industry in Boosting Indonesia's Economic Growth in the New Normal Era)	2021	*LIKUID: Jurnal Ekonomi Industri Halal* (Journal of Halal Industrial Economics)	Google Scholar

Source Author's compilation

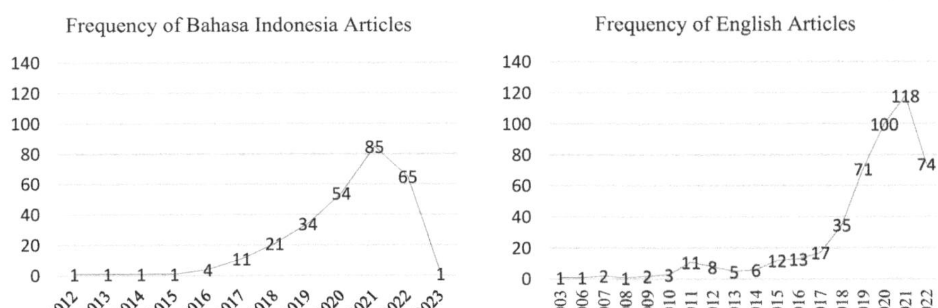

Fig. 8.2 Frequency of Bahasa Indonesia (Left) and English (Right) Articles. *Source* Author's compiled numbers

As a compilation of book chapters published by Springer Link, Habib and Adekunle (2019), Beik et al. (2019a, b), Zulkhibri (2019), Billah (2019a, b, c, d), and Kakkatil discussed on various sides of halal cryptocurrency. The topics cover both conceptual and empirical approaches.

Table 8.2 English papers with highest citation

Cites	Authors	Title	Year	Source	Indexed by
240	Butt and Aftab	Incorporating attitude toward halal banking in an integrated service quality, satisfaction, trust and loyalty model in online Islamic banking context	2013	International Journal of Bank Marketing	Scopus-Q2
206	Bohari et al.	An analysis on the competitiveness of halal food industry in Malaysia: An approach of SWOT and ICT strategy	2013	Malaysia Journal of Society and Space	Web of Science
191	Yousaf and Xiucheng	Halal culinary and tourism marketing strategies on government websites: A preliminary analysis	2018	Tourism Management	Scopus-Q1
186	Nurjuliana et al.	Rapid identification of pork for halal authentication using the electronic nose and gas chromatography-mass spectrometer with headspace analyzer	2011	Meat Science	Scopus-Q1
156	Akyol and Kilinç	Internet and halal tourism marketing	2014	Electronic Turkish Studies	Scopus-Q1

Source Author's compilation

8.4.2 Vosviewer and Atlas.ti Analysis for Articles in Bahasa Indonesia

The clustering result from Bahasa Indonesia articles abstracts is as follows:

Cluster 1 (159 items): Activity, application, area, article, aspect, attention, awareness, challenge, communication, concept, consumer, consumption, country, covid, culture, data collection, data technique, depth interview, development, documentation, economy, effort, effort, era, existence, factor, fashion, food, form, function, government, halal certification, halal food, halal industry, halal product, halal tourism, impact, indicator, information, internet, interview, knowledge, library research, majority, marketing, MSMEs, Muslim population, Muslim tourist, Muslims, number, object, observation, opportunity, order, pandemic, paper, person, population, potential, process, qualitative approach, research, research method, researcher, role, secondary data, sector, service, social medium, stakeholder, strategy, study, technology, tourism, tourism sector, type, use, world (Fig. 8.3).

Table 8.3 Publication with frequent Bahasa articles on halal and digitalization

Source	Indexed by	Authors	Title	Year
Jurnal Ilmiah Ekonomi Islam	Sinta 3	Azizuddin and Ainulyaqin	*Industri Pariwisata Halal: Pendorong Inovasi Untuk Halalpreneurs* (Halal tourism industry: Innovation drivers for halalpreneurs)	2022
		Komariah et al.	*Pengaruh Inovasi dan Media Sosial terhadap Kinerja UMKM Makanan Halal di Kota Bekasi Utara* (The influence of innovation and social media on the performance of halal food MSMEs in North Bekasi City)	2022
		Lubis and Izzah	*Faktor Penentu Gaya Hidup Halal Generasi Z di Sumatera Utara* (Determinants of generation Z halal lifestyle in North Sumatra)	2022
		Ma'rifah and Normasyhuri	*Daya Tarik Halal Tourism Pada Minat Wisatawan di Indonesia Ketika Era Covid-19* (The attraction of halal tourism to tourist interest in indonesia during the COVID-19 era)	2022
		Nugroho and Rahman	*Digitalisasi dan Keberlangsungan UMKM Kuliner Halal Selama Pandemi Covid-19* (Digitalization and sustainability of halal culinary MSMEs during the COVID-19 pandemic)	2022
Sewagati	Sinta 4	Gunawan et al.	*Merajut Ekosistem Industri Halal dalam Menumbuhkembangkan Usaha Kompetitif Melalui Merdeka Belajar* (Linking the halal industry ecosystem in enhancing business competitiveness through independent learning program)	2022

(continued)

Table 8.3 (continued)

Source	Indexed by	Authors	Title	Year
		Savitri and Putra	*Sosialisasi Sistem Jaminan Halal Bagi Pelaku Usaha Mikro, Kecil dan Menengah (UMKM)* (Socialization of halal assurance system for micro, small, and medium enterprises (MSMEs))	2022
		Gunawan et al.	*Pendampingan Berkelanjutan Sistem Jaminan Halal Bagi Usaha Mikro, Kecil, dan Menengah (UMKM)* (Sustainable assistance program for halal assurance system for micro, small, and medium enterprises (MSMEs)	2021
Halal Research Journal	Google Scholar, Moraref	Adinegoro	*Relasi media sosial dan pasar dalam perubahan nilai publik mengenai pengaruh dakwah di media sosial terhadap pandangan gaya hidup halal* (The relationship between social media and the market in changing public values regarding the influence of Islamic preaching on social media on the view of halal lifestyle)	2022
		Qoni'ah	*Tantangan dan strategi peningkatan ekspor produk halal Indonesia di pasar global* (Challenges and strategies to increase exports of Indonesia halal products in the global market)	2022
		Ashari	*Pengembangan sistem logistik produk halal di Indonesia* (Development of halal product logistics system in Indonesia)	2021

Source Author's compilation

Cluster 2 (64 items): *analisis data, berdasarkan, data, deskriptif, di* Indonesia, digital marketing, dsn mui, *dunia*, gmti, *hal, halal*, halal awareness, halal *di Indonesia*, haram, *harga, hasil, hasil penelitian ini menunjukkan, hasil penelitian menunjukkan*, Indonesia, islam, *kata kunci, kedua, konsumen, kualitif, label halal, makanan halal, masyarakat*, medium, *mengenai, metode, minat beli*, model,

Table 8.4 Publication with frequent English articles on halal and digitalization

Source	Indexed by	Authors	Title	Year
Journal of digital marketing and halal industry (JDMHI)	Sinta 3	Peristiwo	Impact of the COVID-19 pandemic on Indonesia halal tourism transportation	2021
		Purwanto et al.	Policy analysis of enforcement of halal product guarantee regulations through the regulatory impact analysis (RIA) approach	2021
		Fataron and Rohmah	Effect analysis of trust, ease, information quality, halal product on online purchase decision of 2016–2018 batch students of Islamic economics study program in UIN Walisongo at Shopee marketplace	2020
Halal cryptocurrency management	Springer Link Book	Habib and Adekunle	Case study of bitcoin and its halal dimension	2019
		Beik et al.	Waqf led halal cryptocurrency model	2019b
		Beik et al.	Zakat standard framework of halal cryptocurrency	2019a
		Zulkhibri	Halal cryptocurrency and financial stability	2019
		Billah	Halal cryptocurrency: Its establishment and operational mechanisms	2019b
		Billah	SWOT analysis of halal cryptocurrency structure	2019d

(continued)

Table 8.4 (continued)

Source	Indexed by	Authors	Title	Year
		Billah	Cryptocurrency? Its halal alternative model	2019a
		Billah	Innovative action of halal cryptocurrency management	2019c
		Kakkattil	Blockchain technology in managing halal cryptocurrency	2019
Annual international conference on Islamic Economics and business (AICIEB)	Google Scholar	Apriliani et al.	The influence of brand image, halal label, and product quality on Kentucky fried chicken (KFC) purchase decisions	2021
		Siswoyo	Strengthening halal industry in increasing competitiveness and economic opportunities in Industrial Revolution era 4.0	2021
		Trimulato	Linkage sharia banking and sharia fintech to support halal industry in Indonesia	2021
		Trishananto et al.	Halal culinary tourism promotion strategy in improving performance of small and medium enterprises through digital marketing in Batam city	2021
		Nurohman and Qurniawati	The intention to use e-money: An empirical study of halal food SMEs in Surakarta	2021

Source Author's compilation

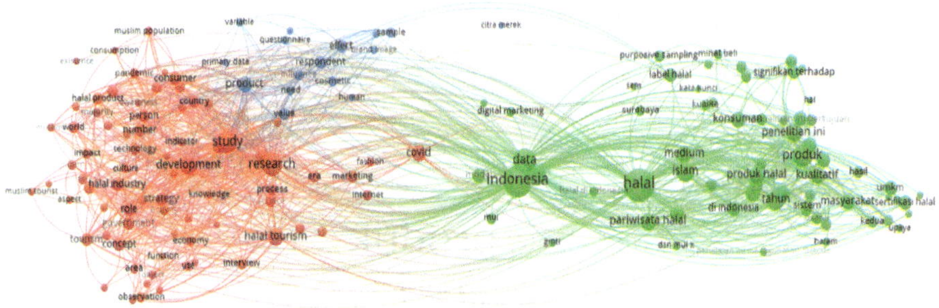

Fig. 8.3 Vosviewer visualization for Bahasa Indonesia articles. *Source* Author's drawn visual

mui, muslim, *pariwisata halal, penelitian ini, penelitian ini bertujuan, penelitian ini dilakukan, penelitian ini menggunakan, penelitian ini menggunakan metode, produk, produk halal*, purposive halal, purposive sampling, *responden*, sem, sertifikasi *halal, signifikan terhadap*, system, *sjh, strategi*, Surabaya, *syariah, tahun, teknologi, tentang jaminan produk halal, uji, umkm, undang undang nomor, upaya, usaha, wawancara, wisata, wisata halal*, Yogyakarta.

Cluster 3 (16 items): Brand image, *citra merek*, cosmetic, effect, halal label, human, influence, need, primary data, product, quantitative approach, questionnaire, respondent, sample, value, variable.

Some of the keywords are in English because some publication also requires authors to provide abstract in English. There are several tourism-related keywords on some clusters such as "halal tourism", "Muslim tourist", "*pariwisata halal*" (halal tourism), "tourism", "tourism sector", "*wisata*" (tourism), and "*wisata halal*" (halal tourism). The third cluster has keywords which indicate that numerous empirical papers use quantitative analysis (Fig. 8.4).

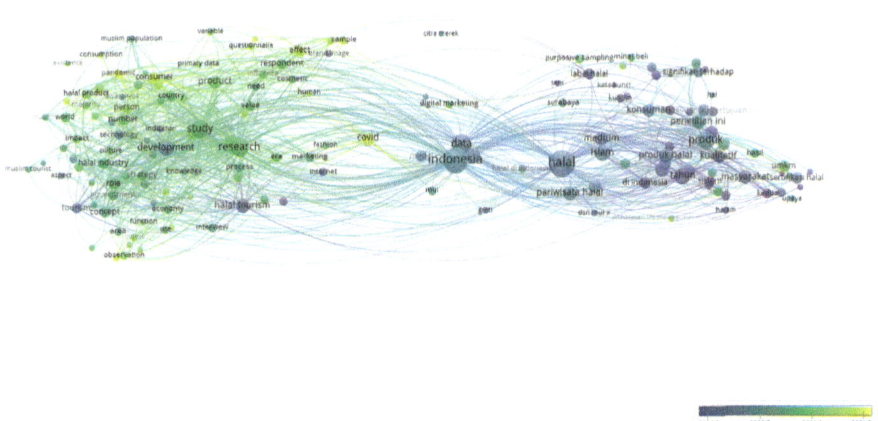

Fig. 8.4 Overlay visualization for Bahasa articles. *Source* Author's drawn visual

Fig. 8.5 Atlas.ti word map for Bahasa articles. *Source* Author's drawn visual

Overlay visualization shows trending keywords on the abstracts in the last four periods. We can see that "covid", "pandemic", and "consumer" are among the keywords highlighted yellow, indicating the major research interest focus on research since the middle of 2021 (Fig. 8.5).

As Vosviewer analyzes abstracts, Atlas.ti word map is resulted from the full texts. From the size of some keywords, it can be concluded that there is a significant amount of research discussing on halal tourism, as the words "*wisata*"/ "*pariwisata*" (tourism), and "tourism" are among the words with the biggest size. As the articles were all in Bahasa Indonesia, it is just natural that Indonesia is the only country mentioned, probably as the study case or as the origin of the author(s).

8.4.3 Vosviewer and Atlas.ti Analysis for Articles in English

The clustering result from English articles abstracts is as follows:

Cluster 1: (89 items) Accordance, addition, adoption, benefit, beverage, blockchain, blockchain technology, business, case study, certification, challenge, chapter, company, competitiveness, concern, consumption, cosmetic, cost, cryptocurrency, demand, distribution, efficiency, emergence, fact, field, food, food industry, food product, framework, gap, global halal industry, goal, growth, guideline, halal, halal certification, halal concept, halal food industry, halal industry, halal market, halal status, halal supply chain, hand, health, identification, implementation, industry, ingredient, innovation, internet, Islam, Islamic law, issue, lack, law, literature review, Malaysia, market, meat, Muslim population, Muslims, overview, paper, perspective, practice, presence, process, product, production, regulation, religion, review, scholar, scope, situation, SMEs, source, state, success, supply chain, system, technology, thing, traceability, transaction, use, view, way, world.

Cluster 2 (70 items): Activity, area, article, aspect, case, communication, community, concept, contribution, country, covid, culture, depth interview, destination, development, digital marketing, economy, education, effort, era, existence, government, halal lifestyle, halal tourism, halal tourism destination, increase, Indonesia, interview, line, literature study, majority, marketing, marketing strategy, Muslim community, Muslim tourist, need, number, object, observation, opportunity, order, pandemic, place, policy, population, potential, promotion, qualitative approach, reason, recommendation, region, research, research method, secondary data, sector, service, social medium, stakeholder, strategy, term, tourism, tourism industry, tourism sector, tourist, trend, type, value, website, year.

Cluster 3 (63 items): Analysis, approach, attitude, author, awareness, brand, consumer, consumer behavior, context, customer, data, decision, design methodology approach, effect, entrepreneur, factor, future, halal awareness, halal cosmetic, halal food, impact, implication, importance, influence, insight, intention, insight, intention, interest, knowledge, limitation, literature, marketer, methodology, model, Muslim, Muslim consumer, organization, originality value, perception, practical implication, primary data, purchase intention, quality, question, questionnaire, regard, relationship, religiosity, research limitation implication, researcher, respondent, role, sample, satisfaction, significant effect, study, subjective norm, survey, theory, topic, total, trust, understanding, variable.

Cluster 4 (27 items): Application, attention, business actor, condition, ease, form, halal certificate, halal label, halal logo, halal product, image, indicator, information, JAKIM, level, part, performance, person, previous study, problem, public, response, solution, technique, time, today, user (Fig. 8.6).

Among the difference between English and Bahasa articles is that there are some keywords related to halal finance issues such as "blockchain", "blockchain technology", and "cryptocurrency". As "Indonesia" and "Malaysia" are the only countries mentioned in all clusters, it seems that studies on halal and digitalization issues are dominated by cases or authors based in Indonesia (Fig. 8.7).

The overlay visualization shows quite a similar result on the increasing trend, the difference is that the trend in Bahasa Indonesia has a lag of approximately one year. What was trending in the middle of 2021 for Bahasa Indonesia articles was actually the trend one year before for English articles. Other than that, English articles have an increasing trend of discussing issues related to cryptocurrency (Fig. 8.8).

The word map shows a similar result with Fig. 8.5 as "tourism" is among the most mentioned word. "Food", "marketing", and "technology" are next on the biggest size word indicating that research on halal and digitalization has

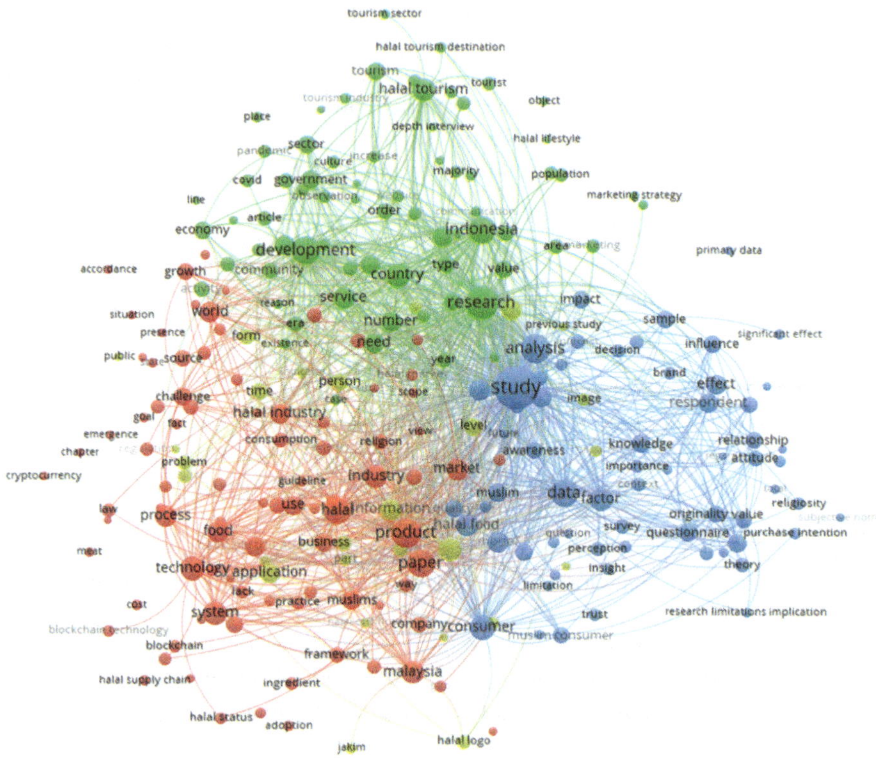

Fig. 8.6 Vosviewer visualization for English articles. *Source* Author's drawn visual

also the significant number of research interest on halal food, digital marketing, and how to use technology to boost the halal industry. There are now more countries mentioned; "Indonesia" and "Malaysia" indicating that case studies or authors affiliations from these two countries lead the publication on halal and digitalization.

8.5 Conclusion and Recommendation

The result shows that the research topics in Bahasa Indonesia are dominated by Muslim-friendly tourism and digital marketing for almost all halal industry sectors. Papers in English have more varied topics including cryptocurrency but are dominated by cases in Indonesia and Malaysia. Therefore, research in Bahasa Indonesia related to halal and digitalization may cover uncovered topics such as halal bitcoin and cryptocurrency. Research in English may adopt cases from the two most mentioned cases, but with implementation in other countries.

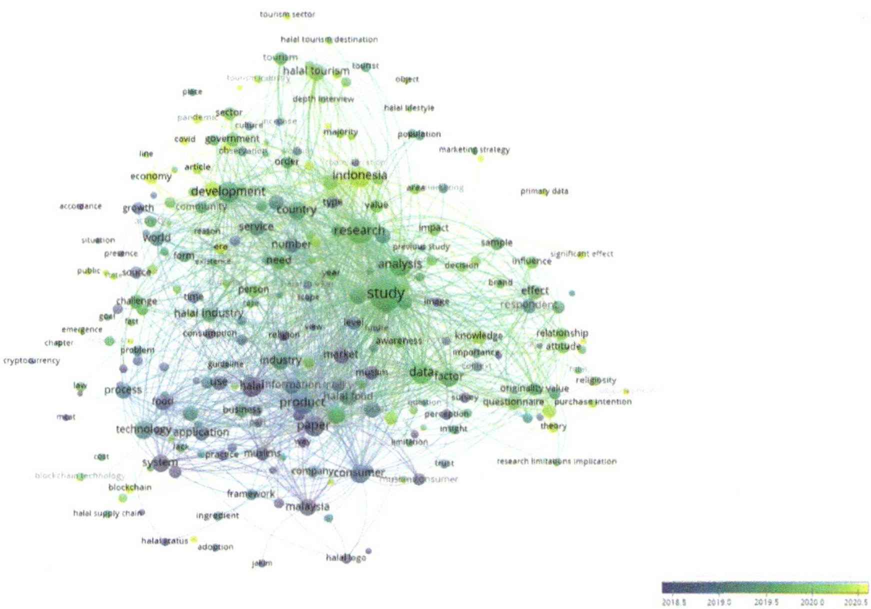

Fig. 8.7 Overlay visualization for English articles. *Source* Author's drawn visual

Fig. 8.8 Atlas.ti word map for English articles. *Source* Author's drawn visual

8.6 Limitation of Study

This study is limited only to articles containing the words, "halal", "digital", and "online". There might be papers with content related to this topic which do not mention the word but instead uses the synonyms. Therefore, it is highly possible that this paper does not cover all papers with similar or interrelated topics.

References

Abdullah KH (2021) Publication trends on halal tourism: a bibliometric review. Halalpshere 1(2):41–53. https://doi.org/10.31436/hs.v1i2.29

Adinegoro KRR (2022). Relasi media Sosial dan Pasar dalam Perubahan Nilai Publik mengenai Pengaruh Dakwah di media Sosial terhadap Pandangan Gaya Hidup Halal. Halal Res J 2(1):1–7. https://doi.org/10.12962/j22759970.v2i1.214

Agrawal N, Tripathi AP (2021) A bibliometric analysis of halal food and certification: existing research & future directions. Delhi Bus Rev 22(2):9–20. https://doi.org/10.51768/dbr.v22i2.222202102

Akyol M, Kilinç Ö (2014) Internet and halal tourism marketing. Electr Turk Stud 9(8):171–186. https://doi.org/10.7827/TurkishStudies.7278

Antonio MS, Rusydiana A, Laila N, Hidayat YR, Marlina L (2020) Halal value chain: a bibliometric review using R. Library Philos Pract 4606

Apriliani P, Tikawati T, Fadhilah N (2021) The influence of brand image, Halal label, and product quality on Kentucky fried chicken (KFC) purchase decisions. Ann Int Conf Islam Econ Bus 1:202–218. https://doi.org/10.18326/aicieb.v1i0.21

Ashari RT (2021) Pengembangan Sistem Logistik Produk Halal di Indonesia. Halal Res J 1(1):8–19. https://doi.org/10.12962/j22759970.v1i1.13

Azizuddin I, Ainulyaqin MH (2022) Industri Pariwisata Halal: Pendorong Inovasi Untuk Halalpreneurs. Jurnal Ilmiah Ekonomi Islam 8(1):106–116. https://doi.org/10.29040/jiei.v8i1.4040

Azmi NAN, Noor NAM, Elgharbawy AA (2021) A bibliometric analysis on halal cosmetics over twenty years. Halalpshere 1(1):53–71. https://doi.org/10.31436/hs.v1i1.9

Bayuny AFR (2022) Bibliometric analysis on management of halal tourism. Rev Islam Econ Finance 5(2). https://doi.org/10.17509/rief.v5i2.52657

Beik IS, Nurzaman MS, Sari AP (2019a) Zakat standard framework of halal cryptocurrency. In: Halal cryptocurrency management. Palgrave Macmillan, Cham, pp 275–284

Beik IS, Zaenal MH, Rizkiningsih P (2019b). Waqf led halal cryptocurrency model. In: Halal cryptocurrency management. Palgrave Macmillan, Cham, pp 285–298

Billah MMS (2019a) Cryptocurrency? Its halal alternative model. In: Halal cryptocurrency management. Palgrave Macmillan, Cham, pp 3–14

Billah MMS (2019b) Halal cryptocurrency: its establishment and operational mechanisms. In: Halal cryptocurrency management. Palgrave Macmillan, Cham, pp 221–234

Billah MMS (2019c) Innovative action of halal cryptocurrency management. In: Halal cryptocurrency management. Palgrave Macmillan, Cham, pp 23–33

Billah MMS (2019d) SWOT analysis of halal cryptocurrency structure. In: Halal cryptocurrency management. Palgrave Macmillan, Cham, pp 15–22

Bohari AM, Hin CW, Fuad N (2013) An analysis on the competitiveness of halal food industry in Malaysia: an approach of SWOT and ICT strategy. Malays J Soc Space 9(1):1–11. http://journalarticle.ukm.my/6055/1/1-11.pdf

Butt MM, Aftab M (2013) Incorporating attitude towards halal banking in an integrated service quality, satisfaction, trust and loyalty model in online Islamic banking context. Int J Bank Mark 31(1):6–23. https://doi.org/10.1108/02652321311292029

Cancino C, Merigó JM, Coronado F, Dessouky Y, Dessouky M (2017) Forty years of computers & industrial engineering: a bibliometric analysis. Comput Ind Eng 113:614–629. https://doi.org/10.1016/j.cie.2017.08.033

Dinar Standard (2022) State of the global Islamic economy report. https://www.salaamgateway.com/specialcoverage/SGIE22#:~:text=The%20State%20of%20the%20Global,Islamic%2Dthemed%20media%20and%20recreation

Fahimnia B, Sarkis J, Davarzani H (2015) Green supply chain management: a review and bibliometric analysis. Int J Prod Econ 162(1):101–114. https://doi.org/10.1016/j.ijpe.2015.01.003

Fataron ZA, Rohmah H (2020) Effect analysis of trust, ease, information quality, halal product on online purchase decision of 2016–2018 batch students of Islamic economics study program in UIN Walisongo at Shopee Marketplace. J Dig Mark Halal Indus 1(1):1–8. https://doi.org/10.21580/jdmhi.2019.1.1.4772

Gunawan S, Juwari J, Aparamarta H, Darmawan R, Rakhmawati NA (2021) Pendampingan Berkelanjutan Sistem Jaminan Halal Bagi Usaha Mikro, Kecil, dan Menengah (UMKM). Sewagati 5(1):8–14. https://doi.org/10.12962/j26139960.v5i1.8120

Gunawan S, Rakhmawati NA, Juwari J, Aparamarta HW, Darmawan R, Pradhana YW, Prabowo A (2022) Merajut Ekosistem Industri Halal dalam Menumbuhkembangkan Usaha Kompetitif Melalui Merdeka Belajar. Sewagati 6(4):1–10. https://doi.org/10.12962/j26139960.v6i4.110

Habib F, Adekunle SS (2019) Case study of bitcoin and its halal dimension. In: Halal cryptocurrency management. Palgrave Macmillan, Cham, pp 235–255

Haleem A, Khan MI, Khan S, Jami AR (2020) Research status in halal: a review and bibliometric analysis. Modern Supply Chain Res Appl 2(1):23–41. https://doi.org/10.1108/MSCRA-06-2019-0014

Handayani DI, Masudin I, Haris A, Restuputri DP (2022) Ensuring the halal integrity of the food supply chain through halal suppliers: a bibliometric review. J Islam Mark 13(7):1457–1478. https://doi.org/10.1108/JIMA-10-2020-0329

Hashim YZHY, Ahmad AN, Samsudin N, Abd Lateef MH, Majlan NS (2022) The evolution of halal research: a bibliometric for 1996–2020. J Fatwa Manag Res 27(3):1–38. https://doi.org/10.33102/jfatwa.vol27no3.443

Hermawan E (2019) Strategi Kementerian Pariwisata Indonesia Dalam Meningkatkan Branding Wisata Halal. Jurnal Ilmu Manajemen dan Akuntansi 7(2):87–95. https://core.ac.uk/download/pdf/270210028.pdf

Hidayat SE, Rafiki A, Nasution MDTP (2021) A bibliometric analysis and review of halal tourism. Signifikan: Jurnal Ilmu Ekonomi 10(2):177–194. https://doi.org/10.15408/sjie.v10i2.20270

Izza NN (2022) Bibliometric mapping of halal fashion research development with Scopus index. J Islam Econ Lit 3(2). https://journals.smartinsight.id/index.php/JIEL/article/view/88

Kakkattil SK (2019) Blockchain technology in managing halal cryptocurrency. In: Halal cryptocurrency management. Palgrave Macmillan, Cham, pp 53–67

Kamila EF (2021) Peran industri halal dalam mendongkrak pertumbuhan ekonomi indonesia di era new normal. LIKUID: Jurnal Ekonomi Industri Halal 1(1): 33–42. https://doi.org/10.15575/likuid.v1i1.12731

Komariah NS, Nursal MF, Rianto MR (2022) Pengaruh Inovasi dan media Sosial terhadap Kinerja UMKM Makanan Halal di Kota Bekasi Utara. Jurnal Ilmiah Ekonomi Islam 8(2):2247–2252. https://doi.org/10.29040/jiei.v8i2.5957

Lubis RH, Izzah N (2022) Faktor Penentu Gaya Hidup Halal Generasi Z di Sumatera Utara. Jurnal Ilmiah Ekonomi Islam 8(1):97–105. https://doi.org/10.29040/jiei.v8i1.3961

Ma'rifah AN, Normasyhuri K (2022) Daya Tarik Halal Tourism Pada Minat Wisatawan di Indonesia Ketika Era Covid-19. Jurnal Ilmiah Ekonomi Islam 8(1):328–334. https://doi.org/10.29040/jiei.v8i1.4594

Mahardika R (2020) Strategi Pemasaran Wisata Halal. Mutawasith: Jurnal Hukum Islam 3(1):65–86. https://doi.org/10.47971/mjhi.v3i1.187

Mas'ud R, Hamim K (2022) Bibliometric analysis of journals, authors, and topics related to halal tourism listed in the database Scopus by Indonesian authors. Khazanah Pendidikan Islam 4(2):71–86. https://doi.org/10.15575/kp.v4i2.19677

Masudin I, Rahmatullah BB, Agung MA, Dewanti IA, Restuputri DP (2022) Traceability system in halal procurement: a bibliometric review. Logistics 6(4):67. https://doi.org/10.3390/logistics6040067

Nugraha YM (2018) Analisis Potensi Promosi Pariwisata Halal Melalui E-Marketing Di Kepulauan Riau. Jurnal Penelitian dan Karya Ilmiah Lembaga Penelitian Universitas Trisakti 3(2):63–68. https://doi.org/10.25105/pdk.v3i2.2990

Nugroho AP, Rahman A (2022) Digitalisasi dan Keberlangsungan UMKM Kuliner Halal Selama Pandemi Covid-19. Jurnal Ilmiah Ekonomi Islam 8(2):1654–1660. https://doi.org/10.29040/jiei.v8i2.4470

Nurjuliana M, Man YB, Che H, Mat D, Mohamed AKS (2011) Rapid identification of pork for halal authentication using the electronic nose and gas chromatography mass spectrometer with headspace analyzer. Meat Sci 88(4):638–644. https://doi.org/10.1016/j.meatsci.2011.02.022

Nurohman YA, Qurniawati RS (2021) The intention to use e-money: an empirical study of halal food SMEs in Surakarta. Ann Int Conf Islam Econ Bus 1:46–56. https://doi.org/10.18326/aicieb.v1i0.8

Omar WMW, Ismail MN, Ismail M, Ahmad GO, Ya MS (2020) Visualizing research literature in halal supply chain: a bibliometric analysis. Malays J Cons Family Econ 24(S2):75–92. https://www.majcafe.com/wp-content/uploads/2020/10/Vol-24-S2-2020-Paper-6.pdf

Oxford Learners Dictionary (n.d) Digital. https://www.oxfordlearnersdictionaries.com/definition/english/digital_1?q=digital

Peristiwo H (2021) Impact of the Covid-19 pandemic on Indonesia halal tourism transportation. J Dig Mark Halal Indus 3(1):19–36. https://doi.org/10.21580/jdmhi.2021.3.1.7814

Priantina A, Sapian SM (2021) Halal food and compliance: a bibliometric analysis. In: e-Proceedings of the 3rd international halal management conference 2021 (3rd IHMC 2021). https://oarep.usim.edu.my/jspui/handle/123456789/15541

Purwanto H, Jati S, Rofiq A (2021) Policy analysis of enforcement of halal product guarantee regulations through the regulatory impact analysis (RIA) approach. J Dig Mark Halal Indus 3(2):149–164. https://doi.org/10.21580/jdmhi.2021.3.2.9345

Qoniah R (2022) Tantangan dan Strategi Peningkatan Ekspor Produk Halal Indonesia di Pasar Global. Halal Res J 2(1):52–63. https://doi.org/10.12962/j22759970.v2i1.246

Rejeb A, Rejeb K, Zailani S (2021) Are halal food supply chains sustainable: a review and bibliometric analysis. J Foodserv Bus Res 24(5):554–595. https://doi.org/10.1080/15378020.2021.1883214

Rusydiana AS, Aswad M, Pratomo WA, Hendrayanti E, Sanrego YD (2021) Halal tourism indicators: a bibliometric study. Library Philos Pract 5738. https://digitalcommons.unl.edu/libphilprac/5738/

Savitri NA, Putra RR (2022) Sosialisasi Sistem Jaminan Halal Bagi Pelaku Usaha Mikro, Kecil dan Menengah (UMKM). Sewagati 6(2):224–230. https://doi.org/10.12962/j26139960.v6i2.217

Siswoyo E (2021) Strengthening halal industry in increasing competitiveness and economic opportunities in industrial revolution era 4.0. Ann Int Conf Islam Econ Bus 1:67–74. https://doi.org/10.18326/aicieb.v1i0.9

Suban SA, Madhan K, Shagirbasha S (2021) A bibliometric analysis of halal and Islamic tourism. Int Hosp Rev. https://doi.org/10.1108/IHR-05-2021-0038

Suryani W, Mohd Subri I, Rafiki A, Tobing F, Hidayat SE (2022) Overview of halal cosmetics in a decade: a bibliometric analysis. Al-Iqtishad Jurnal Ilmu Ekonomi Syariah 14(1). https://doi.org/10.15408/aiq.v14i1.23361

Thesaurus Online Dictionary. Digital. https://www.dictionary.com/browse/digital

Trimulato T (2021) Linkage sharia banking and sharia Fintech to support halal industry in Indonesia. Ann Int Conf Islam Econ Bus 1:138–151. https://doi.org/10.18326/aicieb.v1i0.15

Trishananto Y, Arianti DINr, Maulani ED (2021) Halal culinary tourism promotion strategy in improving performance small and medium enterprises through digital marketing in Batam City. Ann Int Conf Islam Econ Bus 1:433–438. https://doi.org/10.18326/aicieb.v1i0.27

Wahyuni H, Vanany I, Ciptomulyono U (2019) Food safety and halal food in the supply chain: review and bibliometric analysis. J Indus Eng Manag 12(2):373–391. https://doi.org/10.3926/jiem.2803

Yagmur Y, Ehtiyar R, Aksu A (2020) Evaluation of halal tourism in terms of bibliometric characteristics. J Islam Mark 11(6):1601–1617. https://doi.org/10.1108/JIMA-05-2019-0101

Yanti R, Febrianti MA, Sulistio J (2022) Halal blockchain: bibliometric analysis for mapping research. Asian J Islam Manag 72–85. https://doi.org/10.20885/AJIM.vol4.iss1.art6

Yousaf S, Xiucheng F (2018) Halal culinary and tourism marketing strategies on government websites: a preliminary analysis. Tour Manag 68(1):423–443. https://doi.org/10.1016/j.tourman.2018.04.006

Zahrah A, Fawaid A (2019) Halal food di Era Revolusi Industri 4.0: Prospek dan Tantangan. Hayula Indones J Multidisc Islam Stud 3(2):121–138. https://doi.org/10.21009/hayula.003.2.01

Zakaria Z, Ramli SQ, Shoid NSZM, Sulaiman A (2022) Mapping halal cosmetics research: a bibliometric analysis. Malays J Syariah Law 10(1):63–75. https://doi.org/10.33102/mjsl.vol10no1.384

Zulkhibri M (2019) Halal cryptocurrency and financial stability. In: Halal cryptocurrency management. Palgrave Macmillan, Cham, pp 35–49

The Relationship Among Religiosity, Consumption, and Life Satisfaction of Muslim Workers

9

Lola Fitria Sari, Syafruddin Karimi, Endrizal Ridwan, and Mohamad Fany Alfarisi

Abstract

This study investigates the relationships among religiosity, consumption, and life satisfaction of Muslim workers in Padang, West Sumatera, Indonesia. Four hundred and twenty questionnaires were distributed to Muslim workers. The respondents are classified into hijra and non-hijra groups. Ideology, rituals, intellectual capacity, consequential and experimental aspects were used as indicators to quantify the independent variable, namely religiosity. Consumption is a moderating variable, and life satisfaction represents the dependent variable. Consumption and life satisfaction are positive and significantly influenced by religiosity. Data analysis was performed using structural equation modeling (SEM). Overall, empirical evidence explains the significant impact of religiosity on the economic life of Muslim workers.

Keywords

Religiosity • Consumption • Life satisfaction

9.1 Introduction

The world records that nearly 80% of its population follows a religious path to life (PEW Research Center 2008). Approximately 70% of the US population states that religion shapes the behavior of their lives (PEW Research Center 2012). Religiosity is reflected ritualistically and symbolically on human behavior (Khraim 2010).

L. F. Sari (✉)
Sekolah Tinggi Ilmu Ekonomi "KBP", Padang, Indonesia
e-mail: lolafitriasari@akbpstie.ac.id

S. Karimi · E. Ridwan · M. F. Alfarisi
Universitas Andalas, Padang, Indonesia

A. Rafiki (ed.), *Digitalization in Halal Management*, Contributions to Environmental Sciences & Innovative Business Technology,
https://doi.org/10.1007/978-981-99-5146-8_9

The level of religiosity and stance of a person represents religion and beliefs that impact their conduct (Sood and Nasu 1995). Therefore, religion is a long-range manifestation of meaningful concepts for a comprehensive behavior pattern including in consumption and production. Religion will guide a person in making economic decisions which are significantly ideological and cultural elements (Fam et al. 2002). For example, the refusal to eat pork by a Muslim and a Jew demonstrates the effect of religion on consumption preferences.

In sociology and psychology, spiritual values are acknowledged as key elements that impact economic behavior (Allport 1965). Economic behavior reflects the environment's social, political, and business prospects and cultural and religious opportunities (Mathras et al. 2016; Minkler 2004). A person's economic vision manifests itself in his economic decision (Schumpeter 1955).

Many things can influence human beings. Some negative influential things can lead to wrong choices in personal finance, healthcare, credit and even damage the world (Thaler 2008). Group consumption decisions or mood swings influence consumer preferences. Moreover, sometimes, consumers make rational buying decisions, and at the other time, they buy on impulse (Pindyck 2013).

Various literature explains the theories and definitions of religion as a subfield of economics. For economists, the definition is essential to the modeling process. Religious economics are relatively new subject of economics research. Some people believe religion and economics are more closely related. Because of this, more studies should focus on the socioeconomic ramifications of religion. The availability of historical and contemporary data highlights the need to identify the impact of religiosity (Iyer 2016). Religion is an essential part of human existence. Even in a predominantly secular country, it remains a primary topic of discussion (Jawad 2012).

Religion can be defined in various ways, notably its functionality and substance (Iyer 2016). It can be seen as a method of beliefs, rituals, symbols, and practices. Moreover, it is expected to encourage closeness with "the Absolute Power" and build up interpersonal interactions in society with religious beliefs (George et al. 2000). In other words, believing in God should only be followed by a commitment to obey His commands.

Religiosity is regarded as a construct of consumer behavior due to its relationship with lifestyle choice (Wilkes et al. 1986). Essoo and Dibb (2004) explain that the impact of commitment to a particular religion significantly influences the consumption behavior. Moreover, as a cultural element, religion is the universal social institution that substantially influences individual and societal attitudes, values, and human behavior. Additionally, consumer and producer preferences are also influenced by their beliefs (Mokhlis 2009).

Islam is a complete way of life. It contains a set of laws that governs all aspects of life. Furthermore, Islam teaches the vertical relationship between humans and God the Creator and the horizontal ties among all God's creations. In view of this, the Islamic teachings comprise the norms and values about spiritual and social life, political activities, culture, legal, and economic activities.

One of the Islamic tenets applicable to the economy is the teaching about consumption. From the Islamic perspective, a person should maximize maslahah in his consumption activities. Imam Syatibi argues that maslahah is not only about utility or satisfaction. Moreover, it has a deeper meaning than similar terminology in conventional economics (Islahi 1944). Maslahah is the nature or tendency of goods and services to help obtain the primary goal of human life (Amin 2019).

Sharia determines the criteria that govern the life of all humankind. Nevertheless, it is contrary to the Pareto optimum concept, which states that a person's optimal condition might increase at the expense of a decrease in the well-being of others (Islahi 1944).[1] The concept of maslahah is relevant to all economic activities, namely production, consumption, and distribution.

The dynamic of religious understanding in the Muslim ummah recently has manifested in a new phenomenon called "hijra". The meaning of hijra here can be found in the following hadits: "A Muslim saves another Muslim from his tongue and hands, and one who emigrates is one who stays away from what Allah has forbidden" (Sahih Bukhari) (Khan 2009). Hijra, in this context, means somebody has to give up anything that Allah Azza wa jalla does not like (ridha). Furthermore, it can also be associated with the concept of repentance (tawbah), in which the wrongdoer returns to the straight path (Islam). Having spent their entire life engaging in sinful activities, people might choose to make a U-turn and be a better person from the perspective of Allah Azza wa jalla.

In the last few decades, many educated urban Indonesians have gone through a spiritual journey to be devoted Muslims. They have been attracted to learning more about Islam after listening to speeches from famous Islamic scholars and preachers in the country. Thanks to the advancement of information technology that has assisted the spread of Islamic content through video streaming platforms and social media. Meanwhile, to strengthen their faith, some urban Muslims have decided to join Islamic groups such as the Salafi, Tarbiyah, and Jamaah Tabligh (Fajriani and Sugandi 2019). Understanding the Islamic tenets gives a strong spirit for developing the phenomenon of hijra, which attracts large audiences, including public figures and trendsetters in society. In addition, social media also helps to spread the message of Islam, reaching the younger generation, notably millennials and gen Z.

Moreover, Islam is a comprehensive life guide that can shape personal and economic behavior. In other words, those who have migrated (hijra) from the false to the right condition are expected to prefer economic activities under the Islamic law (Sharia). With that in mind, we can expect the hijra groups to strongly advocate the Islamic economic practices such as Islamic banking and halal business.

[1] Since the 1970s, the concept of well-being has significantly changed. It goes beyond the traditional well-being concept that focuses more on material elements. In view of this, well-being characteristics are related to the subjective world of feelings and emotions that ultimately determine the quality of life.

9.2 Literature Review

Religion is a significant cultural power and affects economic behavior (Delener 1990). Moreover, it explains the value of human life reflected in social attitudes. A religious person tends to behave according to his values (Johnson et al. 2001).

According to Ellis et al., religiosity can be defined as how an individual commits to religious values.[2] Religious commitment shows how much a person insists on applying religious values in day-to-day activities (Worthington et al. 2003). Furthermore, it exerts an impact on consumers' choices and preferences. Rehman and Shabbir (2010) argue that the consumers prefer to consume the goods and services that are not contrary to their belief system.

Islam means absolute submission to Allah (God) only. To govern the human's life, Allah has prescribed the law called Sharia, mainly derived from the Al-Qur'an and Sunnah of the Prophet Muhammad (Peace be upon him). The Sharia contains the injunction about halal (permitted things) and haram (forbidden things) (Aydin 2013).

Islam is a comprehensive belief system. There is no separation between religion and worldly affairs (secularism) in Islam. For Muslims, the significance and centrality of Islam are determined by its fundamental values. Furthermore, it is not the result of the evolution of human intellectual understanding. In the Qur'an, religion is referred to as "deen", which regulates all aspects of life (Haneef 1997).

Committing any unlawful economic activities is prohibited (haram), while earning income permissible by the Sharia is considered an act of worship (ibadah). For instance, earning interest on a loan (usury) is strongly prohibited. Islam also forbids obtaining unreasonable profits from business transactions which are considered as excessive wealth accumulation (Nasr 2006).

The Islamic economy is a three-sector economy. It consists of (a) the profit-driven private sector, (b) the philanthropic sector, and (c) the public sector. These three sectors interact within the boundaries of Sharia in realizing several objectives, notably the preservation of private property rights, social responsibility, pricing, and market mechanisms (Bhat 2016).

A religious Muslim displays a personality resulting from the Islamic worldview. Moreover, it is expressed through his nature in dealing with various situations in life. It refers to aspects of personality, life's definition, worldview, motivation, behavior, attitudes, and emotions (Morshidi 2015). The relationship between religiosity, worldview, and their impact on action is essential. The worldview philosophically influences personal religiosity (Krauss et al. 2005).

Moreover, moral obligations and enforcement of brotherhood take precedence over personal satisfaction (Bhat 2016). In Islam, the fulfillment of needs must be in line with the Sharia. In view of this notion, not all instinctive desires become the needs until they pass the maslahah test (Raysuni 2013). The economic activities, namely production and consumption must be based on maslahah consideration. It

---

[2] Religiosity and religious commitment can be used interchangeably.

is mainly directed at obtaining goodness as well as happiness in this world and the hereafter. Thus, its central concepts are related to human welfare (faalah) and the good life (tayyibah), emphasizing brotherhood and socioeconomic justice and ensuring that all material and spiritual needs of humans are met in a balanced way (Chapra 1995).

Economic activities that carry maslahah for people are something that must be achieved. The Islamic perspective on economic activities is more on fulfilling the needs than satisfying the wants. Moreover, Islam does not command the man to accomplish all his desires. Maximizing the utility per se is not in line with Islamic tenets. Nevertheless, it is also contrary to the concept of maslahah and leads people to only focus on materialistic achievement (Haneef 1997; Raysuni 2013).

Islam encourages human beings to achieve both material gains and spiritual (immaterial) well-being. Spiritual well-being means the welfare that contains social relations, protection of individual dignity, and emotional relationships. It also encompasses the success in this world and the hereafter. The emphasis on a spiritual dimension increases the pursuit of meaning in life (Kenny and Kenny 2006; Moss 2011). It is deemed a response to modern society's rising individualism.

The secularization theory views the secular state's welfare as the primary substitute for religious principles. However, religion can play a significant role in connecting the cultural and social welfare. Deep-rooted welfare values and beliefs are urgently needed at the individual level (Van Oorschot et al. 2008). People with a low level of religiosity tend to have lower welfare problem than those with high level of religiosity (Pavolini et al. 2017).

Individuals with high levels of religiosity tend to have lower levels of social insurance than secular individuals. Thus, social spending will be lower in countries with higher levels of religiosity (Scheve 2006). Welfare spending and religious participation by cross-country data indicate that the religious participation is a function of government welfare spending which has a negative and robust relationship between the variables after controlling other aspects of modernization (Gill and Lundsgaarde 2004).

From the standpoint of Islamic economics, an Islamic state must introduce an Islamic social welfare system. Moreover, the vision of Islam is to have a just socioeconomic order in society (Ahmad 2011). Three aspects of economic well-being are based on the Al-Qur'an and Hadith, namely poverty alleviation, distribution of wealth, and environmental preservation. The Islamic approach to justice focuses almost exclusively on the distribution of personal income and then addresses fair levels of income and wages for workers and presents the structure of fair distribution described in Islam (Weiss 2002).

Religiosity explains participation in education as social capital based on the arguments of economic and sociological theories (Coleman 1988). The present study has some interesting implications, which can illustrate significant evidence about religion and education in a dynamic life. The results illustrate that individuals in a steady state will allocate time and resources positively to religious activities (Simon Fan 2008). It further strengthens the existing literature on why religions, which appear unproductive, have existed for so long (Coleman 1988; Simon Fan 2008). Religion is an invaluable investment in developing human resources in children and consumption value (Simon Fan 2008).

The relationship between religiosity and economic behavior has contributed to the development of religious economics study. Religiosity in Islam is a complex construction to measure. The present study, therefore, helps to highlight some of the problems in measuring religiosity among Muslims and their influence on behavior in economic decision-making.

The literature has highlighted the role of religion as an essential aspect of the culture that can cause a difference in economic behavior, particularly consumption behavior. The level of religious understanding or religiosity significantly impacts the economic behavior. Furthermore, it is implemented by Muslim in selecting the permitted economic practices and rejecting the prohibited ones. In view of this notion, it can significantly affect the level of life satisfaction.

From the literature on the topic, the researchers accommodate the incorporation of several variables of consumption that are influenced by the dimension of religiosity. This study also looks at the influence of consumption on life satisfaction in the household with a more comprehensive method.

9.3 Data

The data are processed by statistical modeling techniques, structural equation modeling (SEM), which is most commonly used in behavioral science (Jöreskog et al. 2016). SEM is employed to describe the correlation among the main variables in this study.

To collect the data, this study uses a questionnaire. The questionnaire is designed using a Likert scale with the following score: (1) strongly disagree, (2) disagree, (3) simply agree, (4) agree, (5) strongly agree. The statements of economic behavior use the following scale (1) never, (2) rarely, (3) sometimes, (4) often, (5) always.

The present study shows inferential conclusions for the population to compare the levels of belief and the understanding of religion. The samples were classified into hijra and non-hijra groups. Hijra group data were obtained from several Islamic (Hijra-based) organizations in Padang City. There were 210 samples in the hijra group and 210 in the non-hijra group, so the total sample was 420 respondents (Table 9.1).

Table 9.1 Stratified random non-proportional

Stratified	Amount	Non-proportional
Hijra group	7771	210
Non-hijra group	223,561	210
Total	331,332	420

Stratified random sampling with this non-proportional method is possible in certain strata, the amount of sample is close to the number of stratified. Instead, there are relatively small samples with the number of strata. This procedure, known as oversampling or optimal allocation, is used to analyze the strata with a large enough number of samples (Levy and Lemeshow 1999).

Religion is well-defined as a strong faith in supernatural powers that can determine human intention or a society that expresses a strong belief in the power of God. However, a more detailed explanation is needed to describe this research. Therefore, the operational method used to measure the religiosity agrees with the opinion of Rehman and Shabbir (2010). Operationally, the religiosity is determined by five dimensions: ideological, ritualistic, intellectual, consequential, and experimental (Table 9.2).

The ideological dimension consists of the Islamic fundamental belief system: belief in: the God's existence, the holy books, the Prophets, the destiny, the angels, and the judgment day. The ritualistic dimension includes compulsory acts of worship (ibadah) such as shahadah, five times prayer, fasting, compulsory donation, and pilgrimage. The intellectual dimension is a person's knowledge of his religion. The consequential dimension is an aspect of human relations, while the experience dimension explains the experience of carrying out religious teachings.

The characteristics of research respondents are based on gender, age, and occupation, with a total of 420 respondents, which is described in Table 9.3:

9.4 Result and Interpretation

This study aims at investigating the association between religiosity and economic behavior. This section will describe the results on how religiosity relates to consumption behavior.

From Table 9.4, the z statistic is 20.10, and of the P value is less than 0.05. The finding suggests that the religiosity has a positive and significant impact on consumption. It is in line with the study of Wilkes et al. (1986), which analyzes the role of religiosity in determining consumer lifestyles. Moreover, the results show that consumers with higher religious commitment tend to buy something on credit and prefer national products. From these results, it can be said that the religiosity is feasible to be a construct of economic behavior because (1) religiosity is correlated with lifestyle choice variables; (2) religiosity can contribute directly to the model along with age, gender, and income; (3) religiosity has been operated using repeated measurements (Wilkes et al. 1986). Religiosity is the most influential cultural force in economic behavior. Religious commitment has an impact on consumers' preferences. Consumers will buy a product if it seems consistent with their religious teachings (Rehman and Shabbir 2010).

Culture originating from religion affects behavior and social attitudes. Thus, religion is an invaluable construct in understanding human behavior (Khraim 2010; Mokhlis 2009; Carvalho 2012; Cutler 1991; Lindridge 2005; Schouten and McAlexander 1995).

Table 9.2 Data definitions and summary statistics of the final data sample

Variable	Definition
Religiosity	
Ideological	
x1.11	Believing in Allah (God)
x1.12	Believing in Angels
x1.13	Believing in the Holy Books
x1.14	Believing in the Prophets
x1.15	Believing in the Judgment day
x1.16	Believing in Allah's will and destiny
Ritualistic	
x1.21	Praying at the beginning of time five times a day
x1.22	Fasting a whole month in Ramadhan
x1.23	Reciting Qur'an regularly
x1.24	Giving alms (Zakat)
x1.25	Performing Hajj (Pilgrimage)
Intellectual	
x1.31	Rejecting income from haram sources
x1.32	Trying to avoid minor and major sins
x1.33	Learning more about religious knowledge
x1.34	Obeying Islamic rules
x1.35	Learning religious knowledge by reading books and attending lectures
Consequential	
x1.41	Respecting and giving the rights of others
x1.42	Avoiding any act that will hurt others
x1.43	Helping people in need
x1.44	Being honest and fair to others
x1.45	Avoiding talking about other people's disgraces
Experience	
x1.51	Feeling guilty if doing something against Islamic teachings
x1.52	Fearing of Allah
x1.53	Believing that every sin will be punished
x1.54	Being happy to see other people follow the teachings of Islam
Consumption	
x1.21	Consuming based on needs
x1.22	Seeking information first before consuming goods and services
x1.23	Consuming only lawful goods and services
x1.24	Avoiding consuming unlawful products

(continued)

Table 9.2 (continued)

Variable	Definition
x1.25	Limiting the number of product purchased
Life satisfaction	
y1.41	Life satisfaction
y1.42	Living conditions in the next five years
y1.43	Sufficiency of current living conditions
y1.44	Feeling calm in life
y1.45	Feeling happy in life

Table 9.3 Demographic characteristics of the respondents

Characteristic	Frequency	Percentage	Cum
Gender			
Male	350	83.33	83.33
Female	70	16.67	100.00
Age			
16–24	10	2.38	2.38
25–34	126	30.00	32.38
35–49	234	55.71	88.10
50–64	42	10.00	98.10
> 65	8	1.90	100.00
Job			
Self-employed	224	53.33	53.33
Private employee	70	16.67	70.00
Civil servant	42	10.00	80.00
Professional	28	6.67	86.00
Trader	56	13.33	100.00

Table 9.4 Consumption and religiosity

	OIM					
	Coef	Std. err	z	$P > \|z\|$	(95% Conf interval)	
Structural consumption						
Religiosity	1.311584	0.0652409	20.10	0.000	1.183714	1.439454

Table 9.5 shows that the resulting z statistic is 4.51, with a P value of 0.000, which is less than 0.05. With that result, consumption is considered to impact the life satisfaction significantly. Likewise, religiosity on life satisfaction has a

Table 9.5 Consumption, religiosity, and life satisfaction

	OIM					
	Coef	Std. err	Z	*P* > |z|	(95% Conf interval)	
Structural life satisfaction						
Consumption	0.2302607	0.0510664	4.51	0.000	0.1301724	0.3303491
Religiosity	0.9063989	0.0866639	10.46	0.000	0.7365407	1.076257

significant and positive effect, producing a statistical *z* value of 10.46 and a *P* value of 0.000.

Based on Fig. 9.1, the path diagram shows a covariance between the error terms of each manifest variable, where the influence and the direction of the relationship are shown in the *P* value and the direction of the relationship can be seen in the value of the coefficient.

In the path diagram of the standardized solution results, it can be seen that almost all parameters in the model are significant as seen from the *P* value which is less than 0.05.

It can be seen in Table 9.6, that there is several goodness of fit (GOF) criteria that still need to be met, especially RMSEA. Small or insignificant chi-square values are difficult to fulfill, especially in large samples above 200 (Hair Jr 2014; Hoyle 1994). However, it does not mean that the model is not suitable; for that reason, it is recommended to use other criteria for suitability (Hair Jr 2014). As previously explained, a significant *P*-value does not mean the model is good because many samples influence the *P*-value.

The model is fit if one of the absolute measures (Chi-square, SRMR, RMSEA) and one of the comparative measures (CFI, TLI) meet the criteria. The results of this study found that one fit measure has been met, namely SRMR with a value

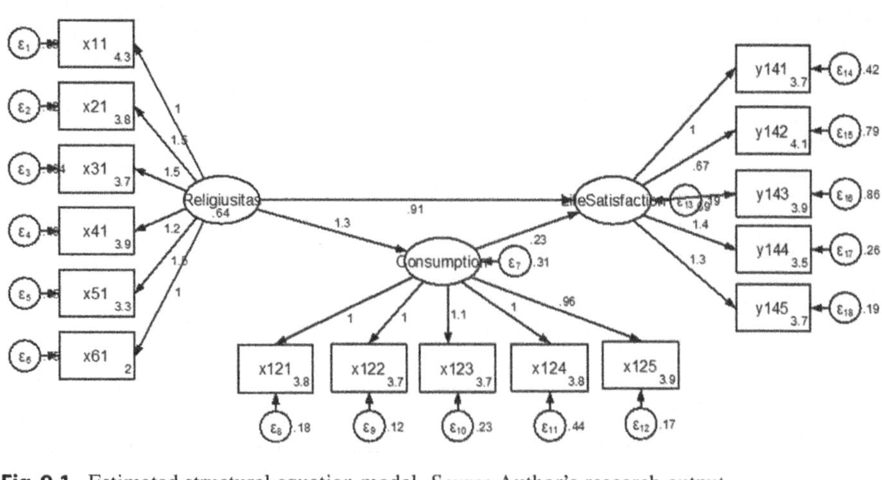

Fig. 9.1 Estimated structural equation model. *Source* Author's research output

Table 9.6 Goodness of fit test results

Fit statistic	Value	Criteria
Likelihood ratio		
chi2_ms(80)	819.013	Not fit
$P >$ chi2	0.000	
chi2_bs(120)	10,039.706	Not fit
$P >$ chi2	0.000	
Population error		
RMSEA	0.141	Not fit
90% CI, lower bound	0.132	
upper bound	0.150	
pclose	0.000	
Information criteria		
AIC	13,097.179	Fit
BIC	13,355.755	Fit
Baseline comparison		
CFI	0.926	Fit
TLI	0.900	Fit
Size of residuals		
SRMR	0.035	Fit
CD	0.989	Fit

that is smaller than 0.08, and for the comparative measure, CFI has a value that is greater than 0.90. Therefore, the model is adequate (Malhotra and Birks 2010).

Based on the covariance value, there is a positive and significant effect between the latent variables of religiosity on life satisfaction. Similar to consumption and life satisfaction, a covariance shows a positive and significant effect. Life satisfaction, which is one of the indicators of welfare from an Islamic perspective, is determined by how well people do in the economy. The vision of Islamic economics is to have a just socioeconomic order for society. The vision of Islamic economics is to have a just socio-economic order for society by using the Al-Qur'an and Sunnah of the Prophet Muhammad (pbuh) as a guideline, three aspects of economic welfare, namely poverty alleviation, distribution of wealth, and the surrounding environment (Ahmad 2011).

This is in line with Pavolini's findings, that people with low level of religiosity do not support of welfare requirements than people who define themselves as religious (Pavolini et al. 2017). In contrast to these findings, Gill and Lundsgaarde (2004) found that the results of welfare spending and participation in religious activities showed a negative relationship, after controlling the modernization factor.

This study finds a positive and significant effect of religiosity and consumption on life satisfaction. This is in line with the perspective of Islamic economics that the state has to improve the life satisfaction by introducing an Islamic social welfare system. This study also reveals how religiosity is reflected in the daily behavior of individuals, including their economic behavior. Every decision on various alternative economic choices is based on an understanding of Islam as an ideology. The emergence of some hijra communities is related with the rising figure of workers and professionals who have a stronger religious commitment. With that in mind, an individual that belongs to the hijra group tends to make economic choices based on the teachings of Islam. As a result, conventional (mainstream) economic practices are increasingly neglected.

A number of experts argue that, when explaining a country's economic growth, we must first assess the deeper role of culture (Landes 1998; Norris and Inglehart 2004). Nevertheless, more studies need to be conducted on evolution of markets, morals, culture, institutions, and religion (Bowels 1998; Friedman 2008).

This study found that religiosity is one of the determinants in deciding various economic choices notably in consumption, saving, investment, and financing. The description of the economic concept of Muslim workers in the city of Padang is clearly illustrated here, and is expected to provide information for stakeholders to further encourage more spiritual-based economic programs.

9.5 Conclusion

This study aims at investigating how religiosity affects the consumption behavior. The findings suggest that religiosity, consumption, and life satisfaction have a positive and significant association. In addition, the economic environment also influences a person's belief, religious and moral choices. In this regard, the economics discipline is not only concerned with the study of markets but also with the study of religion and culture.

Differences in the level of religiosity are related to the differences in understanding not only in the rituals but also in the application of the economic living (muamalah). Religious Muslims adhere to Islamic principles and avoid any unlawful work or income that might promise a more significant material advantage. The empirical finding of this study suggests that the religiosity has a significant impact on consumption. Similarly, the consumption has a significant and positive impact on life satisfaction. In light of this finding, an increase in the level of religiosity and consumption will result in a rise in the life satisfaction.

Religion is described as a multidimensional construct that influences consumers' psychology and behavior. There has been a change in society toward a more spiritual direction; meanwhile, perceived life satisfaction has also increased. This study has contributed in a number of ways: Firstly, the quantitative approach can be used to analyze the role of religion in society. Secondly, the present study highlights non-market behavioral economics, which describes the norms, social capital, values, and spiritual capital that shape human behavior through their belief and actions. Finally, it also helps to explain how Islam and its teachings influence the economic system.

References

Ahmad M (2011) Economic welfare from Islamic perspective. SSRN Electron J 1:1–21. https://doi.org/10.2139/ssrn.1530092
Allport GW (1965) The individual and his religion (Collier-Ma). The Macmillan Company, New York. https://www.pdfdrive.com/the-individual-and-his-religion-a-psychological-interpretation-d187962699.html
Amin H (2019) The Islamic theory of consumer behaviour for ijarah home financing. J Asia Bus Stud 13(4):672–693. https://doi.org/10.1108/JABS-09-2018-0261
Aydin N (2013) Redefining Islamic economics as a new economic paradigm. Islam Econ Stud 21:1–34. http://www.irti.org/English/Research/Documents/IES/024.pdf
Bhat NN (2016) The economic thougt of Khurshid Ahmad. Turk J Islam Econ 3(2):1–1. https://doi.org/10.15238/tujise.2016.3.2.1-11
Bowels S (1998) Endogenous preferences: the cultural consequences of markets and other economic institutions. J Econ Lit 36(March):75–111
Carvalho J-P (2012) VEILING. Q J Econ 1:337–370. https://doi.org/10.1093/qje/qjs045.Advance
Chapra MU (1995) Islam and the economic challenge. In: The Islamic foundation the international institute of Islamic thought, 17th edn. International Islamic Publishing House IIPH. http://eprc.sbu.ac.ir/File/Article/Islam and the Economic Challenge_93524.pdf
Coleman J (1988) Social capital in the creation of human capital author. Am J Sociol 94:95–120. https://doi.org/10.1037/0012-1649.22.6.723
Cutler BD (1991) Religion and marketing: Important research area or a footnote in the literature? J Prof Serv Mark 8(1):153–164. https://doi.org/10.1300/J090v08n01_12
Delener N (1990) The effects of religious factors on perceived risk in durable goods purchase decisions. J Consum Mark 7(3):27–38. https://doi.org/10.1108/EUM0000000002580
Essoo N, Dibb S (2004) Religious influences on shopping behaviour: an exploratory study. J Mark Manag 20:683–712. https://doi.org/10.1362/0267257041838728
Fajriani SW, Sugandi YS (2019) Hijrah Islami Milenial Berdasarkan Paradigma Berorientasi Identitas. Sosioglobal: Jurnal Pemikiran Dan Penelitian Sosiologi 3(5):5–7
Fam KS, Waller DS, Erdogan BZ (2002) The influence of religion on attitudes towards the advertising of controversial products. Eur J Mark 38(5/6):537–555. https://doi.org/10.1108/03090560410529204
Friedman D (2008) Moral and market: an evolutionary account of the modern world, 1st edn. Palgrave Macmillan
George LK, Larson DB, Koenig HG, McCullough ME (2000) Spirituality and health: what we know, what we need to know. J Soc Clin Psychol 19(1):102–116. https://doi.org/10.1521/jscp.2000.19.1.102
Gill A, Lundsgaarde E (2004) State welfare spending and religiosity: a cross-national analysis. Ration Soc 16(4):399–436. https://doi.org/10.1177/1043463104046694
Hair Jr J (2014) Multivariate data analysis, 7th edn. Pearson Education Limited. https://files.pearsoned.de/inf/ext/9781292035116
Haneef MAM (1997) Islam , the Islamic worldview , and Islamic economics. IIUM J Econ Manag 5(1):39–65. http://lib.iium.edu.my/mom2/cm/content/view/view.jsp?admin=1&key=A51jqy6w33wLSxc4XQcJ4NYpGJyYBqbX20070328151304984
Hoyle RH (1994) Introduction to the special section: structural equation modeling in clinical research. J Consult Clin Psychol 62(3):427–428. https://doi.org/10.1037/0022-006X.62.3.427
https://doi.org/10.1016/j.chiabu.2014.03.021
Islahi, A. A. (1944). Thirty years of research on history of islamic economic thought: assessment and future directions, 1937, pp 347–370
Iyer S (2016) The new economics of religion. J Econ Lit 1:1–5. https://www.researchgate.net/publication/281637625_The_New_Economics_of_Religion
Jawad R (2012) Religion and faith based Welfare: from wellbeing to ways of being. Policy Press, Bristol. https://doi.org/10.1111/spol.12022

Johnson BR, Jang SJ, Larson DB, De Li S (2001) Does adolescent religious commitment matter? A reexamination of the effects of religiosity on delinquency. J Res Crime Delinq 38(1):22–44. https://doi.org/10.1177/0022427801038001002
Jöreskog KG, Olsson UH, Wallentin FY (2016) Multivariate analysis with LISREL. J Stat Softw 78(Book Review 1):1–3. https://doi.org/10.18637/jss.v078.b01
Kenny A, Kenny C (2006) Life, liberty, and the pursuit of utility. Imprint Acad. https://doi.org/10.1111/j.1468-2265.2009.00564_26.x
Khan MM (2009) Shahih Bukhari. Shahih Muslim 149:24–21
Khraim H (2010) Measuring religiosity in consumer research from an Islamic perspective. Int J Mark Stud 26(1):52–78. https://doi.org/10.1108/10264116201000003
Krauss SE, Hamzah AHJ, Suandi T, Noah SM (2005) The Muslim religiosity-personality measurement inventory (MRPI)'s religiosity measurement model: towards filling the gaps in religiosity research on Muslims. Social Sci Hum 13(2):131–145
Landes DS (1998) The wealth and poverty of nations: why some are so rich and some so poor, vol 34, issue 2. W. W. Norton. https://doi.org/10.2307/2606732
Levy PS, Lemeshow S (1999) Sampling of populations: methods and applications, 3rd edn. John Wiley & Sons, INC. https://www.pdfdrive.com/sampling-of-populations-methods-and-applications-3rd-ed-wiley-series-in-survey-methodology-e161208020.html
Lindridge A (2005) Religiosity and the construction of a cultural-consumption identity. J Consum Mark 22(3):142–151. https://doi.org/10.1108/07363760510595968
Malhotra N, Birks D (2010) Marketing research: an applied approach. http://capitadiscovery.co.uk/cardiffmet/items/240307
Mathras D, Cohen AB, Mandel N, Mick DG (2016) The effects of religion on consumer behavior: a conceptual framework and research agenda. J Consum Psychol 26(2):298–311. https://doi.org/10.1016/j.jcps.2015.08.001
Minkler L (2004) Religious identity and consumption. Rev Social Econ 37–41 (November 2013). http://www.tandfonline.com/loi/rrse20
Mokhlis S (2009) Relevancy and measurement of religiosity in consumer behavior research. Int Bus Res 2(3):75–84. https://doi.org/10.5539/ibr.v2n3p75
Morshidi A (2015) Relationship between workplace incivility, job attitudes and Muslim religiosity personality among trade union members. Global J Al-Thaqafah 5(December):43
Moss BR (2011) Spirituality , social work education and workplace well-being: towards a critical framework. Staffordshire University, pp 25–27. https://www.semanticscholar.org/paper/Spirituality%2C-social-work-education-and-workplace-%3A-Moss/fe8f500455ae6c874a892f6a0d96098314686447
Nasr SH (2006) Islamic philosophy from its origin to the present. State University of New York Press. https://www.pdfdrive.com/islamic-philosophy-from-its-origin-to-the-present-philosophy-in-the-e16683605.html
Norris P, Inglehart R (2004) Sacred and secular: religion and politics worldwide. Cambridge University Press
Pavolini E, Béland D, Jawad R (2017) Mapping the relationship between religion and social policy. J Int Comp Social Policy 33(3):240–260. https://doi.org/10.1080/21699763.2017.1363801
PEW Research Center (2008) US religious landscape survey: religious beliefs and practices. Pew Research Center. https://www.pewforum.org/2008/06/01/u-s-religious-landscape-survey-religious-beliefs-and-practices/
PEW Research Center (2012) The global religious landscape. Pew Research Center. https://www.pewforum.org/2012/12/18/global-religious-landscape-exec/
Pindyck RS (2013) Microeconomics (eight). Pearson Prentice hall.
Raysuni AA (2013) Imam Al-Shatibi's theory of the higher objectives and intents of Islamic law. The International Institute of Islamic Thought (IIIT). https://www.academia.edu/7120140/Imam_al_Shatibi_s_Theory_of_the_Higher_Objectives_and_Intents_of_Islamic_Law_Ahmad_al_Raysuni
Rehman A, Shabbir MS (2010) The relationship between religiosity and new product adoption. J Islam Mark 1(1):63–69. https://doi.org/10.1108/17590831011026231

Scheve K (2006) Religion and preferences for social insurance. Q J Polit Sci 1(3):255–286. https://doi.org/10.1561/100.00005052

Schouten J, McAlexander J (1995) Subcultures of consumption: an ethnography of the new bikers. J Consum Res. https://doi.org/10.1086/209434

Schumpeter JA (1955) History of economic analysis, vol 22. Routledge. https://doi.org/10.2307/2626853

Simon Fan C (2008) Religious participation and children's education: a social capital approach. J Econ Behav Organ 65(2):303–317. https://doi.org/10.1016/j.jebo.2005.11.002

Sood J, Nasu Y (1995) Religiosity and nationality. An exploratory study of their effect on consumer behavior in Japan and the United States. J Bus Res 34(1):1–9. https://doi.org/10.1016/0148-2963(94)00015-7

Thaler RH (2008) Nudge: improving decisions about health, wealth, and happiness, vol 1, issue 4. Yale University Press

Van Oorschot W, Opielka M, Pfau-Effinger B (2008) The culture of the welfare state: historical and theoretical arguments. In: Culture and welfare state: values and social policy in comparative perspective, pp 1–26. https://doi.org/10.4337/9781848440234.00007

Weiss H (2002) Social welfare in Muslim societies in Africa. Int J Afr Hist Stud 35(1). https://doi.org/10.2307/3097423

Wilkes RE, Burnett JJ, Howell RD (1986) On the meaning and measurement of religiosity in consumer research. J Acad Mark Sci 14(1):47–56. https://doi.org/10.1007/BF02722112

Worthington EL, Wade NG, Hight TL, Ripley JS, McCullough ME, Berry JW, Schmitt MM, Berry JT, Bursley KH, O'Connor L (2003) The religious commitment inventory-10: development, refinement, and validation of a brief scale for research and counseling. J Couns Psychol 50(1):84–96. https://doi.org/10.1037/0022-0167.50.1.84

Halal in Islamic Business

10

Abdul Muizz Abdul Wadud and Khalifah Muhamad Ali

Abstract

The article looks at the possibilities of employing halal regulation and certification to implement Islamic business practices in modern society. The current practices of halal regulation and certification, as well as the literature on Islamic business, were looked at to develop a workable plan for halal Islamic business. Islam recognizes and accepts various degrees of ethical commitment. A multilevel Islamic ethics framework and multilevel halal certification technique are offered to assist the application of Islamic business ethics in a relative setting. Two significant improvements may improve halal corporate practices: harmonizing global standards and governance structures and incorporating responsibility and ethical considerations into halal standards. Halal regulation and certification are effective instruments for assisting Muslims in adhering to their religious obligations in a globalized and diverse world. Disparities in the dependability and functioning of institutions and governance structures in various nations and sectors cause this diversity. Therefore, integrating specific tried-and-true methods from the field of CSR and incorporating ethics and responsibility criteria into halal standards may be used to produce halal regulation and certification. It may be accomplished via a multilevel strategy that provides Muslims with information about items and companies based on ethical and responsibility standards and Halal requirements.

A. M. A. Wadud (✉)
Institute of Islamic Banking and Finance, International Islamic University of Malaysia, Kuala Lumpur, Malaysia
e-mail: wadud.muizz@live.iium.edu.my

K. M. Ali
Department of Islamic Economics, Faculty of Economics and Management, IPB University, Bogor, Indonesia

A. Rafiki (ed.), *Digitalization in Halal Management*, Contributions to Environmental Sciences & Innovative Business Technology,
https://doi.org/10.1007/978-981-99-5146-8_10

Keywords

Islam • Halal • Islamic business • Islamic ethics

10.1 Introduction

Islamic business comprises many sectors, including industry, tourism, and SMEs. The halal business is becoming a world trend. According to a report from the State of The Global Islamic Report (2019), around 1.8 billion Muslim residents are consumers of the halal business. Consumer opportunities in the halal industry increase by 5.2% every year, with total consumer output which reaches USD 2.2 trillion. This number is expected to remain increase gradually every year. The compound Annual Growth Rate (CAGR) states that halal business will increase to 6.2% from 2018 until 2024. The total funds spent by consumers of the halal business will also increase to USD 3.2 trillion by 2024. From this data, we can see that the halal business has very bright prospects in the future.

10.2 Literature Review

10.2.1 The Concept of Halal

A business is declared legal if it complies with the provisions of Allah's law. A work or business is permissible in the Qur'an if it is free of usury, gambling, liquor, corruption, collusion, theft, and others. Halal is defined from a legal perspective as something that is not prohibited, whereas prohibition (*haram*) is defined as an act that ends in sin and the threat of hellfire.

10.2.2 Halal, from an Islamic Perspective

The word "*halal*" is repeated six times in The Holy Quran, two of which are in the condemnation context.

> Ask ⌜the pagans, O Prophet⌝, "Have you seen that which Allah has sent down for you as a provision, of which you have made some lawful and some unlawful?" 1 Say, "Has Allah given you authorization, or are you fabricating lies against Allah?" (10:59)

> Allah has forbidden you only carrion, and blood, and the flesh of swine; also any animal over which the name of any other than Allah has been pronounced. However, whoever eats of them under compelling necessity—neither desiring it nor exceeding the limit of absolute necessity—surely for such action, Allah is Much Forgiving, Most Merciful. And do not utter falsehoods by letting your tongues declare: "This is lawful" and "That is unlawful," thus fabricating lies against Allah. Surely those who fabricate lies against Allah will never prosper (16: 115–116).

According to the two verses above, Allah would condemn people who combined the lawful things (*halal*) and the unlawful things (*haram*). Therefore, only combining lawful (*halal*) and unlawful (*haram*) is prohibited, especially committing and consuming the prohibited things. The four halal things mentioned in the Qur'an contain two similar characteristics, namely

a. The word halal is used in the context of eating (*kuluu*).
b. The word halal is combined with the word "*thayyibah*" (the good stuff).

The Qur'an frequently interprets "eating" as "carrying out all actions." It is likely because eating is the primary source of caloric gain that can result in activity. Consequently, the command to eat in the preceding verse is a command to move. Therefore, not only are its activities halal, but they are also kind. In light of this, the *makruh* does not qualify as halal *al-tayyiban* under Islamic law (halal and good). The Qur'an frequently interprets "eating" as "carrying out all actions." This is likely because eating is the primary source of calories that can promote physical activity.

Consequently, the command to eat in the preceding verse is a command to move. Therefore, not only are its activities halal, but they are also kind. In light of this, the *makruh* does not qualify as halal *al-tayyiban* under Islamic law (halal and good).

10.2.3 Halal in Islamic Business

While the procedures and instruments for implementing Islamic business ethics vary throughout time, the fundamental conceptions of the Islamic ethical system remain constant since they are derived from a transcendent source (i.e., divine revelation). Therefore, Islamic business ethics must adhere to their fundamental principles while attempting to expand their applicability.

According to Islamic beliefs, the caliphate is the foundation of human life and, by extension, their moral duty. A Muslim performs an act of devotion while serving as caliph. The notion of worship is comprehensive and applies to every element of life.

Thus, people are accountable for all their activities. They will be judged hereafter depending on the degree to which they adhere to their purpose on Earth. The message is seen as the adoration of God via the establishment and maintenance of a fair and prosperous existence on Earth (*Imar al-Ard*). Every activity in every element of life carries with it this obligation. In any event, Muslims are obligated to adhere to these teachings, which encompass a variety of facets of life in the form of explicit and precise instructions or basic principles. Sharia is widely used to refer to Islamic doctrines. It is usual in the literature to refer to Sharia as Islamic law, although Sharia's restrictions go much beyond "limited legal frontiers."

While Muslims strive to fulfill their responsibilities as God's heirs on Earth, they face barriers and challenges in various historical and geographical situations.

To address these issues and concerns at various periods and locations, Sharia is dynamic and employs a process to develop laws based on interpretation, inference, and discretion. This approach is referred to as *Usul al-Fiqh.*

Usul Al-Fiqh scholars established maxims for use in formulating Shari'a rulings. The primary objective of these maxims is to increase benefits and decrease harm for both individuals and the general population. The approach of *Usul al-Fiqh* and its maxims may be used to address modern ethical challenges. However, successfully implementing this technique requires a good grasp and awareness of the many factors and dynamics influencing the situation.

10.3 Methodology

This study is qualitative, with paper research as the primary source for examining relevant papers, journal articles, and other published materials. This strategy seeks to assess theoretical data derived from sources such as Hadith and secondary sources such as books and other published publications. The study incorporates modern and classical Shariah perspectives on business, notably the notion of halal Islamic business, to discover and investigate the issues connected to the models utilized to build the conceptual framework of the new models. Based on a literature review technique, this paper investigates and identifies numerous significant features of halal business and its prospective integration with Islamic business principles. The data analysis is carried out by changing and assessing current models, which are either impracticable or lack crucial elements required for effective deployment.

10.4 Result and Discussion

10.4.1 The Concept of Halal Business

The primary issue that every ethical framework face is its implementation. Historically, Muslims have successfully implemented their ethical framework through effective procedures. Markets, for example, were controlled by *Hisba*, an agency that served as a trade standards body (Wilson 2006). However, since then, the corporate climate has evolved, and old systems for implementing Islamic business principles need help to meet the new modern difficulties.

Although religious ethics, particularly Islamic ethics, are commonly seen as idealistic, no substantial difference in idealism or relativism was discovered between Muslims and other religious organizations (Cornwell et al. 2005). Moreover, the fact that Muslims were able to construct a great civilization and dominate international commerce for centuries lends credence to the claim that Islam does not impose an idealized perspective of life and ethics. Instead, Muslims were victorious because Islamic ethics allow for flexibility and adaptation.

Shari'a is dynamic due in part to the adaptability and flexibility of Islamic principles. In addition, they are reinforced by one of the central notions of Islamic theology: religious hierarchy (*Deen*). There are three stages of religion (*Deen*), according to the renowned Hadith known as Hadith Jibril: Islam, Iman, and Ihsan. The first level of Islam is observing Allah's commandments and teachings. The second level of Iman is focused on reinforcing one's core beliefs and ideals. Ihsan's third and final level objective is to live a spiritual experience with God observing every activity. Depending on the complexity and depth of their religious experiences, Muslims have varying degrees of ethical awareness and devotion (level of religion).

According to the theory put forth by Frederick Carney (1983), an obligation, a virtue, and a value are the essential components of any religious ethics. Therefore, "Islam, Iman, and Ihsan represent what Carney calls the obligation, value, and virtue components, respectively, of the Islamic ethic." Moad (2007) proposed the three levels of religion in Islam, "Islam, Iman, and Ihsan" based on this, we can conclude that Muslims can have varying degrees of ethical awareness and commitment depending on the amount of religious experience they have. This approach, which takes into account varying degrees of morality, is supported, as well, by the Shari'a objectives articulated by *Usul Al-fiqh* scholars.

Shari'a goals were put into three groups: those that were essential (*Daroriyyat*), those that were helpful (*Hajiyyat*), and those that were pretty (*Tahsiniyyat*) (Hallaq 2004; Kamali 1991; Abdallah 2010). There is a clear connection between these three tiers of Shari'a goals and the tiers of religion and moral elements.

10.4.2 Regulation of Halal in Certain Countries

The halal status of a product is an essential consideration in Muslim purchasing decisions. As a result, Halal regulations and certifications were created to assist customers in selecting items that adhere to Islamic norms. In the 1970s, attempts were made to establish a system for designating and certifying food items as halal. During this period, western fast-food chains began to invade the rebelling Muslim countries, and naturally, such restaurants were not always adhering to Islamic principles in their food manufacturing. Furthermore, as more multinational food companies entered Muslim markets, the issue of Halal certification became increasingly prominent.

While in the past, the processes were based on trust and shared values among Halal food producers and buyers, this system no longer works in today's commercial world. Muslim customers wanted more information about what was in the product and would not accept simple "Halal" markings, especially when the product came from a country that was not Muslim. Therefore, there was a need for trustworthy Halal certifications, and governments in Malaysia, Indonesia, and other Muslim nations responded by creating governmental organizations to oversee and control the food markets. During this time, institutes like Malaysia's *Jabatan Kemajuan Islam Malaysia* (JAKIM), one of the best Halal institutes today, got their start. Official organizations have also been created in Singapore and Indonesia to oversee and promote the national Halal market. *The Majlis Ulama Islam Singapura* (MUIS) is Singapore's equivalent to JAKIM, whereas the Majlis Ulama Indonesia is an Indonesian institute.

Halal or Sharia-compliant products are already widespread in the food and banking industries, but the concept is increasingly extending to other sectors (i.e., cosmetics, pharmaceuticals, tourism, and entertainment). The most recent example is "Salamworld," a Halal social network developed to offer a Shari'a-compliant communication network for Muslims comparable to Facebook but excludes sensitive content such as pornographic advertisements.

In non-Muslim countries such as France, halal has already exceeded organic and bio-products in market volume (Ahmed 2011). As a result, the global halal market is expected to reach $580 billion yearly, with the halal food industry expanding at an annual rate of 7%.

10.4.3 Improvements in Halal Standards and Regulation

In the middle of modern society with new lifestyles and the consequences of a transforming global economy, new methods are required to enable Muslims to continue to live in line with their fundamental ideals. Applying Islamic business ethics necessitates experiencing several ethical theories and practices, especially in this age of globalization. Keeping this in mind, one of the challenges of adopting Islamic business ethics is adapting them to the many ethical systems and practices around the globe.

Collaboration across critical stakeholders is required to realize the proposed actions for the Halal market's improved and rapid expansion. The Halal business has just recently become aware of the possibilities of merging CSR and corporate responsibility. Currently, companies and certification agencies are mainly concerned with satisfying the product requirements of a Halal (food or finance) product, such as "not containing pork."

Halal certification is already well-established in the nutrition business, at least in Muslim-majority countries. However, the current condition of Halal certification could be more robust in several crucial areas and needs immediate reform. First and foremost, neither in Muslim countries nor internationally is a uniform basis for halal requirements. Around 100 halal certifying agencies worldwide exist, but no mutual recognition mechanism exists. Consequently, the customer must deal with various Halal logos based on different standards. However, these standards are rarely communicated to the customer, making it difficult to determine whether a particular certificate meets one's requirements and level of religious commitment. The halal certification aims to provide a trustworthy signal for *Shari'a*-compliant products; however, this objective needs to be met due to a lack of openness in the certification industry.

In comparison with other businesses, Islamic finance has well-developed and unified standards. International organizations are being formed to standardize and coordinate the governance standards of Islamic financial institutions. The Accounting and Auditing Organization for Islamic Financial Institutions (AAOIFI), the Islamic International Financial Market (IIFM), and the Islamic Financial Standard Board are a few examples of these organizations (IFSB). Such worldwide governing organizations need to improve in the Halal food business and other industries. Instead of certification, several businesses (such as those in hospitality and tourism) employ a rating system.

Halal regulation and certification should create additional criteria in areas other than banking and food to function successfully as a vehicle for implementing Islamic ideals in the long term. These may be incorporated into current halal requirements. Given the variability in Halal legislation and certification, worldwide harmonization is required to improve international commerce and give more clarity to customers. It can only be accomplished with a robust global institutional network and governance system.

Based on an ethical perspective, the Halal business has enormous potential. Halal standards can quickly expand beyond Halal product features, increasing their already-existing ethical component and incorporating corporate social responsibility (CSR) activities and general ethical norms. By doing so, the Halal company may serve as a model for contemporary business growth and contribute to more ethical business practices. At present, Halal commerce and certification are primarily concerned with product qualities, although many believe Halal is more than simply the final product. A more comprehensive approach to halal certification is necessary.

Implementing CSR techniques may assist in incorporating additional criteria and, as a result, establish a more responsible Halal company. By definition, Islamic business ethics and CSR are comparable; both approaches are based on solid principles, responsible conduct, and sustainability. In addition, both topics are founded on the ethical acceptability of items and actions. As a result, CSR can be effectively incorporated into Halal company practices and contribute to a successful future Halal movement.

CSR is a well-established and developed field. CSR employs various techniques, including certification, rating, external audits, mandatory disclosure, and reporting. It has been observed that non-governmental players dominate the corporate responsibility governance framework (governance without government). Self-regulation tools and multi-stakeholder efforts have emerged to assist firms in adopting CSR practices. Halal company operations may benefit from utilizing and adapting well-proven CSR strategies.

It is possible to establish an overall ethical brand by incorporating CSR, based on Islamic business principles, and reinforcing it with product features such as organic, animal welfare, healthy, ecologically friendly, and fair working conditions. By doing so, Halal enterprises can broaden their target market and attract non-Muslims to Halal goods, as long as they provide credible certifications, which is especially important during times of recurring food scandals and the financial crisis with its discredited banking practices. Non-Muslim users of Islamic banks, for instance, linked Islamic financing with an ethical life. They were drawn to the idea that the bank would not invest their money in pornography and weapons.

Furthermore, they accepted the Islamic stance that wealth may only be created through lawful commerce and asset investment rather than by creating money from money. Another potential expansion and development of Halal business practices is the inclusion of animal welfare issues in Halal standards, particularly regarding slaughtering techniques. The International Islamic Fiqh Academy (IIFA), which serves as an ultimate body for Islamic faith and justice, recognized some proper stunning techniques in 1997.

Modern halal businesses should prioritize animal care and use recognized techniques of stunning. It streamlines Halal manufacturing, particularly in non-Muslim nations where Animal welfare regulations formerly impeded halal production. In addition, involvement in animal welfare will soothe animal welfare activists and increase the demand for Halal food among non-Muslims. It might contribute to the global expansion of the halal market.

Halal legislation and certification should be rethought to include CSR and ethical concerns. The multilevel Islamic ethics framework suggested previously in this study may be used to construct this new method. Halal regulatory and certification authorities may establish varying criteria to accommodate the demands of various Muslim communities. All Muslims are required to follow the essential Halal criteria (compulsory component). However, some Muslims have a higher moral awareness and need more information on goods and businesses to ensure they align with their beliefs and not just their responsibilities.

The method by which a multilevel Halal certification may be effectively converted into commercial practice mainly depends on the Halal market's stakeholders' attitudes. Integrating new variables into the Halal certificate evaluation criteria, such as the ethical and social conduct of the manufacturing organization, is a complex process that incurs expenditures. However, as we know from CSR practices, ethically and socially driven investments in business processes have a better chance of being incorporated into a company's strategy if they offer a win–win scenario for both the consumer and the firm.

Consequently, we may anticipate that the Halal business would support and promote a multilevel Halal certification if the consumer thus desires. The primary issue is whether consumers desire a responsible Halal product and a responsible Halal company or if most Halal buyers are content with a Halal-certified product based only on product features. The tendency is toward a heightened awareness of a company's social performance.

While adding more levels of certification, an additional challenge in practical implementation may be that the consumer is overwhelmed by too much information. In addition, adding different levels to the certification might further confuse the buyer, particularly in Europe, where many certification providers administer halal certification. Here, the objective should be to develop a common foundation for halal certification and, once this is accomplished, to enhance halal certification through multilevel certification.

10.4.4 The Role of Halal Business on Economic Growth

The halal business grows annually. This increase unquestionably benefits the economy, especially in Indonesia, reflected in the gross domestic product (GDP). It has been demonstrated that the Islamic economy contributes USD 3.8 billion to the annual Gross Domestic Product (GDP) (Ministry of Finance of the Republic of Indonesia 2019). This contribution to the GDP is reflected in halal export and import activities in Indonesia.

The consumption of halal goods by the Indonesian population is relatively high, among the highest in the world. In 2017, the total consumption of halal products in Indonesia was estimated at USD 200 billion, or more than 36% of the total consumption of households. This sum is also comparable to 20% of Indonesia's GDP (Ministry of Finance of the Republic of Indonesia 2019). On average, the consumption of these halal items will increase by 5.3% every year. As a result, consumption of Indonesian halal goods is anticipated to reach 330.5 billion US dollars in 2025. The bulk of this high overall consumption, according to the Indonesian national development planning body, consists of imported goods (Akbar 2019). However, we can lower the trade imbalance if we enhance output in the halal industrial sector. Therefore, it is easy if we can also enhance halal exports.

In terms of exports, the halal business offers a significant possibility to enhance Indonesia's export value. The export value of the halal business might range between USD 5.1 billion and USD 11 billion each year. In 2018, the halal market produced $7.6 billion in revenue. Globally, Indonesia has a 3.8% export opportunity for halal goods. This number may be enhanced by improving the quality of halal items exported abroad. Moreover, competitive pricing is deemed essential so that our products can compete with those of other nations. If halal products from Indonesia can compete in the global market, then Indonesia may become the halal industry's epicenter.

Humans can stimulate a rise in the value of foreign exchange reserves by utilizing the full potential of the halal business. We could accomplish this in a variety of ways:

a. *First*, the export value of the halal business should be increased. Halal cuisine and Muslim apparel have export potential in Indonesia. In order to raise the export value, it is necessary to expand commodity production in conjunction with quality enhancement and a competitive price, as previously indicated.
b. *Second*, expanding halal tourism's potential Halal tourism is an industry with tremendous potential to increase foreign currency reserves. In addition, Indonesia has won several accolades in halal tourism. Therefore, we must design on-target measures to enhance the quality of Indonesia's halal tourism industry. We can implement this plan by enhancing the quality of human resources and bolstering package and integrated tourism.
c. *Third*, strengthening the significance of the Islamic finance industry in order to achieve financial inclusion. Through financing, the Islamic finance industry may facilitate capital access for Micro, Small, and Medium-Sized Enterprises

(MSMEs) operating in the halal industry. This capital can be utilized to expand the firm. Consider that production will grow if many MSMEs expand their operations. This rise in manufacturing can facilitate the worldwide commerce of halal products from Indonesia.

In addition, the financial inclusion initiative supports MSMEs with streamlined payment processes. Therefore, if an increasing number of MSMEs conduct business through the Islamic finance sector, the Islamic finance industry's market share will rise. Consequently, the Islamic finance industry will grow and have a more significant influence on the well-being of society.

In order to expand the halal business, people need the assistance of all stakeholders, including the government and the community. In addition to halal certification, the government may provide support in the form of rules governing halal goods and sectors. Halal rules and certification aid business players, particularly MSMEs, so these items can enter the export market. Meanwhile, community support consists of initiatives to improve halal knowledge and purchase halal goods from local firms.

10.5 Conclusions and Implications

Several nations have tremendous potential to enhance halal business. This potential is in part attributable to the large Muslim community. In addition, the world is beginning to acknowledge the expansion of numerous halal industrial sectors, particularly the banking, tourism, and fashion industries.

Halal regulation and certification provide a practical framework for Muslims to satisfy their religious obligations while living in a globalized and diverse society. However, different countries and industries have different ways of doing halal business due to differences in how reliable their institutions and government systems are and how well they work.

External constraints, such as the number of competing nations and the need for globally recognized halal certifications, would pose obstacles for diverse countries. Meanwhile, internal constraints include the community's need for halal understanding, difficulties in applying halal product assurance requirements, and a lack of public awareness to compete. Through this study, many people will become more aware of the significance of the halal sector and will support efforts to promote the halal industry.

References

Abdallah S (2010) Islamic ethics: an exposition for resolving ICT ethical dilemmas. J Inform Commun Ethics Soc 8:289–301

Ahmed A (2011) Halal food's growing market

Cornwell B, Cui CC, Mitchell V, Schlegelmilch B, Dzulkiflee A, Chan J (2005) A cross-cultural study of the role of religion in consumers' ethical positions. Int Mark Rev 22:531–546

Kula E (2001) Islam and environmental conservation. Environ Conserv 28(1):1–9
Moad EO (2007) A path to the Oasis: Sharī'ah and reason in Islamic moral epistemology. Int J Philos Relig 62(3):135–148
Pattberg P (2006) The influence of global business regulation: beyond good corporate conduct. Bus Soc Rev 111(3):241–268
Possumah BT, Ismail AG, Shahimi S (2013) Bringing work back in Islamic ethics. J Bus Ethics 112(2):257–270
Riaz MN, Chaudry MM (2004) The value of halal food production—Mian N. Riaz and Muhammad M. Chaudry define what Halal and kosher foods are, describe why they are not the same thing, and what is required of processors and. Inform Int News Fats Oils Related Mater 15(11):698–701
Rice G (1999) Islamic ethics and the implications for business. J Bus Ethics 18(4):345–358
Wilson JAJ, Liu J (2010) Shaping the halal into a brand? J Islam Mark 1:107–123

11 Comparative Analysis of Digitally-Enabled Community in Supporting the Halal Industry in Muslim Majority Countries in the ASEAN, Central Asia, and Maghreb Region

Sutan Emir Hidayat and Khairunnisa Musari

Abstract

The halal industry has been acknowledged as one of the notable spheres in the world which develops fast and influences as well as transforms the insight into how business should be performed. The halal industry is no longer just a religious symbol but has developed into a universal symbol of quality, hygiene, health, and safety. This is certainly an opportunity to boost economic growth for countries with a majority Muslim population by taking a part in the halal industry. In the Association of Southeast Asian Nations (ASEAN) region, several member states stir globally the halal industry although digitalization in this region is currently in the early phase of development. By considering the eminence of digitalization in transforming the halal ecosystem and unleashing the halal industry growth, this chapter maps the digitally enabled community in supporting the halal industry in Muslim-majority countries in ASEAN, Central Asia, and the Maghreb Region. Central Asia countries cover Kazakhstan, Tajikistan, Kyrgyzstan, Uzbekistan, and Turkmenistan. Then, the Maghreb region consists of Algeria, Morocco, Tunisia, Libya, and Mauritania. All of those countries are Muslim-majority countries which a high potential to take part in the global halal trade. The halal industry is hoped to make remarkable support to the economic development of these regions. The outcome of this chapter is to encourage the information and communication technologies (ICT) field to have

S. E. Hidayat (✉)
Gunadarma University and National Committee for Islamic Economy and Finance (KNEKS), Jakarta, Indonesia
e-mail: sutan.emir@kneks.go.id

K. Musari
Kiai Haji Achmad Siddiq State Islamic University (UIN KHAS), Jember, Indonesia

A. Rafiki (ed.), *Digitalization in Halal Management*, Contributions to Environmental Sciences & Innovative Business Technology,
https://doi.org/10.1007/978-981-99-5146-8_11

a crucial part in enabling the digitally enabled community to support the digitally enabled halal trade, which in turn inflicts the economic growth of the Muslim-majority countries in these regions.

Keywords

Digitalization • Digital transformation • Halal ecosystem • Halal trade • Halal digital • Muslim-majority countries

11.1 Introduction

The halal industry has been acknowledged as one of the notable spheres in the world which develops fast and influences as well as transforms the insight into how business should be performed. Halal products today not only reach Muslim consumers but also appeal to other non-Muslim consumers who are concerned and aware of quality, hygiene, health, and product safety. This is certainly an opportunity to boost economic growth for countries with a majority Muslim population by taking a part in the halal industry.

State of the Global Islamic Economy (GIE) 2022 reported that the worldwide expenditure of 1.9 billion Muslims is predicted equal to US$2 trillion over six real-economy sectors, namely the food, cosmetics, fashion, travel, media, and pharmaceutical sectors. The obligation to consume halal in the teachings of Islam is a basis for Muslims and an inspiration for non-Muslims, so the need for consumption of halal products grow 8.9% year-on-year from 2020. Despite the continued uncertainty after the pandemic, the expenditure by global Muslims in 2022 is projected still to grow by 9.1% for the sectors embraced in this report. All the expenditures of these sectors have reverted to the pre-pandemic levels by the end of 2021, exception for the travel sector. In 2025, at a cumulative annual growth rate (CAGR) for 4 years, Muslim expenditure is predicted to grow by 7.5% or equal to US$2.8 trillion.

In line with the increase of the worldwide Muslim population which is forecasted to attain 2.2 billion in 2030 and 2.6 billion in 2050, the need for halal products pursues to increase to fulfill the Muslim market. The worldwide Muslim population reflects almost 30% of the world population. In the Association of Southeast Asian Nations (ASEAN) region, several member countries stir globally the halal industry. ASEAN is highly regarded as the motor in the global halal trade (Musari and Zaroni 2021; Musari and Hidayat 2022; Musari 2022).

By considering the eminence of digitalization in transforming the halal ecosystem and unleashing the halal industry growth, this chapter compares the digitally enabled community in Muslim-majority countries in the ASEAN, Central Asia, and Maghreb, in supporting the halal industry. Central Asia countries cover Kazakhstan, Tajikistan, Kyrgyzstan, Uzbekistan, and Turkmenistan. Then, Maghreb region consists of Algeria, Morocco, Tunisia, Libya, and Mauritania. All of those countries are Muslim-majority countries which a high potential to increase global halal trade. The halal industry is hoped to make remarkable support to the economic development of these regions.

Therefore, this chapter would like to map the digitally enabled community in supporting the halal industry in countries with majority Muslim population in the ASEAN, Central Asia, and the Maghreb. The outcome of this chapter is to encourage the information and communication technologies (ICT) field has a crucial part in enabling the digitally enabled community to support the digitally enabled halal trade, which in turn inflicts the economic growth of the Muslim-majority countries in these regions. These countries should not only act as consumers in the halal industry but also act as major players by becoming producers of halal products. The ICT sector advancement becomes an enabler to accelerate the wider cross-border and domestic trade of halal products.

11.2 Literature Review

11.2.1 Halal Industry

Halal products have gone global and circulated easily on five continents. Codex Alimentarius, an institution recognized by the World Trade Organization (WTO), has even provided standardization for halal products. In France, commercial halal labels are available. Since the 1990s, international halal trade has grown intensively so the Codex Alimentarius adopted guidelines on the use of halal claims to prevent differences in the concept of halal becoming a global trade barrier on July 17, 1997 (Bergeaud-Blackler 2004).

As a reputable community that engines the international halal industry, ASEAN can be the best practice of a region in developing a halal conscious community. As a place for 42% of Southeast Asian Muslims and 25% of worldwide Muslims, ASEAN can promote the halal industry in Asia. Then, Asia can work to influence other countries worldwide, leastwise in the Muslim countries, as Asia is a place to 65% of the Muslims in the world and most of them are emerging countries (Musari and Zaroni 2021; Hidayat and Musari 2022; Musari 2022).

Based on the State of the GIE Report 2022, Malaysia, Indonesia, and Singapore became a representative of ASEAN member states in the GIE Indicators as the Top 15 countries. Moreover, based on the previous Report in 2020/2021, there were also Brunei in Media and Recreation and Halal Food sector as Top 10 Countries, as well as Thailand in Muslim-Friendly Travel sector as Top 10 Countries. Despite Thailand is not a country with a Muslim-majority population, Thailand is also promoting halal travel, particularly for visitors from its Muslim-majority neighbors, and implementing incentives to encourage investors in the food industry to enlarge the halal food industry (DinarStandard 2020, 2022).

Table 11.1 shows the 15 countries with the highest scores in GIE Indicators since 2014/2015. GIE Indicators are comprehensive picture of countries that are currently best positioned to address the multitrillion-dollar global halal economy opportunity. In 2013, the State of the GIE Report 2013 still did not use the term

GIE Indicators to assign the Top 15 countries. But, at that time, the 15 countries with the highest scores had been determined as the Top Centers of the Islamic Economy through a similar method as GIE Indicators.

Apart from ASEAN, another region in the world that has a community of countries with a majority Muslim population, namely Central Asia. Certainly, countries in this region also have the opportunity to lead a halal hub in Asia and the world. Moreover, the growth of halal trade in this region has increased, and there is a country included in the Top 15 countries in GIE Indicators. As Table 11.1 shows, Kazakhstan is included in the 15 countries with the highest GIE Indicators in 2022. Salama (2018) reported that the Muslims of Kazakhstan are 70.2% of the 17.8 million people in 2016. Kazakh consumers spent around US$22 billion on food and beverages, and about US$12.4 billion of that was spent by Muslims.

In addition, the Report noted that Kazakhstan is also the Top 10 the Organisation of Islamic Cooperation (OIC) importers of Halal Fashion with a trade value of US$0.97 billion by 2020. The Kazakhstan government and Malaysian eCommerce company Aladdin will cooperate on building bilateral trade through the Aladdin platform. To increase the bilateral collaboration to improve the fashion trade, a memorandum of understanding (MoU) has been signed between the UK-based University of Central Lancashire (UCLan) Preston Fashion Department and the Central Asian Tashkent Institute of Textile and Light Industry to help boost relationships between universities and the creative industry in both countries (DinarStandard 2022).

Apart from Kazakhstan, other countries in Central Asia also have great potential to become leaders in the halal industry. For example, Uzbekistan has a chance for the Muslim-Friendly Travel sector. Uzbekistan is promoted as a halal destination with awesome and inspiring Islamic monuments and mosques, delicious halal culinary tours through its local cuisine, unique historical tourist attractions by bringing them to streets, a birthplace of ancient Islamic knowledge, magnificent culture and heritage, experience breathtaking and baffling nature, architectural crossroads of the world, bazaars and shopping galore, friendly and welcoming locals, as well as surprising ease of travel and visa exemptions (Danik 2019; Aniqoh and Hanik 2021). Hence, all the countries in Central Asia actually have opportunities to become the leading Muslim Friendly-Travel sector.

Then, all the countries in Central Asia have opportunities also to lead the Halal Food sector. Considering that five Central Asian countries are all Islamic countries that only have halal food (Sahida 2021), the people in this region are an available large market to absorb their halal products. In Turkmenistan, to enlarge the production of halal meat, European Bank for Reconstruction and Development (EBRD) gave a loan US$4.5 million to Taze Aye, a meat producer. Through the funding, this Turkmen company will generate its daily production of halal meat to five times the 3–4 tons into 15 tons (Euro Meat News 2022). The first international halal certificate in this country which regularly organizes halal events, namely Ashgabat Halal Expo, has been issued to Hasar. Hasar is the leader in the production of confectionery products in Turkmenistan. Halal certificate of Hasar

Table 11.1 The 15 countries with the highest scores in GIE Indicators (2013–2022)

No	2013	2014/2015	2015/2016	2016/2017	2017/2018	2018/2019	2019/2020	2020/2021	2022
1	*Malaysia*	*Malaysia*	*Malaysia*	*Malaysia*	*Malaysia*	*Malaysia*	*Malaysia*	*Malaysia*	*Malaysia*
2	UAE	UAE	UAE	UAE	UAE	UAE	UAE	Saudi Arabia	Saudi Arabia
3	Saudi Arabia	Bahrain	Bahrain	Bahrain	Saudi Arabia	Bahrain	Bahrain	UAE	UAE
4	Turkey	Oman	Saudi Arabia	Saudi Arabia	Bahrain	Saudi Arabia	Saudi Arabia	*Indonesia*	*Indonesia*
5	*Indonesia*	Saudi Arabia	Pakistan	Oman	Oman	Oman	*Indonesia*	Jordan	Turkey
6	Iran	Qatar	Oman	Pakistan	Pakistan	Jordan	Oman	Bahrain	Bahrain
7	Qatar	Kuwait	Kuwait	Kuwait	Qatar	Qatar	Jordan	Kuwait	*Singapore*
8	Kuwait	Jordan	Qatar	Qatar	Kuwait	Pakistan	Pakistan	Pakistan	Kuwait
9	USA	Pakistan	Jordan	Jordan	*Brunei*	Kuwait	Kuwait	Iran	Iran
10	UK	*Indonesia*	*Indonesia*	*Indonesia*	Jordan	*Indonesia*	Qatar	Qatar	Jordan
11	*Singapore*	*Brunei*	*Singapore*	*Singapore*	*Indonesia*	*Brunei*	*Brunei*	Oman	Oman
12	Bahrain	Sudan	Sudan	*Brunei*	Iran	Sudan	Sudan	Turkey	Qatar
13	Pakistan	*Singapore*	Egypt	Sudan	*Singapore*	Iran	Turkey	Nigeria	UK
14	Nigeria	Turkey	Iran	Iran	Sudan	Bangladesh	Iran	Sri Lanka	*Kazakhstan*
15	Bangladesh	Australia	Bangladesh	Bangladesh	Bangladesh	Turkey	Bangladesh	*Singapore*	Pakistan

Source Reuters (2013), Hidayat and Musari (2021) and DinarStandard (2020, 2022)

will impact Turkmen products more widely accepted by the international market (Arzuw News 2021; Société Générale de Surveillance [SGS] Caspian 2021).

In Tajikistan, the halal standard for food has been developed by the Government since 2011 with support from the Council of Ulema as the highest Islamic institution in this country, the Committee for Religious Affairs, the Ministry of Health and Social Protection of Population, and State Veterinary Inspection Service under the Ministry of Agriculture (Salama 2013b). Likewise, Kyrgyzstan launched a voluntary certification and labeling of products and services to meet the requirement of halal standards. Kyrgyzstan becomes the second country in Central Asia to implement halal standards (Salama 2013a, 2018).

Regarding the Maghreb, there are no countries in this region that becomes the Top 15 countries in GIE Indicators from 2013 until 2022. This could be because countries in the region still have unresolved problems. To succeed in establishing a halal industry, all Maghreb countries must work together and work for hand in hand. Indeed, the journey of the Maghreb today has not lived up to the expectations of its founders in the past. The Maghreb is a flawless sample of an economic block whose member countries cannot be well integrated. The trading performance of this alliance is categorized as the poorest in the world. This is reflected in the trade between Maghreb members only 4.8% of their trading volume. Nevertheless, despite the Maghreb's existence now resulting in an asleep status, this union remains solid even to this day (Hamza 2017; Al Ghweel 2022; Musari and Sayah 2022).

Furthermore, the countries in the Maghreb region in fact have the prospect to enhance themselves as major halal players, for example, Libya. This country establishes a commission for halal food products and commodities. The commission develops a consolidated procedure for controlling the products and commodities of halal food to verify their fulfillment of the requirement of Islamic law. In addition, the commission also regulates the process of import, examination, traceability, and accreditation for the halal certification bodies (Libyan News Agency [LANA] 2022).

11.2.2 Digitally Enabled Community in Supporting the Halal Industry

Starting from the study by Hidayat and Musari (2022) examined the relationship between the ASEAN digitally enabled community and the occasion of this region to be a hub of global halal logistics. By measuring the Internet subscribers per user per 100 persons and cellular/mobile phone per 100 persons, then measuring the gross domestic product (GDP), total trade in goods, international air passenger traffic, and total air cargo in this region, the study proved that the digitally enabled community and the occasion for ASEAN to be a global halal logistics statistically have a positive significant relationship.

Indeed, digital transformation is arriving more rapidly in all fields. Pandemic COVID-19 contributed to accelerating the transformation. Despite digitalization does not resolve all matters, it presents more effective and excellent data, management, automation, optimization, efficiency, digitization and can attain the wider community (Musari and Zaroni 2021; Musari and Sayah 2022). Likewise, small micro-enterprises (SMEs) that integrated the Internet into their businesses created twice as many jobs as the average (Du Rausas et al. 2011). Similarly, fintech as a portmanteau for "financial technology" has emerged as an appliance to drive financial literacy in the era of growing uncertainty, risks, and opportunities (Sala 2022).

Through digital technologies, if companies in the emerging market might grow as global rivals for established companies from developed countries and strengthen the inclusive growth in emerging economies (McKinsey Global Institute [MGI] 2016; Stonehouse and Konina 2020), certainly digital technologies provide opportunities also to Muslim-majority countries in ASEAN, Central Asia, and the Maghreb Region become global leading halal fabrication and powering inclusive growth economy to these regions.

Then, if much evidence that digitalization rapidly may fuel expansion and scale up the entrance for financial services, can boost the headway agenda to attain the Sustainable Development Goals (SDGs), also generate the perpetual resilience of social and economic (MGI 2016; United Nations Secretary-General's Special Advocate for Inclusive Finance for Development [UNSGSA], the Better Than Cash Alliance, the United Nations Capital Development [UNCDF], and the World Bank 2018; Osafo-Kwaako et al. 2018), certainly digitalization provide opportunity also to the halal industry can be expanded, be scaled up the access, be ignited faster the progress, and be created the long-lasting.

Referring to the State of the GIE Report 2022, digital connectivity and the arise of electronic-commerce (e-commerce) and mobile-commerce (m-commerce) also become the driver of the Islamic economy. In the year 2020, several OIC countries from Southeast Asia and the Middle East and North Africa (MENA) had a higher internet penetration than the global average of 59%, whereas the Gulf Cooperation Council (GCC) countries had a penetration of almost 99%. Then, the Internet penetration for countries in the African region grows by nearly 21% more than in pre-COVID-19 years, which is also the highest growth globally. Furthermore, Malaysia, Indonesia, and Saudi Arabia are several of the OIC countries which become the adopter of the highest global e-commerce and m-commerce, more than each global average of 55.4 and 76.8% per December 2020. In the post-COVID era, the trend is also realized in most Muslim-majority countries and is expected to keep on its continue for the growth of the virtual business (DinarStandard 2022).

Referring to Global Islamic FinTech (GIFT) Index 2022, the transaction volume of Islamic fintech within OIC countries reached US$79 billion in 2021 and will grow to US$179 billion at the compound annual growth rate 17.9% in 2026. This growth forecast is higher compared to the growth forecast of conventional fintech for the same period at a rate of 13.5%. GIFT is a composite index consisting of 32 indicators with five categories, namely regulation, capital, talent, infrastructure, and Islamic fintech market and ecosystem. Even though OIC countries master the ten highest scores, the ecosystems that develop the fastest are non-OIC countries (DinarStandard and Elipses 2022).

Table 11.2 shows the country index of the 60 Islamic fintech markets within the GIFT Index. There are four countries from ASEAN that are included as the Top 20. Even Malaysia and Indonesia as Muslim-majority countries in ASEAN are included as the Top 3. Then, there are countries from Central Asia and the Maghreb which are in the Top 40, namely Kazakhstan, Tunisia, and Tajikistan. In the Top 60, there are Thailand and the Philippines from ASEAN; Algeria, Morocco, Mauritania, and Libya from Maghreb; then Kyrgyzstan and Uzbekistan from Central Asia. Overall, Muslim-majority countries in ASEAN are ahead of the Muslim-majority countries in Central Asia and the Maghreb region in terms of Islamic fintech globally.

Therefore, as resumed by Ahmedov (2020), the digital field becomes a primary space for global trade. In international commerce, the collaboration between digitalization and the global economy is rising fast. The primary assignment of digital economy development is to make certain economic growth and accelerate its integration in the worldwide forum. Thus, if the halal industry in Muslim-majority

Table 11.2 GIFT Index

Rank	Countries	Rank	Countries	Rank	Countries
1	*Malaysia*	21	Switzerland	41	Cameroon
2	Saudi Arabia	22	Canada	42	Afghanistan
3	*Indonesia*	23	Luxembourg	43	*Thailand*
4	UAE	*24*	*Kazakhstan*	44	Maldives
5	UK	25	Germany	*45*	*Algeria*
6	Bahrain	26	China	*46*	*Mauritania*
7	Kuwait	27	France	*47*	*Morocco*
8	*Singapore*	28	Japan	48	Palestine
9	Qatar	29	Egypt	49	Iraq
10	Hong Kong	30	Sri Lanka	*50*	*Philippines*
11	Pakistan	31	Russia	*51*	*Kyrgyzstan*
12	Iran	32	Spain	52	Azerbaijan
13	USA	33	India	53	Albania
14	Bangladesh	*34*	*Tunisia*	54	Syria
15	Jordan	35	Lebanon	*55*	*Uzbekistan*
16	Turkey	36	Kenya	56	Burkino Faso
17	Oman	37	Bosnia and Herzegovina	57	Senegal
18	Australia	38	South Africa	*58*	*Libya*
19	Nigeria	*39*	*Tajikistan*	59	Ivory Coast
20	*Brunei*	40	Sudan	60	Mali

Source DinarStandard and Elipses (2022)

countries in the ASEAN, Central Asia, and Maghreb is hoped to make a meaningful ante to the economic development of these regions, the ICT sector has a role as an enabler to construct the digitally enabled community. Then, the digitally enabled community becomes a lubricant to drive the digitally enabled halal trade. In other words, by transforming the halal industry and digitalization toward digital halal management, they will engine the economic growth in the Muslim-majority countries in these regions.

11.3 Discussion

A study by Moi et al. (2016) found the most competitive country in the halal industry in the Maghreb region is Mauritania, come after by Morocco, then Tunisia, while the lowest competitiveness countries are Algeria and Libya. Nevertheless, the previous study by Noor and Moi (2015) found that Mauritania has no prospect to develop the halal industry. Only Algeria, Libya, Morocco, and Tunisia have the prospect of halal industry development. The main consideration is its positive net export.

Furthermore, as given by Table 11.3, as in the ASEAN and Central Asia, there are three countries in the Maghreb region that are included in the Global Muslim Travel Index (GMTI) 2022 ranking. Among 138 countries accounted for in this report, Morocco in 12th rank, Tunisia in 17th rank, and Algeria in 21st rank. Likewise Kazakhstan and Kyrgyzstan are ranked 22nd, Tajikistan is ranked 32th, and Uzbekistan is ranked 9th as well as the highest rank among Central Asian countries. All indicate that countries in this region actually have the potential to become halal tourist destinations and have the opportunity to become the leading Muslim-Friendly Travel sector.

Table 11.3 GMTI 2022 ranks for the Muslim-majority countries in ASEAN, Central Asia, and Maghreb region

ASEAN	GMTI	Central Asia	GMTI	Maghreb Region	GMTI
Malaysia	1	Uzbekistan	9	Morocco	12
Indonesia	2	Kazakhstan	22	Tunisia	17
Singapore	9	Kyrgyzstan	22	Algeria	21
Brunei	12	Tajikistan	32	Libya	–
Thailand	28	Turkmenistan	–	Mauritania	–
Philippines	37				
Viet Nam	68				
Cambodia	76				
Laos	129				
Myanmar	–				

Source CrescentRating (2022)

Many studies have proven that digitalization can help empower the economic development of the region and state. In order to support the halal industry to boost the economic growth of Muslim-majority countries in ASEAN, Central Asia, and the Maghreb region, a digitally enabled community become a requirement to create a digitally enabled halal economy that is secure, innovative, inclusive, transformative, and sustainable. Digitally enabled community is marked of them with high Internet users, high Internet penetration—Internet users per population—and high social media user.

Table 11.4 shows the internet users, internet penetration, and Facebook users in Muslim-majority countries in ASEAN, Central Asia, and the Maghreb region. This table indicates the average Internet penetration rate in Muslim-majority countries in ASEAN is the highest, followed by the Maghreb, then Central Asia. This situation is also not much different from the average Facebook subscribers in ASEAN which is the highest, followed by the Maghreb, and then Central Asia. However, the highest growth of internet users in 2022 than 2000 is in Uzbekistan and Tajikistan as representatives of countries in Central Asia. This is surely understandable because the number of Internet users in 2000 was very low compared to other countries in Central Asia, especially compared to countries in ASEAN and the Maghreb region.

In Central Asia, the leading destination for fintech is Kazakhstan. In other countries in this region, fintech is still underdeveloped (Mynbayev and Aushev 2020). In Central Asia, most countries are hurt by expensive Internet connectivity and poor quality due to limited use of ICT, limited regional integration, incomplete policy and regulatory environments, global internet traffic bypassing Central Asia, landlocked, and low private investment level (Navas-Sabater 2019).

However, Central Asia has immense potential for fintech because the utilizes of devices for mobile communication and Internet continue to increase, particularly with the appearing of digital payments, better support initiatives of government, and the economic recovery after pandemic. Anyway, the fact that financial innovation and fintech opportunities are underutilized in Central Asia is an indication of weak literacy and implementation. To emerge fully, there is a strong need for a comprehensive national development plan that addresses issues related to financial inclusion and how to successfully introduce practical fintech solutions (Fintech Bulletin 2021).

Regarding the Maghreb region, in terms of Public Policy Economics, the existence of the Maghreb is still not generating the optimal economic outcomes for its members. The impact of public policies on resource allocation and income distribution in the economy in this region is still far from Pareto Optimum as well as Al-Ghazali/Al-Shatibi Optimum (Musari and Sayah 2022). Therefore, higher education institutions have strategic roles in serving and driving the unified innovation system in this region. Regrettably, the political nexus to power in the Maghreb, both internally and in relation to the central government, obstructs a wider expansion of this integrative part, despite all the endeavors embedded to present a new impulse to research. Power-sharing and decision-making decentralization are still hard to realize (Djeflat 2009a, b).

Table 11.4 Digitally enabled community of the Muslim-majority countries in ASEAN, Central Asia, and Maghreb Region

Countries	Population (2022 estimation)	Internet users (December 31, 2000)	Internet users (2021/2022)[a]	Internet penetration (%)	Facebook subscriber[b]
ASEAN					
Brunei Darussalam	444,336	30,000	531,800	119.7	508,300
Malaysia	33,049,551	3,700,000	30,987,000	93.8	30,167,100
Indonesia	278,268,685	2,000,000	212,354,070	76.3	202,200,000
Central Asia					
Kazakhstan	19,146,252	70,000	16,465,777	86.0	13,954,600
Kyrgyzstan	6,703,015	51,600	3,683,700	55.0	3,590,900
Uzbekistan	34,271,815	7500	17,161,534	50.1	7,003,400
Tajikistan	9,898,203	2000	3,013,256	30.4	1,587,300
Turkmenistan	6,177,955	2000	1,562,794	25.3	321,800
Maghreb					
Libya	7,024,811	10,000	6,658,900	94.8	6,658,900
Algeria	45,150,879	50,000	37,836,425	83.8	26,291,400
Morocco	37,344,795	100,000	25,589,581	68.5	21,730,000
Tunisia	11,935,766	100,000	8,170,000	68.4	8,170,000
Mauritania	4,775,119	5000	969,519	20.3	927,300

Source Internet World Stats (2022a, b)
[a]For Asia, per July 31, 2022; and for Africa, per December 31, 2021
[b]For Asia, per July 31, 2022; and for Africa, per April 31, 2022

In the Maghreb region, the financial inclusion level driven by Islamic microfinance in this region is lagging behind Southeast Asia. The market has not yet visible the full potential of the Islamic microfinance industry in driving socioeconomic impacts at scale for people within the "bottom of the pyramid" in the Maghreb as in Southeast Asia. Rooted in sharia values, Islamic finance should have eligibility to lead microfinance in contributing the financial inclusion in the Maghreb region which consist of Islamic countries (Aaminou 2022; Musari and Sayah 2022). Whereas, Islamic finance can be an ideal opportunity for financing infrastructure and development projects in North Africa (Daoud and Kammoun 2020). Therefore, coupled with the Islamic finance industry, ICT as enabler can integrate Islamic finance institutions, the halal industry, and all halal stakeholders in this region to develop the halal economy ecosystem.

In Central Asia, to meet the various needs of the Muslim society, Central Asian governments have also introduced Halal Banking and Finance in the market. Indeed, Islamic reform in Post-Soviet Central Asia has various effects in many fields. In general, the influence of Islamic cultural and educational features is growing in all parts of Central Asia. After gaining independence from Russia in 1991, the Islamic heritage becomes a potential alternative to fill the political and social void in the Central Asian states (Tatari and Shykhutdinov 2010; Malik 2015). These things that are owned by Muslim-majority countries in Central Asia are the biggest social capital in developing the halal industry.

It is undeniable, related to the halal industry and digitally enabled community, Muslim-majority countries in ASEAN are ahead of Central Asia and the Maghreb. Brunei, Malaysia, and Indonesia have succeeded in encouraging non-Muslim-majority countries in the region to work together to make the halal industry a booster for regional economic growth. Through the ASEAN Working Group on Halal Food (AWGHF), ASEAN member countries agreed to design a Plan of Action (POA) during 2017–2020 to be implemented in the 2021–2025 period for the cooperation in halal food. It shows that the halal industry is attracted by all ASEAN member countries, particularly to halal food.

Hence, to realize a digitally enabled community in supporting the halal industry in Central Asia and the Maghreb region, partnerships with Muslim-majority countries in one region are a necessity. Particularly in the Maghreb region, the countries should work together and work hand in hand. By using the recommendations of Hamza (2017) and a study by Musari and Sayah (2022), at least there are two steps that must be done in the short term to realize them. **First,** integrated political economy. The countries need to collaborate and generate a common social economy policy. The most important thing that must be implemented soon is the free movement of goods, people, services, and capital by opening up borders between neighbors. **Second,** cooperation among cyber security and technology providers. The interchange of data and information is expected to create a safer area to encourage investment and tourism. The two of them will drive collectively the real economy which in turn will also drive the halal sector.

11.4 Conclusion

Indonesia, Malaysia, and Brunei as ASEAN state members, also Kazakhstan, Tajikistan, Kyrgyzstan, Uzbekistan, and Turkmenistan in Central Asia, then Algeria, Morocco, Tunisia, Libya, and Mauritania in the Maghreb region, are Muslim-majority countries that have occasion to be major players in the halal industry. To realize the halal industry in the Muslim-majority countries in these regions bring remarkable support to the economic development of the countries and region, the ICT sector has a role as an enabler to construct the digitally enabled community. As a lubricant to drive the digitally enabled halal trade, the digitally enabled community will hasten the growth of global halal trade through digital integration. No doubt, the intraregional economy and digitalization become major areas for global halal commerce. Therefore, national ownership and government partnership in each region are essential to encouraging and sustaining a lasting halal industry at scale.

In the short term to realize the digitally enabled community for supporting the halal industry in Muslim-majority countries, particularly in Maghreb Region, the first step that must be implemented is to establish a new foundation for political integration. All state members must end the long-standing competition as well as unaccomplished tenseness among them. All state members must wake up from their long deep sleep and must support each other for the success of the halal economy in each country and the region. This is the first as well as the key requirement of Public Policy Economics to generate optimal outcomes for the economic regional development.

References

Aaminou MW (2022) Towards a better financial inclusion in Maghreb countries: insights on the role of Islamic microfinance and fintech (research report no. 22). Research Center for Islamic Economics (IKAM), İstanbul

Al-Ghwell H (2022) Arab Maghreb Union needs a whole new foundation. Arab News, Opinion. https://www.arabnews.com/node/2091501

Aniqoh NAFA, Hanik U (2021) The potential of halal tourism industry based on local wisdom in Uzbekistan. J Dig Mark Halal Indus 3(2):111–120. https://doi.org/10.21580/jdmhi.2021.3.2.8634

Arzuw News (2021) Hasar products from Turkmenistan have been awarded the international halal certificate. https://arzuw.news/en/produkcija-hasar-iz-turkmenistana-otmechena-mezhdunarodnym-sertifikatom-halal

Bergeaud-Blackler F (2004) Social definitions of halal quality: the case of Maghrebi Muslims in France. In: Harvey M, McMeekin A, Warde A (eds) Qualities of food. Manchester University Press, Manchester, pp 94–107

Crescent Rating (2022) Global Muslim travel index (GMTI) 2022. CrescentRating and Mastercard Asia/Pacific, Singapore

Dania S (2019) 10 reasons Uzbekistan is the Islamic destination to visit in 2019. Have Halal Will Travel. https://www.havehalalwilltravel.com/reasons-uzbekistan-islamic-destination-2019

Daoud Y, Kammoun A (2020) Islamic banking in North Africa: emergence, growth, and prospects. In: Rafay A (ed) Growth and emerging prospects of international Islamic banking. IGI Global, Hershey, PA, pp 78–98. https://doi.org/10.4018/978-1-7998-1611-9.ch005

DinarStandard (2020) State of the global Islamic economy report 2020/2021. DinarStandard, Dubai Islamic Economic Development Center, and Salaam Gateway, Dubai

DinarStandard (2022) State of the global Islamic economy report 2022. DinarStandard, Dubai Islamic Economic Development Center, and Salaam Gateway, Dubai

Djeflat A (2009a) Building systems for innovation "take off" in African economies. Afr J Sci Technol Innov Dev 1(1):53–75

Djeflat A (2009b) Universities and scientific research in the Maghreb states: power politics and innovation systems. Int J Technol Manage 45(1/2):102–113. https://doi.org/10.1504/IJTM.2009.021522

DinarStandard & Elipses (2022) Global Islamic fintech report 2022. DinarStandard, Elipses, and Saalam Gateway, Dubai

Du Rausas MP, Manyika J, Hazan E, Bughin J, Chui M, Said R (2011) Internet matters: the Net's sweeping impact on growth, jobs, and prosperity. McKinsey Global Institute (MGI), McKinsey & Company

Euro Meat News (2022) Taze Ay receives $4.5m from EBRD to increase halal meat production. https://www.euromeatnews.com/Article-Taze-Ay-receives-%244.5m-from-EBRD-to-increase-halal-meat-production/237

Fintech Bulletin (2021) The importance of fintech in Central Asia. Bull Bus Finance Consul. https://bulletins.bfconsulting.com/en/the-importance-of-fintech-in-central-asia/

Hamza WA (2017) The Maghreb union is one of the world's worst-performing trading blocs. Here are five ways to change that. World Economic Forum. https://www.weforum.org/agenda/2017/06/five-ways-to-make-maghreb-work/

Hidayat SE, Musari K (2022) ASEAN towards a global halal logistics through the digitally-enabled community. Int J Asian Bus Inform Manag 13(2):1–15. https://doi.org/10.4018/IJABIM.20220701.oa1

Internet World Stats (2022a) Internet users statistics for Africa. https://www.internetworldstats.com/stats1.htm

Internet World Stats (2022b) Internet 2022b usage in Asia. https://www.internetworldstats.com/stats3.htm

LANA (2022) Ministry of economy forms a committee to monitor halal food products. Libyan News Agency (LANA). https://lana.gov.ly/post.php?lang=en&id=252510

Malik BA (2015) Halal banking in post-soviet central Asia: antecedents and consequences. Mark Brand Res 2:28–43

MGI (2016) Digital finance for all: powering inclusive growth in emerging economies. Executive summary, celebrating 25 years of insight McKinsey Global Institute (MGI), McKinsey & Company

Moi MR, Noor MAM, Ismail MA (2016) Competitiveness of halal industry in Maghreb countries. Global J Al-Thaqafah 6(1):61–67

Musari K (2022) ASEAN toward circular economy and sustainable energy in tackling the climate change: Islamic green financing taxonomy for sustainable finance. In: Ordóñez de Pablos P (ed) Handbook of research on building greener economics and adopting digital tools in the era of climate change. IGI Global, Hershey, PA, pp 210–230. https://doi.org/10.4018/978-1-6684-4610-2.ch011. https://www.igi-global.com/chapter/asean-toward-circular-economy-and-sustainable-energy-in-tackling-climate-change/309805

Musari K, Hidayat SE (2022) Digital financial knowledge and behavior of generation Z in Indonesia: a survey of Islamic fintech literacy toward digital financial inclusion. In: Anshari M, Almunawar MN, Masri M (eds) Fintech development for financial inclusiveness. IGI Global, Hershey, PA, pp 96–117. https://doi.org/10.4018/978-1-7998-8447-7.ch007

Musari K, Sayah F (2022) Encouraging digital financial inclusion ecosystem in Maghreb region through Islamic helix approach. A paper was presented at the 3rd international conference of Islamic finance and business (ICIFEB) held by Faculty of Economics and Business (FEB) of Syarif Hidayatullah State Islamic University (UIN), Jakarta—Indonesia

Musari K, Zaroni Z (2021) Reverse logistics in the age of digital transformation for circular economy and halal logistics through the leadership of Asia. In: Ordoñez de Pablos P, Zhang X, Almunawar MN (eds) Handbook of research on disruptive innovation and digital transformation in Asia. IGI Global, Hershey, PA, pp 83–103. https://doi.org/10.4018/978-1-7998-6477-6.ch006

Mynbayev B, Aushev K (2020) Overview of fintech development in Central Asia: Kazakhstan, Kyrgyzstan, Uzbekistan, Tajikistan. KPMG Tax and Advisory of Kazakhstan. https://assets.kpmg/content/dam/kpmg/kz/pdf/2020/12/Overview-of-Fintech-Development-in-Central-Asia.pdf

Navas-Sabater J (2019) World Bank regional digital programs in Central Asia: the example of the Digital CASA regional program. A presentation material at United Nations Economic and Social Commission for Asia and the Pacific (UNESCAP) Expert Group Meeting on the Asia-Pacific Information Superhighway in North and Central Asia held by Digital Development Global Practice, World Bank, in Almaty—Kazakhstan

Noor MAM, Moi MR (2015) Potential of halal industry for Maghreb countries. J Islam Bank Finance 32(1):82–93

Osafo-Kwaako P, Singer M, White O, Zouaoui Y (2018) Mobile money in emerging markets: the business case for financial inclusion. Global Banking Practice, McKinsey & Company

Reuters T (2013) State of the global Islamic economy report 2013. Thomson Reuters, DinarStandard, and Dubai Islamic Economic Development Center, Dubai

Sahida (2021) Central Asia—Islamic halal countries. Sah World Wide Food & Travels. https://sahworldwidefood.com/central-asia-islamic-halal-countries/
Sala C (2022) How fintech can foster financial inclusion and literacy. Forbes, Leadership Strategy. https://www.forbes.com/sites/esade/2022/11/02/how-fintech-can-foster-financial-inclusion-and-literacy/?sh=f289a2964153
Salama (2013a) Kyrgyzstan to develop halal standards. Halal Focus. https://halalfocus.net/kyrgyzstan-to-develop-halal-standards/
Salama (2013b) Tajikistan certifies halal food. Halal Focus. https://halalfocus.net/tajikistan-certifies-halal-food/
Salama (2018) Central Asia countries to increase exports of halal products. Halal Focus. https://halalfocus.net/central-asia-countries-to-increase-exports-of-halal-products/
SGS Caspian (2021) Hasar confectionery company receives halal certificate. Société Générale de Surveillance (SGS) Caspian. https://www.sgs-caspian.com/en/news/2021/06/hasar-halal-cert
Stonehouse GH, Konina NY (2020) Management challenges in the age of digital disruption. In: Proceeding of 1st international conference on emerging trends and challenges in the management theory and practice (ETCMTP 2019), series advances in economics, business and management research, vol 119. Atlantic Press. https://doi.org/10.2991/aebmr.k.200201.001
Tatari E, Shaykhutdinov R (2010) State response to religious revivalism in post-Soviet central Asia. Eur J Econ Poli Stud 2(3):85–110
UNSGSA, Better Than Cash Alliance, UNCDF, and the World Bank (2018) Igniting SDG progress through digital financial inclusion. United Nations Secretary-General's Special Advocate for Inclusive Finance for Development (UNSGSA), Better Than Cash Alliance, the United Nations Capital Development (UNCDF), and the World Bank, NY

Analysis of Motives and Satisfaction with the Use of Islamic Website Media on Halal Food for Islamic Economics Students in Jabodetabek, Indonesia

12

Riskayanto, Amri Dunan, Yanuar Farida Wismayanti, Bambang Mudjiyanto, Ahmad Budi Setiawan, Karman, Djoko Walujo, and Ari Cahyo Nugroho

Abstract

As the country with the largest Muslim population in the world, Indonesia does not occupy the top position in the world's halal industry. It will focus on the halal food sector as one of the sectors in the halal industry and the fact that Indonesia ranks at the top of the mobile e-commerce adoption by millennials. This phenomenon becomes very interesting to analyze. The authors conducted a descriptive study on the target audience of the Islamic economics students in Indonesia and Islamic social media as a digital communication medium. The researchers chose republika.co.id and eramuslim.com as objects with the assumption that they are currently the most popular digital media among Muslim students. The authors' analysis drew on the uses and gratification theory, among others, to observe the gap between motives (gratification sought) and satisfaction (gratification obtained) in the use of social media. They found that the students' highest expectation for using the website was on the motive of self-identity and information, whereas they considered eramuslim.com more desirable. However, the high value of expectations for eramuslim.com makes the gap in the motives of personal identity negative. According to the percentage criteria, the level of fulfillment of the motives expected from utilizing the two websites is high. In this criterion, the percentage level of fulfillment of self-identity motives with republika.co.id is higher than with eramuslim.com.

Riskayanto
Master's Program in Communication Studies, Gunadarma University, Depok City, Indonesia

A. Dunan (✉) · Y. F. Wismayanti · B. Mudjiyanto · A. B. Setiawan · Karman · D. Walujo · A. C. Nugroho
National Research and Innovation Agency (National Research and Innovation Agency-BRIN), Central Jakarta, Indonesia
e-mail: amri003@brin.go.id

A. Rafiki (ed.), *Digitalization in Halal Management*, Contributions to Environmental Sciences & Innovative Business Technology,
https://doi.org/10.1007/978-981-99-5146-8_12

Keywords

Halal food • Digital social media • Uses and gratification • Communication gap

12.1 Introduction

For the Indonesians, who are predominantly Muslim, the awareness to use something following sharia or halal is very common. The tendency for legal consumption is increasing from year to year. Assurance of the halalness of a product is essential for Muslims for safety, comfort, and perfection in carrying out religious law. The development of the Islamic finance industry, initially triggered by developments in the Islamic banking industry, also encourages industrial development of other sharia-based sectors, including food, finance, travel, cosmetics, fashion, and arts and culture sectors. Maftuhah (2018) stated that the potential for halal industry players in Indonesia is enormous, but it needs to be optimized. Indonesia has 175.4 million (64%), potential Internet users, and 160 million people are social media users (Riyanto 2020). The use of social media shows that, in Indonesia, Internet users search most portals, including YouTube, WhatsApp, Facebook, Instagram, and Twitter.

Several facts motivated the authors' interest in the topics of this descriptive study; among others, they are: (1) Maftuhah's (2018) findings about the connection between a very large use of social media in Indonesia in six sectors of halal industry; (2) Agustin's (2019) findings revealing that some non-Muslim countries appreciated the opportunity of the halal industry, as well as the conclusion that Indonesian people need to become more aware of the importance of choosing domestic halal products; (3) Karman's (2013) study of usage media selection for the fulfillment of the needs of audiences who experience development from traditional to Internet media; (4) the fact that Indonesia's performance is not top ranked in the halal food industry; (5) the residents are productive in Indonesia, with the highest use of online transactions by a mobile device.

Regarding halal food, the authors chose the category conservative as a reference, and found eramuslim.com and VOA-islam.com as the most popular websites. According to the rating from Alexa.com, in July 2020, IBTimes.id released a ranking list of 100 Islamic websites in Indonesia and globally. The results of a research involving 27 FISIP lecturers at Muhamadyah University Jakarta-UMJ in 2019 showed that the Internet is preferred as a means to learn Islam among Muslim students. Surveys from several studies show that respondents mostly access YouTube (31.94%), follow Muslim pages (17.02%), then eramuslim (10.21%), and Nahdlatul Ulama-NU pages (6.81%), while the others learn Islam directly with ustadz (Reaktor 2019).

The main idea of the research arose from the many problems Muslims and Indonesians face concerning Internet media or websites (Adiandirgantoro et al. 2017). Therefore, this chapter addresses the following research questions:

1. What is eramuslim.com and republika.co.id media users' expectation (i.e., GS) when looking for information about halal food?
2. How satisfied are eramuslim.com and republica.co.id media users (i.e., their GO) looking for information about halal food?
3. What is the difference between GS and GO of eramuslim.com and republika.co.id media users looking for information about halal food?

Based on these research questions, the authors focused on the following objectives:

1. Describing and analyzing eramuslim.com and republica.co.id media users' GS in looking for information about halal food.
2. Describing and analyzing eramuslim.com and republika.co.id media users' GO in looking for information about halal food.
3. Describing and analyzing the gap between eramuslim.com and republica.co.id media users' GS and GO in looking for information about halal food.

This study is focused on the issues related to halal food only, as one area of the halal industry. Restrictions on the industry of halal food only are, of course, very interesting to study, considering Indonesia is a populated country, the most prominent Muslim in the world, and ever claimed as a prosperous country. In 1984, the new reigning order's aim was to achieve food self-sufficiency, even though, at the time, a bowl of imported rice still was the actual availability. Besides the focus on halal food, the authors limited the subject of their research to Islamic economics students in five Indonesian cities, namely Jakarta, Bogor, Depok, Tangerang, and Bekasi (Jabodetabek). The authors based this choice on the consideration that a significant part of new media users surfing the virtual world is made by the teenagers who are still attending high school. The last limitation the authors posed in the study was related to their methodology. The authors investigated only the development of the world of halal food associated with the development of new media.

12.2 Literature Review

This section will describe the theoretical basis of the authors' discussion topics in this chapter, including new media, Islamic websites (i.e., eramuslim.com and republika.co.id), and uses gratification (U&G) theory.

12.2.1 Development in the Use of New Media

"New media" is a term for explaining convergence among technology, computerized digital communication, and connection to the Internet. New media can channel information (i.e., they are an intermediary) from its source to its receiver (McQuail as cited in Efendi et al. 2017).

New media have some specific features, according to the following experts (Karman 2013):

1. McQuail (2010) identified the following five characteristics of new media:
   a. Digitization and convergence of all aspects of media.
   b. Increased interactivity and network connectivity.
   c. Mobility and delocation of sending and receiving.
   d. Adaptation of publication and audience roles.
   e. Appearance of diverse new forms of media gateways.
2. Rogers (1986) identified new media according to three criteria:
   a. Interactivity, where with level the interactivity of the Internet is approaching the level of interpersonal communication, so that the participants can communicate more accurately, effectively, and satisfactorily.
   b. Demassification, where the message is memorable, enhances individual exchange among the participants.
   c. Asynchronicity, which allows to send and receive the communication whenever the user wants.

The term "new media" is often used to explain the emergence of digital, computerized, and networked media. Since its introduction in 1972, the ability of the Internet to facilitate any activity has increased enormously (Masse 2017). Through the Internet, everyone can access the global world to get information in various fields and influence behavior (Rusno 2010).

12.2.2 Islamic Sites

One growing phenomenon, fast in line with the development of technology, is the appearance of many sites in the cyberspace. These websites are used for various needs by various parties, with different aims and objectives. Some surveys have been conducted to rate them (Reaktor 2019). For this study, the authors chose two Islamic websites, namely eramuslim.com and republika.co.id, which are very popular among Islamic teenagers and students.

Eramuslim.com
Eramuslim.com was founded in 2000, during the period of rampant use of the Internet, by Indonesians. It was not conceived as an Islamic website that focuses solely on the news but also offers consulting.

In order to ease visitors' search for news or information, eramuslim.com provides some rubrics and columns, as follows:

1. The *News* is a rubric that offers the latest domestic and overseas news, in several columns, namely:
   a. *Editorial.*
   b. *Islamic world.*

   c. *Palestine.*
   d. *Friendship.*
   e. *National.*
   f. *People info.*
   g. *Editor's notes.*
2. *Oasis of faith* is a column that offers an article by a post-reader about a life experience that can benefit others.
3. *Civilization* is a column that serves articles about everyday life in the divided Qur'an and Sunnah. It includes the following subcolumns:
   a. *Sharia economics.*
   b. *Islamic thought.*
   c. *Reflecting on the Salaf.*
   d. *Sirah thematic.*
   e. *Miracles of the Qur'an and Sunnah.*
   f. *Tafsir Fii Zilalil Qur'an.*
4. *Focus* is a rubric; it includes the following columns:
   a. *Analysis.*
   b. *Special report.*
   c. *Did You Know?*
   d. *Talk.*
5. *Muslim* is a column that offers life-divided Muslims. It includes the following subcolumns:
   a. *Talking woman.*
   b. *Women's inspiration.*
   c. *Quranic parenting.*
   d. *Muslimah lifestyle.*
   e. *Culinary.*
6. *Wisdom* is a column that serves rubric-related contemplation and wisdom from everyday life. It offers two discussion subcolumns:
   a. *Contemplation.*
   b. *Business is jihad.*
7. *Education and Family* is a rubric with the following three columns of discussion:
   a. *Education.*
   b. *Family.*
   c. *Education and family consultation.*
8. *Business* is a column that serves business information. It includes the following parts:
   a. *Business info.*
   b. *Review book.*
   c. *Product info.*
   d. *Review.*
   e. *E-market.*
   f. *E-plaza.*

9. *Chaplain* is a column where experts in the field raise and discuss problems about morality, prayer, and others. This column is divided into:
   a. *The Chaplain answers.*
   b. *Ulama's advice.*
   c. *Friday sermon.*
10. *Voice Sky* is a column that serves light articles that have deep meaning as well as various stories or possible events related to reward and pleasure of Allah. It includes the following parts:
    a. *Penetration ideology.*
    b. *Lightweight.*
    c. *Going to true Life.*
    d. *Invitation to paradise.*
11. *Consultation* is a column that offers help to website visitors to solve their ongoing problems. The inner rubric column is divided as follows:
    a. *Behind conspiracy.*
    b. *Consultation family.*
    c. *Ribbon Nabawi.*
    d. *Consultation architecture.*
    e. *Zakat consultation.*
    f. *Waqf.*

The majority of eramuslim.com contains information about news, articles, and opinions. Besides, eramuslim.com displays some links to *zakat* and *waqf* websites. Eramuslim.com also uses Twitter and Facebook to spread news or the most recent articles.

Republika.co.id

Republika.co.id is a news portal presenting information by text, audio, and video based on technology hypermedia and hypertext. With progress information and development in social media, republika.co.id has a new feature that mixes digital media communication. The information is always updated, making it a trustworthy news portal.

Republika.co.id website encompasses several rubrics and columns:

- *Treasures,* containing news about the Islamic world.
- *Islamic Digest*, which contains stories about the prophet, Al-Qur'an, *sunnah*, and learning contents about Islam.
- *International*, containing news about international.
- *Republikbola*, which contains news about football competitions, score updates, and information about football teams.
- *Hajj Journal*, which contains information and news about *hajj*.
- *Leisure*, which contains news about style, halal life, and tourism.
- *In Pictures*, contains news with high resolution images.
- *Videos*, contains news program with audio visual about current affair and music.
- *Publications*, containing news written by the public (i.e., citizen journalism).

- *The Economy*, which contains economy news.
- *English*, containing news in English.
- *The Sheath*, which contains news in the Islamic perspective.

12.2.3 Uses and Gratification Theory

The U&G theory was first proposed by Katz and Blumler (1974), in their study *The Uses of Mass Communication: Current Perspectives on Gratification Research* on the use of media to fulfill the chosen audience's expectations and needs.

In relation to the use of media, Karman (2013) argued that the U&G theory was applied with an emphasis on assessment of what audiences do with the media they use. In other words, Karman observed that the development of the U&G theory tended to emphasize the concept of gratuity.

This theory reveals that media users are active parties in the communication process. Because of their uses, media users try to find the best media sources to meet their needs. It can be assumed that users have choices to meet their needs (Nurudin 2003). The essence of the U&G theory is that the audience uses the mass media based on certain motives. The media are considered to be trying to fulfill the audience's motives. If this motive is fulfilled, the audience's needs will be fulfilled. In the end, the media that can meet the audience's needs will be called effective media. According to Ebersole (2006), the U&G theory is one of the media theories that have repeatedly been able and relevant to research new media such as the Internet. Instead, Katz et al. (1973) stated that the U&G theory includes the following five basic assumptions:

1. The audience is active and goal-oriented.
2. Initiative in linking need satisfaction to specific media choices is available to audience members.
3. Media compete with other sources for the satisfaction of needs.
4. People have sufficient self-awareness of their media use, interests, and motives to give researchers an accurate picture of their use.
5. The audience can only judge the assessment of media content.

12.2.4 Gratification Sought (Motive)

According to Palmgreen (1984), GS is the motive sought and encourages someone to consume media. GS originates from a person's beliefs about what the media can provide and evaluations of media content. GS is the expectation of satisfying user needs from media exposure converted into active involvement to achieve a certain level of satisfaction (Hardyanti 2017). It was found that there was a significant relationship between activity and motives for using media. There is a positive relationship between pre-activity with post-activity actions.

12.2.5 Gratification Obtained (Satisfaction)

GO is a person's absolute satisfaction after consuming a specific media type. According to Palmgreen (1984), gratification is obtained by questioning specific things about what is obtained after using the media by mentioning specific events or rubrics. GO is a preference for favorite material presented by certain mass media, which is considered to have no difference from one individual to another (Hardyanti 2017).

12.2.6 Motives for Using Media

The study of media use always explains how individuals use the media. Regarding the use of media by individuals, McQuail (1972) suggested the following four motives for the use of media by individuals:

1. Diversion, i.e., escape from daily routines or problems.
2. Personal relationships, which happen when people use media instead of friends.
3. Personal identity, which is a way to strengthen individual values; e.g., many students are better able to learn if accompanied by music from the radio.
4. Supervision, namely information on how the media helps individuals achieve something.

12.2.7 Previous Relevant Studies

As the authors started to address the problem they described in the introduction of this chapter, they analyzed previous relevant studies related to the application of the U&G theory on communication media. Table 12.1 provides an outline of the main references for their research.

12.3 Methodology

This section describes the framework and methodology the authors used in the study. Based on the description in the background and this research problem, the population of this study is made of school or college teenagers. The authors chose this target because they are rated as a very relevant generation for this study. Indeed, they are very dependent on the Internet world daily; they are a valuable example of Internet users for socializing, for more formal needs in support of

Table 12.1 Previous relevant studies

No.	Author and year	Title	Methodology	Results/conclusion
1	Adiandirgantoro et al. (2017)	*Descriptive Analysis of Gratification Sought and Gratification Obtained on Online News Site (a Comparative Study of Online News Site Readers detik.com and kompas.com)*	Descriptive, quantitative	Detik.com readers are interested in information, and detik.com managed to provide the highest satisfaction in this regard. Meanwhile, at kompas.com, the highest expectation is also in the information aspect, but kompas.com gives the highest satisfaction in the entertainment aspect
2	Karman (2013)	*Research on Media Use and Current Developments*	Quantitative	The U&G theory is commonly used in research about media usage. The advantage of the U&G theory is that it can guide getting an explanation of what audiences do with the media by adopting the socio-psychological side of audiences (often referred to as paradigms, perspectives, and schools)
3	Rianto (2016)	*New Media, Active Audience Vision, and Media Literacy Urgency*	Participation, observation	In this research, through participation and observation, the authors found that individuals actively search and share information in a group-customized conversation with needs, values, and beliefs
4	Arifin (2013)	Competition Seven News Portals Online Indonesia by Analysis, Uses, and Gratifications	Descriptive, quantitative	Based on research, the author concluded that the news portal audience is an individual who is already good enough in media literacy

(continued)

Table 12.1 (continued)

No.	Author and year	Title	Methodology	Results/conclusion
5	Hardyanti (2017)	*Discrepancies in Reader Satisfaction in Islamic online media: A Comparative Study of Gaps in Reading Satisfaction on the eramuslim.com Site among Lecturers and Students in Malang*	Quantitative with survey method	The research results show that the education level of the course influences the readers' discrepancy satisfaction. The higher the readers' level of education, the higher the level of discrepancy
6	Wardani (2015)	*Motives and Audience Satisfaction of Ramadan Programs on National Television, Majoring in Islamic Communication and Broadcasting*	Survey method (single cross-sectional surveys)	The results strongly support the U&G theory. This means that there is a correlation between the motive gap and satisfaction. Thus, the Tausiah, Sinetron, Vartiety programs, shows, and features are considered to fulfill student motives on the dimensions of social integration and entertainment

Source Author's compilation

their studies, and for being actively related to religion. The authors chose them as respondents to a survey on searching the Internet for information on halal food, especially on eramuslim.com and republica.co.id.

12.3.1 Population and Sample Study

The population is combined from whole-shaped elements, events, things or people with characteristics seen as universe research (Sugiyono 2017; Arikunto 2021). The study's population is Muslim teenagers, especially school or college teenagers. In contrast, the sample is a part of the population and can represent its characteristics (Sugiyono 2017). In this study, the sample consisted of Islamic economics students at Gunadarma University and several academic institution in Jabodetabek, including the University of Indonesia, the Bogor Agricultural Institute, and Syarif Hidayatullah State Islamic University.

12.3.2 Data Collection Techniques

The authors collected data from the participants' answers to their online questionnaires. The authors' choice of online data collection was due to several reasons: first of all, the COVID-19 pandemic, which limited mobility in all physical activity; secondly, the availability of limited time to complete the study; finally, the possible effectiveness of digital technology and the Internet in getting the desired data quickly and with minimal cost.

The researchers communicated with the sample of Islamic Economics students via social media, especially WhatsApp groups of Islamic Economics classes at Gunadarma University. The authors distributed their online questionnaires they had designed by the snowball technique.

12.3.3 Approach Study

In this study, the authors used a quantitative descriptive approach, which is based on the audience and the use of massive numeric analysis and interpretation. As Prasetyo and Jannah (2005) stated, quantitative research is an approach that relies on calculations, numbers, and statistics as a central tool for support analysis. As to the method, the authors used a questionnaire to collect the data. A more comprehensive study is a study with a descriptive survey. Rachmat (2008) mentioned that the explanative survey method is carried out to know why one situation occurs and what affects happening something. In several cases, this method is chosen to explain a connection or comparison among two or more variables. The survey describes the study conducted for explaining the connection between factors of expectation or motive (i.e., GS) and factors of satisfaction (i.e., GO) for users of Islamic websites searching for certain information.

12.4 Research Design

Time Scale

The authors measured some elements simultaneously in different samples. In other words, the authors assessed their research variables across different objects simultaneously. Thus, the measurement of the object or the same sample may give different results in different periods (Prasetyo 2011). The period the authors considered for distribution of their questionnaire, data collection, and data analysis was March–April 2021.

Variables and Operationalization Variable Study

Quantitative descriptive research for analyzing connection or comparison basically consists in comparing quantities in the research sample. The authors identified two variables: (1) a person's motive to use media, or GS, and (2) a person's absolute satisfaction with the use of a specific type of media, or GO. As something abstract, the measurement of satisfaction in using a medium requires an operationalization so that results from the sample could reflect socialization information or appropriate marketing with needs studies. The authors operationalized several necessary concepts for the measurement in the study, namely the use of media, the motive for their use (i.e., GS), and the satisfaction obtained after use (i.e., GO). Operationalization of the concept refers to Wardani (2015) (Table 12.2).

McQuail (as cited in Kriyantono 2008) defined motive as sought satisfaction or wanted user moment to consume a particular medium. Based on this definition, in this case, the motive for media consumption is distinguished into four categories:

Table 12.2 Operationalization concepts and variables

Draft	Variable	Dimension	Indicator	Scale
Media usage	Exposure to media and media content	Intensity	1. Observed level of media exposure with observed outpouring time for a browse 2. User's attention level to the website content	1. Ordinal 2. Interval
GS	Level of satisfaction sought	1. Cognitive 2. Affective 3. Personal integrative 4. Social integrative 5. Tension release	1. Getting more knowledge about something 2. Getting wisdom from self-learning 3. Applying religion in building connections with other people in daily life 4. Making browsed material as ingredient talks daily 5. Getting entertainment and getting rid of boredom	Interval
GO	Level of satisfaction obtained	1. Cognitive 2. Affective 3. Personal integrative 4. Social integrative 5. Tension release	1. Obtaining more knowledge about something 2. Getting wisdom from self-learning 3. Applying religion in building connections with other people in daily life 4. Making browsed material as ingredient talks daily 5. Getting entertainment and eliminating boredom	Interval

Adapted from Wardani (2015)

(1) information motives, (2) personal identity motives, (3) social integration and interaction motives, and (4) entertainment motives. Subsequent satisfaction is interpreted as the amount of satisfaction an individual gets as a result of the fulfillment of their specific needs after consuming the concerned media. As for the motive, the satisfaction obtained as a consequence of consuming media is identified into four categories: (1) satisfaction for information, (2) satisfaction for personal identity, (3) satisfaction for social integration and interaction, and (4) satisfaction for entertainment. In the study, the authors used grid operationalization variables as guidelines for developing a questionnaire as a data collector (Table 12.3).

12.4.1 Method Measurement

In data analysis, the authors used the method applied for nominal or ordinal scale variables. Referring to previous research, they used a five-point Likert scale (Table 12.4). In social studies, the Likert scale is commonly used for measuring attitudes, opinions, and perceptions of an individual or a group of people about a social phenomenon.

12.4.2 Stages of Analysis

The procedure for deep data analysis generally includes the following stages:

1. **Editing.** The editing process of the collected data is conducted to confirm the required data or follow framework research and discard inappropriate data.
2. **Scoring.** This is a coding response in the form of numbers by a Likert scale. In other words, this is a quantification process on respondents' answers.
3. **Tabulation.** The tabulation process concludes the handling of respondent data. This process of working for a group or compiling response data regularly to be more adherent to the users' needs continued the data processing.

12.4.3 Instrument Test Data Collector

Validity Test

The validity test is needed for the study with primary data sources such as respondent data collected from the questionnaires. The authors conducted this test to confirm whether a questionnaire is a response collector to the question meant by the research concept (Ghozali 2009). A questionnaire is valid if the questions can disclose something that the researcher will measure by it by their research variable. According to Arikunto (2021), validity is a condition that describes the ability of the tool used as data collector to measure what will be measured. The authors

Table 12.3 Definitions of grid operationalization variables

Variable	Operational definition	Indicator
Motive	Information	The user aims to: 1. Know various related events and conditions with society 2. Get various information about related events with the environment 3. Know more about something
	Personal identity	1. Want to find supporting values related to personal access 2. Identify themselves with other values in the media 3. Want to get added values
	Social integration and interaction	1. Get knowledge related to halal food 2. Want to find material for conversation and social interaction with other people around him 3. Be close to other people
	Entertainment	1. Release themselves from problems 2. Relax and fill up free time 3. Get entertainment and fun
	Information	1. Know various related events and conditions with society 2. Know various information about related events in the environment 3. Obtain more knowledge about something
Satisfaction	Personal identity	1. Want to find supporting values related to personal access 2. Identify themselves with other values in the media 3. Want to get added values
	Social integration and interaction	1. Obtain knowledge related to halal food 2. Want to find material for conversation and social interaction with other people around him 3. Have a connection with other people
	Entertainment	1. Release themselves from problems 2. Relax and fill up free time 3. Get entertainment and fun

Adapted from Syarifah (2017). The end questionnaire followed the abovementioned grill

conducted this test by comparing score r_{count} with r_{table} in degrees free (DB) $= n - 2$. If the score r_{count} is more significant than the r_{table}, then details question as an indicator for the research variable could be confirmed as valid (Ghozali 2009).

Table 12.4 Likert scale for the measurement

Rank	Score
Strongly agree	5
Agree	4
Neither agree nor disagree	3
Disagree	2
Strongly disagree	1

Reliability Test

As for the validity test, the reliability test is a precise process required to confirm response data from the sample study. Fulfill conditions can be processed more continuously. The authors conducted this test to confirm or not the questionnaire was reliable as gauge indicator of the research variable. Reliable questionnaires have as a characteristic that participants' responses are consistent from time to time (Ghozali 2009). In the test procedure, this consistency is indicated by the value coefficient. The Cronbach alpha must be more than 0.60.

Data Analysis Techniques

Data analysis required in the study determined the existence of discrepancy between motive (i.e., GS) and satisfaction (i.e., GO) in media use. Subsequently, the authors classified and interpreted the data on a five-point Likert scale. They identified and analyzed the differences between the score of the audience's GS and the score of their GO after using a medium. The researchers completed this analysis using Palmgreen's idea (Oktriyanto and Mursito 2013), which is stated in the following formula:

$$D = \frac{\sum n \cdot i \cdot j}{\sum \sum_{i \neq j}^{i \neq j} n \cdot i \cdot j} \tag{12.1}$$

where D is the gap (discrepancy), n is the sample number, i is GS, and j is GO.

The measurement gap uses the above formula implemented by cross tabulation on the application data processor. The application data processor is generally a software such as MS Excel and SPSS (i.e., spreadsheets). This method intersects items from the motive indicator with similar items from the satisfaction indicator. Data comparison allows to determine the gap level in media use for each indicator. Table 12.5 gives the criteria for satisfaction with media usage based on the gap Bekti (2018) and Hardyanti (2017) reported in their studies.

Adiandirgantoro et al. (2017) implemented the method for identifying the existence of the discrepancy between GS and GO. This method is applied only to compare the mean value of the scale Likert score of each indicator from GS and GO. Based on Arifin's (2013) work, the authors drew the following conclusions from data processing and descriptive analysis with measurement method of the gap above could:

Table 12.5 Satisfaction categories in media usage

No.	Fulfillment satisfaction	Percentage gap (%)
1	Low	21–30
2	Moderate	11–20
3	High	0–10

Source Bekti (2018)

1. If the mean score is more than big than the mean score satisfaction or GS > GO, then, for the relevant indicator, discrepancy occurs where the media could be rated as not satisfying the audience.
2. If the mean score of the motives is the same as the mean score of satisfaction or GS = GO, then no gap occurs for the relevant indicator. The media could fulfill the audience's desired needs.
3. If the mean score is smaller than the mean score of satisfaction or GS < GO, then, for the relevant indicator, discrepancy occurs where the media could be rated as successful in satisfying the audience.

12.5 Results and Discussion

The study population was made of students in studies related with Islamic or sharia economics. We employ purposive sampling, selecting participants based on specific characteristics, to align with our research focus on investigating the utilization of online communication media, widely regarded as the predominant medium on the Internet. Indonesian millennials' dependency on the Internet as a communication medium is very high, even the highest in the world. Thus, two main factors can explain the reason of the authors' choice for sampling by purposive research, namely:

1. Generation Z and the millennial generation in Indonesia are active users of digital media. Therefore, Islamic economics or sharia economics becomes the population in this research.
2. The issue of the halal industry is one of the country's current economic issues. The topic of halal food related to Islamic digital media is the focus of the problem in this research.

In no more than 2 months, the authors obtained responses from 99 respondents who came from students in study programs related with Islamic economics or sharia economics. Table 12.6 gives the respondents' details. It is assumed that students are studying in a program related to Islamic economics, where the curriculum always involves a related subject—the halal industry as one of the elements of the economy. Islam in Indonesia is expected to support the hope that the government realizes the country as one of the world centers of Islamic finance and economy.

Table 12.6 Respondents' details based on origin region or institution

No.	Institution	Amount	Percentage
1	Gunadarma University, Depok	67	67.7
2	Higher Education in Greater Jakarta	32	32.3
Total		99	100

Source Bekti (2018)

12.5.1 Instrument Tests for Data Collection

Instrument tests for data collection include validity tests and reliability tests. They are used to confirm the level of validity of the questions from the questionnaire and how reliable the data collected by a questionnaire could be.

Validity Test

The authors conducted a validity test to compare values r_{count} with the r_{table} on formulation product moments. The items of the questionnaire are considered valid if $r_{\text{count}} > r_{\text{table}}$. The test use level significance of 5% and the amount of data n as much as 43 with degrees free $df = n - 2 = 43 - 2 = 41$, then used $r_{\text{table}} = 0.301$ for comparison. Tables 12.7 and 12.8 give test results.

Tables 12.7, 12.8, 12.9 and 12.10 give the validity test results. They confirm that the questions the authors used as an indicator for the motive variable (i.e., GS) and satisfaction variable (i.e., GO) are significantly valid for use.

Reliability Test

For testing the reliability questionnaire, the authors used the Cronbach's alpha, where the criteria reliability indicator obtained is a score of Cronbach's alpha 0.6. Tables 12.11, 12.12 and 12.13 give the results.

Tables 12.11, 12.12, 12.13 and 12.14 give the reliability test results. They confirm that the questions the authors used as indicators for the motive variable (i.e., GS) and satisfaction variable (i.e., GO) are significantly reliable for use as indicator.

Eramuslim.com Site Gap

The authors defined the gap as the difference between the audience's GS and GO from the use of eramuslim.com (Table 12.15).

The summary of the processed data in Table 12.15 highlights that the public's use of media related with halal food and gratification by general rated could fulfill an audience's needs. The biggest expectation, as eramuslim.com's audience, is from teenagers who are Islamic economics students. The halal food is for the fulfillment needs aspect of identity private, while the smallest is for the fulfillment needs aspect of entertainment.

Based on Syarifah's (2017) study, can mean that sharia students in implementing halal food have the most significant motive for finding aspects that can become means for them to actualize themselves.

Table 12.7 Validity test results for the motive variable in eramuslim.com

Question	r_{count}	r_{table}	Information
1.1	0.838	0.301	Valid
1.2	0.792	0.301	Valid
1.3	0.715	0.301	Valid
1.4	0.760	0.301	Valid
1.5	0.917	0.301	Valid
1.6	0.854	0.301	Valid
2.1	0.760	0.301	Valid
2.2	0.867	0.301	Valid
2.3	0.945	0.301	Valid
2.4	0.975	0.301	Valid
3.1	0.773	0.301	Valid
3.2	0.869	0.301	Valid
3.3	0.882	0.301	Valid
3.4	0.718	0.301	Valid
4.1	0.835	0.301	Valid
4.2	0.817	0.301	Valid
4.3	0.823	0.301	Valid
4.4	0.885	0.301	Valid

Source Research output

However, Table 12.16 also gives the opposite from GS because of fulfillment needs. The authors found the gap with the most significant positive (+) precisely on the indicator "entertainment," and the gap with the most negative (-) on the indicator "private identity." The description above indicates that the only fulfilled needs from the use of eramuslim.com media by teenage students in Islamic economics programs with non-halal food could be on "private identity."

Table 12.16 gives the rating gaps, from the most significant positive (+) to the most negative (–), on eramuslim.com.

Republika.co.id Website Gaps

The authors defined gap as the difference between republika.co.id's audience GS and GO (Table 12.17).

Table 12.17 gives that republika.co.id is a medium that teenagers who are students in the Islamic economic program use and that turns out to fulfill their needs or give satisfaction in all aspects. However, the gap in Table 12.17 gives the same pattern as in the case of ermuslim.com.

The authors found the most significant positive (+) gap in the aspect of entertainment, and the gap positive (+) most diminutive in the aspect of personal identity. This fact, at a time, implies that, for the audience of teenage students

Table 12.8 Validity test results for the motive variable in republika.co.id

Question	r_{count}	r_{table}	Information
1.1	0.884	0.301	Valid
1.2	0.739	0.301	Valid
1.3	0.723	0.301	Valid
1.4	0.888	0.301	Valid
1.5	0.903	0.301	Valid
1.6	0.860	0.301	Valid
2.1	0.783	0.301	Valid
2.2	0.894	0.301	Valid
2.3	0.927	0.301	Valid
2.4	0.954	0.301	Valid
3.1	0.802	0.301	Valid
3.2	0.885	0.301	Valid
3.3	0.886	0.301	Valid
3.4	0.773	0.301	Valid
4.1	0.826	0.301	Valid
4.2	0.813	0.301	Valid
4.3	0.814	0.301	Valid
4.4	0.879	0.301	Valid

Source Research output

in the field of Islamic economics, the use of Islamic media such as eramuslim.com and republika.co.id in the context of halal food could give satisfaction if they seek entertainment. The opposite result emerged on fulfillment needs on aspects such as personal identity.

These findings could be interpreted as odd, the respondent sample is made of students of educational programs related to sharia economics. There is a relation between character and behavior. Generation millennials are being viewed as having orientations different from the previous generation. Discussion carries on about this; naturally, it will be clarified through advanced research in more appropriate fields, i.e., Islam and psychology (Table 12.18).

Motive Fulfillment Rate

The analysis of the audience's use and gratification showed that the level of users' fulfilled needs on the media depend on the motive. With base classification as used by Bekti (2018) and Hardyanti (2017), the analysis gap between GS and GO can also be reviewed from side-level fulfillment (satisfaction) on the use of media by the audience relative to each of the underlying motives (Tables 12.19 and 12.20).

In context-level fulfillment, the higher percentage gap could be interpreted as a low fulfillment level of motives (GS). As an Islamic medium for audience

Table 12.9 Validity test results for the satisfaction variable in eramuslim.com

Question	r_{count}	r_{table}	Information
1.1	0.797	0.301	Valid
1.2	0.767	0.301	Valid
1.3	0.896	0.301	Valid
1.4	0.901	0.301	Valid
1.5	0.824	0.301	Valid
1.6	0.881	0.301	Valid
2.1	0.921	0.301	Valid
2.2	0.909	0.301	Valid
2.3	0.938	0.301	Valid
2.4	0.857	0.301	Valid
3.1	0.827	0.301	Valid
3.2	0.931	0.301	Valid
3.3	0.812	0.301	Valid
3.4	0.886	0.301	Valid
4.1	0.804	0.301	Valid
4.2	0.805	0.301	Valid
4.3	0.704	0.301	Valid
4.4	0.861	0.301	Valid

Source Research output

teenagers who go through education in sharia economic programs, eramuslim.com and the republika.co.id have a minimal percentage gap between GS and GO. In other words, the level of fulfillment and the motives for using these media are very high. The authors found the highest fulfillment rate in private identity motives, while the lowest fulfillment was in the motive of entertainment. Although in this study the authors used Bekti's (2018) and Hardyanti's (2017) categorization of fulfillment level, data processing in this study would be equal to a high level of fulfillment according to other references.

12.6 Discussion

The population of this study is the millennials generation, who are active as students in sharia economics programs. Furthermore, in the context of the U&G theory, this population developed many relationships with the Internet world to access Islamic media, e.g., related access to problem halal industry in the sector lots of halal food found the intended audience is very often to access to Islamic media, especially eramuslim.com and republika.co.id.

From the criteria gap or level fulfillment, the audience's use of Islamic media showed the same pattern. Both on eramuslim.com and republika.co.id, the authors

Table 12.10 Validity test results for the satisfaction variable in republika.co.id

Question	r_{count}	r_{table}	Information
1.1	0.789	0.301	Valid
1.2	0.794	0.301	Valid
1.3	0.889	0.301	Valid
1.4	0.885	0.301	Valid
1.5	0.841	0.301	Valid
1.6	0.841	0.301	Valid
2.1	0.907	0.301	Valid
2.2	0.873	0.301	Valid
2.3	0.906	0.301	Valid
2.4	0.827	0.301	Valid
3.1	0.817	0.301	Valid
3.2	0.942	0.301	Valid
3.3	0.786	0.301	Valid
3.4	0.906	0.301	Valid
4.1	0.823	0.301	Valid
4.2	0.801	0.301	Valid
4.3	0.739	0.301	Valid
4.4	0.848	0.301	Valid

Source Research output

Table 12.11 Reliability test results for the motive variable in eramuslim.com

Indicator	Cronbach value	Threshold limit	Information
Information motive	0.896	0.6	Reliable
Identity motive personal	0.910	0.6	Reliable
Motives for integration and social interaction	0.817	0.6	Reliable
Entertainment motive	0.860	0.6	Reliable

Source Research output

Table 12.12 Reliability test results for the variable motive in republika.co.id

Indicator	Cronbach value	Threshold limit	Information
Information motive	0.911	0.6	Reliable
Personal identity motive	0.913	0.6	Reliable
Motives for integration and social interaction	0.852	0.6	Reliable
Entertainment motive	0.848	0.6	Reliable

Source Research output

Table 12.13 Reliability test results for the satisfaction variable in eramuslim.com

Indicator	Cronbach value	Threshold limit	Information
Satisfaction with information	0.915	0.6	Reliable
Personal identity satisfaction	0.992	0.6	Reliable
Satisfaction with integration and social interaction	0.885	0.6	Reliable
Entertainment satisfaction	0.793	0.6	Reliable

Source Research output

Table 12.14 Reliability test results for the satisfaction variable in republika.co.id

Indicator	Cronbach value	Threshold limit	Information
Satisfaction with information	0.912	0.6	Reliable
Personal identity satisfaction	0.899	0.6	Reliable
Satisfaction with integration and social interaction	0.883	0.6	Reliable
Entertainment satisfaction	0.806	0.6	Reliable

Source Research output

Table 12.15 Gaps fulfillment needs in the use of the eramuslim.com

Indicator	GS	GO	Difference
Information	3.89	3.95	0.061
Personal identity	3.90	3.88	–0.015
Social integration and interaction	3.61	3.69	0.077
Entertainment	3.42	3.63	0.215

Source Research output

Table 12.16 Gaps fulfillment needs in the usage of eramuslim.com according to the rating

Indicator	GS	GO	Difference
Entertainment	3.42	3.63	0.215
Integration and interaction social	3.61	3.69	0.077
	3.61	3.69	0.061
Personal identity information	3.90	3.88	–0.015

Source Research output

found that the highest gap is in the aspect of entertainment. On the other hand, they found the lowest gap in the aspect of private identity. The gap between GS and GO, which became the base measurement here, related to the satisfaction level on the underlying motive of the use of media by the intended audience.

Table 12.17 Gaps fulfillment needs in the use of the republika.co.id

Indicator	GS	GO	Difference
Information	3.73	3.83	0.10
Personal identity	3.74	3.75	0.01
Social integration and interaction	3.61	3.69	0.08
Entertainment	3.37	3.54	0.17

Source Research output

Table 12.18 Gaps fulfillment needs in the use of republika.co.id according to the rating

Indicator	GS	GO	Difference
Entertainment	3.37	3.54	0.17
Social integration and interaction	3.61	3.69	0.08
Information	3.73	3.83	0.10
Personal identity	3.74	3.75	0.01

Source Research output

Table 12.19 Level of compliance needs on media use in eramuslim.com

Type needs	Percentage gap (%)	Fulfillment rate
Information motive	1.54	High
Personal identity motive	–0.38	High
Motives for integration and social interaction	2.08	High
Entertainment motive	5.93	High

Source Research output

Table 12.20 Compliance level needs on media use in republika.co.id

Type needs	Percentage gap (%)	Fulfillment rate
Information motive	2.59	High
Personal identity motive	0.33	High
Motives for integration and social interaction	2.08	High
Entertainment motive	4.89	High

Source Research output

According to the satisfaction level, eramuslim.com and republika.co.id provide the highest level of audience's satisfaction on the motive of private identity. The results show that the GS score as the audience's underlying motive to access Islamic sites. The authors found that both eramuslim.com's and republika.co.id's audience access these websites primarily for information and private identity. The results is in line with Adiandirgantoro et al.'s (2017) and Hardyanti's (2017) findings.

This finding is significant because the audience of reference is young students in sharia economics programs. It might be they access these sites as students. This hypothesis needs more studies to lead to more correct and appropriate results.

As to the evaluation of the two Islamic sites the authors studied in their research, it is necessary to make a more acceptable comparison based on gap criteria and level of fulfilled needs or satisfaction. Based on this consideration, then, simply republika.co.id is more refined because there is no valuable negative (–) gap.

12.7 Conclusion

In the end, below, the authors provide conclusions and recommendations based on their findings:

1. In the audience made of teenagers with a background in the Islamic economics study program, the Islamic websites eramuslim.com and republika.co.id are expected to be the largest to fulfill personal identity motive on halal food.
2. For teenagers with a background in Islamic economics study program, the Islamic sites eramuslim.com and republika.co.id provide the highest satisfaction on the motive of information on halal food.
3. For teenagers with a background in the Islamic economics study program, the use of the Islamic websites eramuslim.com and republika.co.id provides the highest gap in the entertainment aspect.

As a result, the authors put forward the following suggestions:

1. The Indonesian government incessantly spurred Islamic economics in the development of Indonesian economy as one of the centers of Islamic world finance. Islamic sites such as eramuslim.com and republika.co.id can play a role in the achievement of policy targets for digital economy.
2. The Internet media period has become a mainstay for the millennials generation to develop themselves in every aspect of life. Islamic media, such as eramuslim.com and republika.co.id, should also realize the role of supporting the development of human resources in Indonesia.
3. In the time of the COVID-19 pandemic, the role of sites such as eramuslim.com and republika.co.id should be more developed to stabilize the community's activity in all aspects, by deploying appropriate information with the website's specialties, including in the halal food industry. This is in line with findings from other studies, according to which information is one of the highest motives or satisfaction for users.

References

Adiandirgantoro DW, Mayangsari ID, Apriyanti A (2017) Analisis deskriptif gratification sought dan gratification obtained situs berita online (studi komparasi pembaca situs berita online detik.com dan kompas.com). *e-Proc Manage* 4(2):206–229

Arifin P (2013) Persaingan tujuh portal berita online Indonesia berdasarkan analisis uses and gratifications. *J Ilmu Komunikasi* 10(2):353

Arikunto S (2021) *Penelitian tindakan kelas: Edisi revisi*. Bumi Aksara

Agustin I (2019) *Perkembangan produk halal di Indonesia*. https://www.kompasiana.com/itaagustin/5e020ec8097f364e3c34ad12/perkembangan-produk-halal-di-indonesia. Accessed 7 Feb 2021

Bekti BK (2018) *Motif dan kepuasan menggunakan jejaring sosial Instagram* (unpublished thesis master). Program Studi Ilmu Komunikasi, Fakultas Komunikasi dan Informatika, Universitas Muhammadiyah, Surakarta

Katz E, Blumler JG (1974) The uses of mass communications: current perspectives on gratifications research. Pub Opin Q 37(4):509

Ebersole S (2006) Uses and gratifications of the Web among students. J Comput-Mediat Commun. https://doi.org/10.1111/j.1083-6101.2000.tb00111.x

Efendi A, Astuti PI, Rahayu NT (2017) Analisis pengaruh penggunaan media baru terhadap pola interaksi sosial anak di kabupaten Sukoharjo. J Penelitian Humaniora 18(2):12–24

Ghozali I (2009) Aplikasi analisis multivariate dengan program SPSS. Universitas Diponegoro

Hardyanti, W. (2017). Diskrepansi kepuasan pembaca media online Islam: Studi komparatif kesenjangan kepuasan membaca situs eramuslim.com di kalangan dosen dan mahasiswa di Malang. J Sospol 3(1):180–201

Karman (2013) Riset penggunaan media dan perkembangannya kini. J Stud Komunikasi dan Media 17(1):93–111

Katz E, Blumler JG, Gurevitch M (1973) Uses and gratifications research. Pub Opin Q 37(4):509–523

Kriyantono R (2008) Teknik praktis riset komunikasi—disertai contoh praktis riset media, public relations, advertising, komunikasi organisasi, komunikasi pemasaran. Kencana Predana Media Group

Maftuhah M (2018) Seberapa besar peluang Indonesia di antara produsen halal kelas dunia? https://www.kompasiana.com/ceu_meta/5a548be5dd0fa87ec662cca2/seberapa-besar-peluang-indonesia-diantara-produsen-halal-kelas-dunia?page=all

Masse MR (2017) Internet Dan Penggunaannya (Survei di kalangan masyarakat Kabupaten Takalar Provinsi Sulawesi Selatan). J Stud Komunikasi Dan Media 21(1):13–24

McQuail D (2010) McQuail's Mass Communication Theory. SAGE Publications Ltd

McQuail D (1972) The television audience: a revised perspective. Sociol Mass Commun 1972:35–165

Nurudin (2003) *Komunikasi massa*. Cespur

Oktriyanto DD, Mursito DM (2013) Kesenjangan kepuasan dalam membaca surat kabar harian lokal. https://adoc.pub/kesenjangan-kepuasan-dalam-membaca-surat-kabar-harian-lokal.html

Palmgreen P (1984) Uses and gratifications: a theoretical perspective. Ann Int Commun Assoc 8(1):20–55

Prasetyo B, Jannah LM (2005) Metode Penelitian Kuantitatif: Teori dan Aplikasinya. PT Raja Grafindo Persada, Jakarta

Rachmat M (2008) Pengaruh kepuasan, kepercayaan merek, loyalitas merek pada ekuitas merek. Doctoral dissertation, Universitas Gadjah Mada

Rianto P (2016) Media baru, visi khalayak aktif dan urgensi literasi media. J Komunikasi Ikatan Sarjana Komunikasi Indonesia 1(2):90–96

Riyanto AD (2020, July 6) Digital data Indonesia 2020. https://andi.link/hootsuite-we-are-social-indonesian-digital-report-2020/. Accessed 7 May 2021

Reaktor (2019, July 31). Mayoritas Anak Muda Belajar Islam Melalui Internet. https://reaktor.co.id/mayoritas-anak-muda-belajar-islam-melalui-internet/. Accessed 5 May 2022
Rogers EM (1986) Communication technology. Free Press
Rusno R (2010) Pengaruh penggunaan Internet sebagai salah satu sumber belajar terhadap prestasi mahasiswa pendidikan ekonomi Universitas Kanjuruhan Malang. J Ekonomi Modernisasi 6(2):161–172
Sugiyono (2017) *Metode penelitian kuantitatif, kualitatif, dan R&D*. Penerbit CV Alfabeta
Syarifah AZ (2017) Hubungan Antara Motif Dan Kepuasan Penonton Pada Tayangan Drama Korea (Uses And Gratification Pada Mahasiswa Jurusan Komunikasi Dan Penyiaran Islam Iain Surakarta) (Unpublished Tesis) IAIN Surakarta
Wardani NPK (2015) Motif dan kepuasan penonton program Ramadan di televisi nasional. Unpublished Thesis, UIN Syarif Hidayatullah, Jakarta

Non-Muslim Acceptance of Halal Products in the Context of Business Development

13

Koeswinarno, B. J. Sujibto, M. Fakhruddin, and Siti Atieqoh

Abstract

This study explores the relationship between non-Muslim business actors and halal product regulation. The Halal Product Guarantee (HPG) was implemented in Indonesian state law to create the awareness and acceptance of non-Muslim business actors. To increase their economic value, Muslim and non-Muslim business actors should produce, distribute, and sell products which have test-passed halal criteria. This political-economic paradigm is essential for non-Muslim business actors being trapped in a sectoral and ideological framework. The qualitative approach adhered to the triangle parameter and was applied as an individual analysis unit for non-Muslim business actors in Yogyakarta and Semarang, Indonesia. Data were collected through in-depth interviews, observations, and documentation. This study identified a common thread between halal issues and their status as non-Muslim business actors. First, non-Muslim business actors do not interpret halal products as beneficial socio-economic interests for enhanced economic circulation. Second, they follow a sectoral and ideological framework cantered solely on Islamic interests, causing social issues. These two findings emphasize the state's role in socialization and negotiations, specifically with these business actors. Therefore, the implementation of Halal Product

Koeswinarno (✉) · M. Fakhruddin · S. Atieqoh
The National Research and Innovation Agency/Badan Riset Dan Inovasi Nasional (BRIN), Jakarta, Indonesia
e-mail: koes2008@gmail.com

M. Fakhruddin
e-mail: fakh004@brin.go.id

S. Atieqoh
e-mail: siti081@brin.go.id

B. J. Sujibto
Sunan Kalijaga State Islamic University, Yogyakarta, Indonesia
e-mail: bj.sujibto@uin-suka.ac.id

A. Rafiki (ed.), *Digitalization in Halal Management*, Contributions to Environmental Sciences & Innovative Business Technology,
https://doi.org/10.1007/978-981-99-5146-8_13

Guarantee Law was justified as common ground for enhancing economic value and community welfare.

Keywords

Non-Muslim acceptance • Halal products • Business development

13.1 Introduction

The development of halal products in the global market has increased in the past two decades (Awan et al. 2015; Briliana and Mursito 2017; Jumani and Sukhabot 2020). The increased consumption of halal products for Muslims promotes companies in non-Muslim countries to adopt halal regulations (Afroniyati 2014; Astuti and Asih 2021; Haque et al. 2015). Lies Afroniyati reported that non-Islamic countries in recent years labelled halal products, such as Australia, New Zealand, Singapore, Thailand, France, Japan, Canada, and the USA (Afroniyati 2014). In fact, some data showed that non-Muslim entrepreneurs have gained interest in halal certification. A well-known Indonesia mass media, Republika, reported that the non-Muslim entrepreneurs in Bali Island were highly enthusiastic about certifying their products as halal. Institute for the Study of Food, Drugs, and Cosmetics of The Indonesian Ulema Council (ISFDC-IUC) issued 338 halal certificates. Bali had 60–70% for non-Muslim entrepreneurs (Baraas 2015, accessed 1 February 2021). This shows that the non-Muslim parties' acceptance is increasingly affirming the trend and awareness of the halal products demand from consumers.

The non-Muslim business actors' acceptance of halal products is not homogeneous despite being implemented into Law Number 33 of 2014 on Halal Product Guarantee (HPG). The state rules for halal certification submission and process are systematically promulgated in the HPG Law for business actors regardless of their sectoral background (Kamsari et al. n.d.). Thus, the regulation not only serves the interests and fulfilment of Muslim rights as a formal system of the state structure, but also binds overall citizens and also cover all business actors' interests. The regulation has turned into public discourse with legislation justification. Hence, it takes efforts to accommodate non-Muslim community issues on perceptions and perspectives.

The current study result on halal products related to non-Muslim groups has two tendencies. First, some studies showed that non-Muslim group has limited knowledge of neither halal products and certification nor the process (Ab Rashid and Bojei 2019; Mubarok and Imam 2020). Second, they rely on popular media, including product labels, for information on halal products and certification (Istanti et al. 2019; Pujiono et al. 2018; Rahman 2019). These two tendencies have reinforced non-Muslims into general understanding that halal products are safe for consumption because they are clean and hygienic (Tjitroresmi and Suhodo 2014; Latif et al. 2014; Nurrachmi 2017; Nurrachmi et al. 2020; Rasul et al. 2020; Rejeb et al. 2021).

This study argues that non-Muslim small business actors lack knowledge of Halal Product Guarantee despite being regulated through Law Number 33 of 2014. The lack of information and socialization on this issue allows misinterpretations of halal products definition. However, some non-Muslim small business actors' understanding of Halal Product Guarantee is affected by sectoral aspirations of Muslims than citizens in the political-economic structure. As a result, those type of small business actors ignore the regulation. They prioritize Food and Drug Supervisory Agency (FDSA) certification needed in the health aspect than halal certification, which they disregard as non-Muslims, whereas they cannot oppose to the halal certification implementation.

This study aimed to define halal products' interpretation expressed by words and symbolic expressions by non-Muslim sellers in Yogyakarta and Semarang. The study is based on three main issues: how is the knowledge of non-Muslim small business actors (food and beverage entrepreneurs) about halal products? How non-Muslim small business actors (food and beverage entrepreneurs) interpret halal products? And how is the relationship of halal products for business progress, according to their perception?

13.2 Literature Review

Halal Products

Currently, Muslim and non-Muslim communities are aware of what they consume, specifically halal labelled products (Khattak et al. 2011). According to Belhaj (2018), halal products are items or services which materials and processes follow the provisions of Islamic law. Based on this context, halal is anything permitted, allowed, approved, and valid following the Qur'an and Hadith (Mukhtar and Butt 2012). The Qur'an and Hadith accept halal products that fulfil the following criteria; (1) no content of haram/forbidden (*najis*) ingredients; (2) no exploitation of labour and the environment; (3) not endanger health; and (4) provide quality and clean materials. The issuance of halal certification was released by an entrusted Islamic certification institutions such as the Indonesian Ulema Council endorse halal products and services (Hanzaee and Ramezani 2011; Karoui and Khemakhem 2019; Yunus et al. 2014). Halal product labelling builds consumer confidence in purchasing decisions (Rasul et al. 2020; Nurrachmi et al. 2020; Yeo et al. 2016).

Halal product provisions and benchmarks are determined according to important concepts of halal in the Qur'an and Hadith. Erol (2021) stated that determination of halal products is based on *maslahah* (sharia goals) and m*aqashid* (welfare) concept to achieve the pious and to eliminate the evil in a product. More specifically, Mannan (1987) explained that *maslahah* and *maqashid* contain five principles such as (1) justice; (2) cleanliness; (3) simplicity; (4) generosity; and (5) morality applied in the three main production elements (consumers, producers, and products), to achieve halal based on Islamic sharia. Therefore, permitted materials do not guarantee halal and quality products, instead, the processes of production also need to be incorporated with the Sharia requirements (Rahman et al. 2020;

Othmana et al. 2018). Thus, halal products can be interpreted as clean, consumable items with various benefits to the Islamic Sharia objectives. Therefore, the application of the *maslahah* and *maqashid* concepts promote non-Muslim consumers to purchase and use halal products due to the guarantee of the maintained quality and cleanliness (Mathew et al. 2014).

Regulations on Halal Certification

Halal certification promulgated in the Halal Product Guarantee (HPG) Law, Number 33 of 2014, followed by Government Regulation number 31 of 2019. The regulation was enforced in October 2019 in which the requirement was changed, from voluntary to mandatory (obligation). In a clause number 3, halal product regulation aims (a) to provide comfort, security, safety, and certainty of the halal products available to the consumers and (b) to increase value for business actors to produce and sell halal products. Furthermore, a clause number 4 concerning HPG explains that products entering, circulating, and trading in Indonesia must be certified as halal (Faridah 2019).

The ratification of HPG Law will overcome consumer concerns on halal products, whereas clause number 4 of HPG Law requires food and beverage business actors to certify their products as halal. The halal certification in Indonesia has recently shifted. The Institute for the Study of Food, Drugs and Cosmetics of the Indonesian Ulema Council (ISFDC-IUC) handled the issue as non-governmental organization, and after the Halal Product Guarantee Agency (HPGA). However, IUC authority remains to determine halal and haram, but HPGA controls the formal process, including scientific examination and certificates issuance (Hartati 2019).

The HPG Law stated that implementing clause number 4 of the HPG Law must include the government and the community to enforce these rules. This will be achieved by establishing a Halal Inspection Agency (HIA), assisting HPGA to conduct inspection and halal product testing. Based on clause number 13 of the HPG Law, HPG is formed by the community, proposed by an Islamic Religious Institution, and legal entity. Based on Law Number 33 of 2014, in establishing the agency requires a laboratory owned by the institution and a large capital (Hijriawati et al. 2018).

The halal certification process at the Ministry of Religion's BPHPG is as follows:

1. Applicants register by attaching the required documents (data on business actors, names and types of products, used products and materials, and product processing processes).
2. HPGA examines the submitted documents.
3. Applicants determine the inspected products (sold or eaten).
4. HIA conducts the inspection or product testing (sold or eaten) and submits the results to the IUC to obtain a halal fatwa. HIA is the institution that conducts inspection and testing activities on the halal product.
5. IUC determines whether the products are halal using a halal fatwa session.

6. HPGA issues halal certificates and labels based on the IUC fatwa results.

Non-Muslim Acceptance

The adoption of halal product by companies in non-Muslim countries to adopt halal regulations increases acceptance for halal labelled products among non-Muslims (Chong et al. 2021; Kawata et al. 2018; Wibowo and Ahmad 2016). Based on this context, the acceptance of halal products in the non-Muslim market is influenced by two factors, including (1) the social and work environments, such as establishing relationships and friendships with Muslims (Shariff and Lah 2014; Nurrachmi et al. 2020) and (2) advertisements that positively affect non-Muslim understanding in interpretation and practice of the halal concept (Rezai et al. 2012). However, safety, convenience, and high-quality products also influence non-Muslims to consume these products. Halal emphasizes the process of slaughtering animals, the product quality and the cleanliness (Ismail and Nasiruddin 2014; Tjitroresmi and Suhodo 2014; Latif et al. 2014). Therefore, it is understandable that many non-Muslims consume halal products due to the perception that they are healthy, clean, and nutritious options as implied by Quantaniah et al. (2013).

The development of halal products has various impacts and responses on non-Muslim communities. Soesilowati (2018) stated that non-Muslims have accepted halal as the main consumed product. They positively responded that it affected their lives, specifically understanding that halal products have guaranteed quality and cleanliness (Teng et al. 2017). This shows that understanding, habits, and knowledge of halal products significantly affect sustainable consumption behaviour (Billah et al. 2020). Understanding halal products also impacts the business and employment sectors, as shown in Thailand, where the increased demand in the halal market promoted businesses and productions through exports, for gaining economic benefits (Mohd Nawawi et al. 2020; Abdul 2014; Taghipour 2021); the increasing halal products demand promotes non-Muslim dominant countries such as South Korea to open opportunities with halal tourism. Halal tourism indirectly promotes halal products and creates jobs for non-Muslim communities (Han et al. 2019).

Business Development

The high demand for halal consumption in the market develops the business sector sustainably (Akhtar et al. 2020; Garg and Joshi 2018; Karia and Asaari 2016; Vizano et al. 2021). According to Yeo et al. (2016), halal is a religious consideration which is able to increase sales and competition among business people. In fact, Jabar et al. (2014) argued that any business could grow by adopting halal products labelling. The non-Muslim's awareness of quality and healthy product consumption increases the global demand for halal products (Divianjella et al. 2020).

Halal products generate substantial growth in the halal industry valued at USD 2.1 trillion, which is expected to grow by 7% annually (Pahim et al. 2012; Fathi et al. 2016; Shah Alam and Mohamed Sayuti 2011). Halal products are currently common in the food and financial industries, but the concept has expanded to new sectors such as cosmetics, pharmaceuticals, tourism, entertainment, marketing, logistics, packaging, branding, and financial services. Therefore, these products are beneficial and generate businesses in various sectors (Ismaeel and Katharina 2012; Azam and Abdullah 2020). Based on this context, religious values and business interests are intertwined (Weber 2001).

Several supporting factors influence business success. Muflih and Juliana (2020) stated that beliefs and spirituality increase the global demand for halal products. It is also influenced by a good image where halal products are branded as high quality and clean (Al Abdulrazak and Gbadamosi 2017). Consumer satisfaction was also proven to promote the sustainable purchase of halal products (Olya and Al-Ansi 2018; Yeo et al. 2016; Al-Ansi et al. 2019). Therefore, the demand for halal products directly affects business development, specifically those requiring halal certificates and product logos as labels (Adham et al. 2020; Supaat and Nizam 2019; Borzooei and Asgari 2013). Study conducted by Ariff et al. (2021) who explored Small and Medium Enterprises (SMEs) in Malaysia also showed that most SMEs supported halal products have improved their business. This is because public knowledge of halal principles facilitates products marketing to consumers (Ariff et al. 2021).

13.3 Research Method

This study was conducted on thirteen (13) informants being from non-Muslim business actors in Yogyakarta (8 informants) and Semarang (5 informants) to explore halal product interpretation. The locations were selected based on background similarity in rich historical aspects, religious diversity, ethnicity, socio-politics, and culture.

The study used the qualitative approach method and collected data through in-depth interviews, observations, and documentation techniques. The in-depth interviews were conducted with non-Muslim business actors in the beverage and food sectors. The data were sourced through primary and secondary methods. The primary data were directly obtained through interviews with non-Muslim business actors. In contrast, the secondary data were gathered from interviews with non-Muslim business actors and the Regency/City Ministry of Religion Office officials. Observations and documentation were applied to non-Muslim business actors in halal product certification and other supporting activities for instance their survival strategy.

The observations involved the study object and their social world, while the interviews were conducted with key stakeholders and first experienced business actors in their social world. The sample was non-Muslim food and beverage sellers in Yogyakarta and Semarang. The random sampling (graded random sampling) and purposive (snowball) methods were used. Interview informants were coded as BIS,

MY, APG, A, NLP, GR, EF, and YN (from Yogyakarta City) and EB, S, MSN, DAU, and BC (from Semarang City).

Data were analysed qualitatively following Miles Matthew et al. (2014), which involved four main data analysis activities: reducing data, presenting, organizing, and drawing conclusions. Meanwhile, data reduction was made after the collection and segregation of relevant data to answer study questions. Data classification was based on the study questions, then presented, organized in theory discussions, and concluded.

13.4 Results

13.4.1 Non-Muslim Responses (Business Actors) Regarding Consumption of Halal Products

According to the interview results, it was found that non-Muslim business actors accepted the implementation of halal certification. The informant named MY stated that halal products and certificates issued by MUI still caused public discourse. Similarly, a honey reseller in Yogyakarta named A admitted that he was aware and understood the discourse of halal products. However, his understanding and knowledge were not comprehensive, specifically on the halal certification process. He continued to state that halal products are obtained and processed in a halal manner based on Islamic religious principles, which are healthy and good for consumption. An informant, named GR, reinforced this by explaining that halal certification products are halal labels or permits issued by IUC. His views on halal products for business values were that these products are halal guaranteed from the processed ingredients to the processing.

Based on the observations and interview data, non-Muslim small business actors' responses and knowledge on halal products were not homogeneous. There were different perceptions and knowledge, as shown by the different reactions or responses of non-Muslim small entrepreneurs about halal products. Yogyakarta and Semarang had two trends categorized by field data analysis. The first group was aware of halal products and certification but did not understand the process. The second group understood the meaning of halal products and certification from popular media, including the product labels.

13.4.2 Relationship of Halal Products for Business Progress

The groups' response and knowledge about halal products indicated that they had not integrated the halal certification into their business. As one of informants, GR, an online food trader, stated:

> "*The halal certification process has a business nuance; in short, MUI halal certification is a business field*" (GR, an online food trader)

Therefore, GR's socio-anthropological view is indirectly related to the business world. Further explanation, GR stated:

> *"Certification and halal products are important from a business perspective, considering that the largest consumer in Indonesia is Muslim."*

Thus, GR only considers halal as a way to gain more financial benefits as Muslims become the largest consumers in Indonesia. The non-Muslim group that is not traders or entrepreneurs does not benefit from the halal products and certifications because it does not affect their religious beliefs. Based on GR's perspective, non-Muslim entrepreneurs following the regulations of halal product are not motivated by religion but solely for business interests. Therefore, the certification process for these entrepreneurs is to fulfil the Muslim consumers' interests solely, while it is not applied for non-Muslim consumers. Meanwhile, in relation to the business responsibility, his perspective was that halal certification only benefited non-Muslim entrepreneurs, including MSMEs, as GR further stated:

> *"Certification of halal products is not that important."*

13.5 Discussion

Halal product certificates are a legal requirement for all business actors. It has become the standard for Muslims to target non-Muslim groups (Khattak et al. 2011) as business actors and consumers (Chong et al. 2021; Shariff and Lah 2014). Therefore, halal products are essential considerations for business actors without factoring in sectoral and ideological aspects such as the fulfilment of Muslim consumers. Business actors have slowly accepted the benefits of halal products.

According to the findings, it is showed that some non-Muslim small business actors in Indonesia were not aware of nor understand Halal Product Guarantee (HPG) even though it was regulated by Law Number 13 of 2014. The lack of knowledge affected their articulation and understanding of halal products from a personal perspective before the structural aspect. Most of them were unaware that the issue of halal products was regulated through the law. The lack of knowledge and socialization on halal certification affected their perspective in understanding and social action on the discourse of halal products in their business.

The business actors minimized or ignored the HPG in their practice and view halal products as a sectoral problem or aspiration of Muslims, not as a structural for all citizens in the business world. Meanwhile, previous studies provided evidences that Halal products and certifications were able to support economic activities, business, or other forms of commercialization as it can boost financial gains while also meet religious aspects. Some people view business and religious activities separately, where business pursues financial gains (profit). At the same time, religion teaches people to distance themselves from materialistic worldly interests.

Lack of Knowledge About Halal Products
According to the study finding, non-Muslim small business actors involved in this study associated halal issues with the state system, and the lack of knowledge affected their perception of dominant Muslim countries. The evidenced perceptions of several non-Muslim small business actors showed that Indonesia is inhabited by Muslim and Non-Muslim groups, and the use of Food and Drug Supervisory Agency (FDSA) labelling food products should be sufficient. Non-Islamic countries also apply halal labels to products such as food, beverages, medicines, or cosmetics.

The discourse on the state regulation of halal products in the business world was reflected by Weber (2001) on the tension between ideas and interests in Indonesia, which is not an Islamic country. The tension of ideas and interests in the Weberian context is because Indonesia is not an Islamic state; hence, it is contradictory to regulate halal products into the law. However, based on economic interests, the trade, and distribution of halal products, the state considers it as a strategy to develop the industry and trade in the products. The Director of Global Food Research and Advisory, Irfan Sungkar, in Kuala Lumpur, stated that the global halal industry is valued at more than 600 billion US dollars with a Muslim market population (captivate market) of 1.6 billion people (Afroniyati 2014).

Based on the increased distribution of halal products from an economic perspective, the Indonesian government responded with a postulate to oversee the halal certification process with stricter rules in the HPG Law. Discourse and certification process cannot be understood by exploring the issue solely with religious teachings. Hence, it should consider other contexts such as the political-economic and business. This paradigm can follow the perceptions, views, and reflections of non-Muslim small business actors in Yogyakarta and Semarang City. Furthermore, based on social and economic aspects, Indonesia is a potential market for halal products trading because of Muslim domination. A survey by the Institute for the Study of Food, Drugs, and Cosmetics of the Indonesian Ulema Council (ISFDC-IUC) in 2010 showed increased awareness of people for halal products (Afroniyati 2014).

Non-Muslims Response (Actors), Consumption of Halal Products
This study showed that some groups of small non-Muslim entrepreneurs interpreted HPG personally and not structurally. Most of them were unaware that this issue was regulated through Law no. 33 of 2014 concerning Halal Product Guarantee. The lack of knowledge and socialization on halal certification affected their perspective in understanding. Social actions of the discourse on halal products in their business, where (1) they minimized or ignored the HPG in business practice, and (2) they viewed halal products in the Indonesian context as a sectoral problem or aspiration of Muslims, and not as structural for all citizens in the business world.

These sectoral and ideological aspects should be explored to change their understanding to attain common interests for economic progress. The perception that IUC has a business agenda on certifying halal products affects their chances of obtaining certification. The sociological tendency that most consumers are Muslim is a consequence of a pluralistic nation such as Indonesia. However, the economic aspect, namely solely for business interests, should be the basis for certification. This means the Food and Drug Supervisory Agency (FDSA) label and halal certificate are of common interest for the nation's economic progress. Additionally, halal products are a religious issue and for business and trade (Borzooei and Asgari 2013).

Relationship of Halal Products for Business Progress

Business advancement must be supported by various factors and an adequate support system, including halal product certification. This certification is a structural understanding by the state for the benefit of people who fulfil the requirements. The lack of knowledge and socialization on halal certification affects their perspective in understanding. Social actions of the discourse on halal products in their business, where (1) they minimize or ignore the HPG in business practices, and (2) they view halal products in the Indonesian context as a sectoral problem or aspiration of Muslims and not as structural for all citizens in the business world.

The structural concept of HPG should be separated from piety (ascetic/holy) impulses in the religious sense. This action complies with state regulations as the motive is to follow state laws and not religious teachings. Therefore, the awareness of non-Muslim business actors for halal certification is not individually driven by external forces such as political and economic structures in the state laws. The structuralist paradigm is non-Muslim small business actors following the HPG management. It can also be interpreted as awareness of Islamic values about halal products. However, the consequences of citizens as business actors are bound by the state legal system. Non-Muslim actors are promoted by external factors such as laws to follow the HPG process for economic and not religious interests.

Non-Muslim community recognizes that the value of halal certification is for businesses. However, they do not consider HPG essential outside of business, although HPG reflects pietism and commercialization structurally, it can be accepted when only applied to Muslim groups. This concept does not apply outside Islamic groups because non-Muslim entrepreneurs do not recognize halal teachings in Islamic teachings. The social actions by Islamic groups on the HPG issue are motivated to comply with state laws and not religious teachings (Islam) for business interests. The interests of halal products are projected to those of consumers and not sellers. This is because the sellers, as non-Muslims, do not follow any religious regulations. Some non-Muslims have no prohibited consumed products by their religion.

13.6 Result Implication

Muslims take benefit from the government's policy through the HPG Law. However, the halal certification policy also affects non-Islamic parties, specifically business owners. Therefore, this study recommended several factors affecting government policies such as (1) socializing the HPG Law policy to all citizens, specifically business owners for small and large scale, Muslims and non-Muslims. Several non-Muslim small business actors do not understand the discourse of halal products and halal certification process; (2) there is an understanding among non-Muslim small business actors that the value of halal certification is solely for business, thereby having a cynical view of HPG's policy. Therefore, the HPG Law should be packaged as an aspirational space for all entrepreneurs (small and large), regardless of religious background, emphasizing material benefits from trading halal products; and (3) halal certification management should be simplified and financially inexpensive. Non-Muslim business actors expect that halal certification process is not complicated. The implementation of HPG must be tactical and efficient for business interests.

13.7 Conclusion

This study opened a new discourse on (1) how the terms and meanings of halal are understood by various groups outside the religious majority (non-Muslims); (2) understanding of sectoral and ideological halal certification; and (3) the importance of socializing in all levels of Indonesian agencies in a multicultural society. The understanding of halal meaning and products by non-Muslim groups is increasing.

Academicians can use this study to contribute to the weakness of halal product certification in the country. Meanwhile, there are limited publications related to studies on halal products in Indonesia, and most are descriptive-evaluative and less theoretical.

As this study only involved a few informants, it is recommended to collect more data from more non-Muslim business actors. The data limitation was due to the consideration of only non-Muslim business actors. Some of non-Muslim business actors practised Christianity and Catholicism, hence, did not represent all religions to be interviewed.

References

Ab Rashid N, Bojei J (2019) The relationship between halal traceability system adoption and environmental factors on halal food supply chain integrity in Malaysia. J Islamic Market 11(1):117–142. https://doi.org/10.1108/JIMA-01-2018-0016

Abdul M (2014) Perceptions on Halal food certification in Hat Yai, Thailand. Int J Econ Manage 8(1):178–194

Adham KA, Muhamad NS, Said MF (2020) Diagnosing the Halal Industry of Taiwan: a viable system model approach. J Pengurusan 58:169–179. https://doi.org/10.17576/pengurusan-2020-58-14
Afroniyati L (2014) Analisis Ekonomi Politik Sertifikasi Halal Oleh Majelis Ulama Indonesia. J Kebijakan Admin Publik 18(1):37–52
Akhtar N, Jin S, Alvi TH, Siddiqi UI (2020) Conflicting halal attributes at halal restaurants and consumers' responses: the moderating role of religiosity. J Hosp Tour Manag 45(August):499–510. https://doi.org/10.1016/j.jhtm.2020.10.010
Al-Ansi A, Olya HGT, Han H (2019) Effect of general risk on trust, satisfaction, and recommendation intention for halal food. Int J Hosp Manag. https://doi.org/10.1016/j.ijhm.2018.10.017
Al Abdulrazak RM, Gbadamosi A (2017) Trust, religiosity, and relationship marketing: a conceptual overview of consumer brand loyalty. Soc Bus Rev. https://doi.org/10.1108/sbr-03-2017-0014
Ariff ALZ, Nursalwani M, Amizi AM (2021) SME performance towards marketing halal products. AIP Conf Proc. https://doi.org/10.1063/5.0051510
Astuti Y, Asih D (2021) Country of origin, religiosity and halal awareness: a case study of purchase intention of Korean food. J Asian Fin Econ Bus 8(4):0413–0421. https://doi.org/10.13106/jafeb.2021.vol8.no4.0413
Awan HM, Siddiquei AN, Haider Z (2015) Factors affecting halal purchase intention—evidence from Pakistan's Halal food sector. Manag Res Rev. https://doi.org/10.1108/mrr-01-2014-0022
Azam MSE, Abdullah MA (2020) Global Halal industry: realities and opportunities. *Int J Islam Bus Ethics*. https://doi.org/10.30659/ijibe.5.1.47-59
Belhaj A (2018) Halal matters: Islam, politics and markets in global perspective. *Islam Christian Muslim Relat*. https://doi.org/10.1080/09596410.2018.1495807
Billah A, Rahman MA, Hossain MT (2020) Factors influencing Muslim and non-Muslim consumers' consumption behavior: a case study on halal food. J Foodserv Bus Res 23(4):324–349. https://doi.org/10.1080/15378020.2020.1768040
Borzooei M, Asgari M (2013) The Halal brand personality and its effect on purchase intention. Interdiscip J Contemp Res Bus 5(3):481
Baraas A (2015) Sertifikat halal di bali 70 persen Dipegang non-muslim. Republika, https://khazanah.republika.co.id/berita/dunia-islam/islam-nusantara/15/01/13/ni4044-70-persen-pengusaha-nonmuslim-bali-bersertifikasi-halal?. Accesed May 2023
Briliana V, Mursito N (2017) Exploring antecedents and consequences of Indonesian Muslim youths' attitude towards halal cosmetic products: a case study in Jakarta. Asia Pac Manag Rev 22(4):176–184. https://doi.org/10.1016/j.apmrv.2017.07.012
Chong SC, Yeow CC, Low CW, Mah PY, Tung DT (2021) Non-Muslim Malaysians' purchase intention towards halal products. J Islamic Market. https://doi.org/10.1108/JIMA-10-2020-0326
Divianjella M, Muslichah I, Ariff ZHA (2020) Do religiosity and knowledge affect the attitude and intention to use halal cosmetic products? Evidence from Indonesia. *Asian J Islamic Manag (AJIM)*. https://doi.org/10.20885/ajim.vol2.iss2.art1
Erol A (2021) Genetically modified foods from Islamic law perspective. J Agric Environ Ethics. https://doi.org/10.1007/s10806-021-09845-4
Faridah HD (2019) Halal certification in Indonesia; history, development, and implementation. J Halal Prod Res 2(2):68. https://doi.org/10.20473/jhpr.vol.2-issue.2.68-78
Fathi E, Zailani S, Iranmanesh M, Kanapathy K (2016) Drivers of consumers' willingness to pay for halal logistics. Br Food J. https://doi.org/10.1108/BFJ-06-2015-0212
Garg P, Joshi R (2018) Purchase intention of "Halal" brands in India: the mediating effect of attitude. J Islamic Market. https://doi.org/10.1108/JIMA-11-2017-0125
Han H, Al-Ansi A, Olya HGT, Kim W (2019) Exploring halal-friendly destination attributes in South Korea: perceptions and behaviors of Muslim travelers toward a non-Muslim destination. Tour Manage. https://doi.org/10.1016/j.tourman.2018.10.010

Hanzaee KH, Ramezani MR (2011) Intention to halal products in the world markets. *Interdiscip J Res Bus*

Haque A, Sarwar A, Yasmin F, Tarofder AK, Hossain MA (2015) Non-Muslim consumers' perception toward purchasing halal food products in Malaysia. J Islamic Market. https://doi.org/10.1108/JIMA-04-2014-0033

Hartati R (2019) Peran Negara dalam Pelaksanaan Jaminan Produk Halal. ADIL J Hukum. https://doi.org/10.33476/ajl.v10i1.1066

Hijriawati M, Putriana NA, Husni P (2018) Upaya Farmasis Dalam Implementasi Uu No. 33 Tahun 2014 Tentang Jaminan Produk Halal. Farmaka 16(1):127–132

Ismaeel M, Katharina B (2012) Practitioner contribution toward applied Islamic business ethics: responsible halal business. Tourism Manage

Ismail FRH, Nasiruddin K (2014) Perception of non-Muslim consumers towards halal products in Malaysia. Int J Account Bus Manage 2(1):128–133

Istanti LN, Pratikto H, Agustina Y, Churiyah M, Basuki A (2019) Peningkatan Mutu Produk UKM Melalui Sertifikasi Jaminan Produk Halal. J KARINOV 2(3):198. https://doi.org/10.17977/um045v2i3p198-203

Jabar FA, Ishak M, Johar N, Wahid SNS (2014) A study on the relationship between awareness and knowledge of Muslim retailers toward halal cosmetic products. In: Proceedings of the international conference on science, technology and social sciences (ICSTSS) 2012. https://doi.org/10.1007/978-981-287-077-3_44

Jumani ZA, Sukhabot S (2020) Behavioral intentions of different religions: Purchasing halal logo products at convenience stores in Hatyai. J Islamic Market. https://doi.org/10.1108/JIMA-07-2018-0112

Kamsari DHA, Bidang K, Halal S, Registrasi P, Penyelenggara B, Produk J, Kementerian Agama H (n.d.) Mekanisme Pengajuan Sertifikasi Halal dan Fasilitasi Halal Bagi UMK

Karia N, Asaari MHAH (2016) Halal business and sustainability: strategies, resources and capabilities of halal third-party logistics (3PLs). Prog Indus Ecol. https://doi.org/10.1504/pie.2016.10003063

Karoui S, Khemakhem R (2019) Factors affecting the Islamic purchasing behavior—a qualitative study. J Islamic Market. https://doi.org/10.1108/JIMA-12-2017 0145

Kawata Y, Htay SNN, Salman SA (2018) Non-Muslims' acceptance of imported products with halal logo: a case study of Malaysia and Japan. J Islamic Market. https://doi.org/10.1108/JIMA-02-2016-0009

Khattak JZK, Mir A, Anwar Z, Wahedi HM, Abbas G, Khattak HZK, Ismatullah H (2011) Concept of Halal food and biotechnology. *Adv J Food Sci Technol*

Latif IA, Mohamed Z, Sharifuddin J, Abdullah AM, Ismail MM (2014) A comparative analysis of global halal certification requirements. J Food Prod Market 20(S1):85–101. https://doi.org/10.1080/10454446.2014.921869

Miles MB, Huberman AM, Saldana J (2014) Qualitative data analysis. SAGE Publications

Mannan MA (1987) Islamic economic principles: theory and practice. Westview Press, Boulder

Mathew VN, Abdullah AMR, Binti A, Ismail SN, Binti M (2014) Acceptance on Halal food among Non-Muslim consumers. *Proc Social Behav Sci*. https://doi.org/10.1016/j.sbspro.2014.01.1127

Mohd Nawawi MSA, Abu-Hussin MF, Faid MS, Pauzi N, Man S, Mohd Sabri N (2020) The emergence of halal food industry in non-Muslim countries: a case study of Thailand. J Islamic Market 11(4):917–931. https://doi.org/10.1108/JIMA-05-2018-0082

Mubarok FK, Imam MK (2020) Halal industry in Indonesia; challenges and opportunities. J Dig Market Halal Indus 2(1):55. https://doi.org/10.21580/jdmhi.2020.2.1.5856

Muflih M, Juliana J (2020) Halal-labeled food shopping behavior: the role of spirituality, image, trust, and satisfaction. J Islamic Market. https://doi.org/10.1108/JIMA-10-2019-0200

Mukhtar A, Butt MM (2012) Intention to choose Halal products: the role of religiosity. J Islamic Market. https://doi.org/10.1108/17590831211232519

Nurrachmi I, Setiawan, Saripudin U (2020) Motivation for purchasing halal products: the influence of religiosity, trust, and satisfaction. *Human Soc Sci Rev*. https://doi.org/10.18510/hssr.2020.8520

Nurrachmi R (2017) The global development of halal food industry: a survey Halal industry is the latest trend in the world market. Tazkia Islam Fin Bus Rev 11(1):41–56

Olya HGT, Al-ansi A (2018) Risk assessment of halal products and services: Implication for tourism industry. Tour Manage. https://doi.org/10.1016/j.tourman.2017.10.015

Othmana K, Hamdania SM, Sulaimanb M, Mutaliba MM, Ramlya R (2018) A philosophy of Maqasid Shariah underpinned muslim food consumption and the Halalan Toyyiban Concept. J Islamic Social Sci Human 13:75–86

Pahim KM, Jemali S, Mohamad SJANS (2012) The importance of training for Halal logistics industry in Malaysia. In: SHUSER 2012—2012 IEEE symposium on humanities, science and engineering research. https://doi.org/10.1109/SHUSER.2012.6268847

Pujiono A, Setyawati R, Idris I (2018) Strategi Pengembangan Umkm Halal Di Jawa Tengah Dalam Menghadapi Persaingan Global. Indonesia J Halal. https://doi.org/10.14710/halal.v1i1.3109

Quantaniah NA, Noreina, Nurul S (2013) Selecting halal food: a comparative study of the Muslim and non-Muslim Malaysian student consumer. In: 2nd international conference on technology management, business and entrepreneurship

Rahman NAA, Ahmad F, Rahim SA, Mayor-Vitoria F (2020) Advancing theory on halal food supply chain in aviation: current issues and future research. In: Test engineering and management

Rahman UA (2019). Strategi Promosi dan Pengembangan Produk Kopi Mahkota Raja BLend Doa Menurut Pemasaran Perspektif Islam di Pondok Pesantren Mukmin Mandiri Waru Sidoarjo. Ekonomi Syariah

Rasul T, Ashaduzzaman M, Jebarajakirthy C (2020) How do Muslim consumers arrive at Halal purchase decisions in online food consumption context?. In: 20th Australia and New Zealand Marketing Academy Conference (ANZMAC)

Rejeb A, Rejeb K, Zailani S, Treiblmaier H, Hand KJ (2021) Integrating the Internet of Things in the halal food supply chain: a systematic literature review and research agenda. Internet of Things 13:100361. https://doi.org/10.1016/j.iot.2021.100361

Rezai G, Mohamed Z, Shamsudin MN (2012) Non-Muslim consumers' understanding of Halal principles in Malaysia. J Islamic Market. https://doi.org/10.1108/17590831211206572

Soesilowati E (2018) Business opportunities for halal products in the global market: Muslim consumer behaviour and halal food consumption. *J Indonesian Soc Sci Human*. https://doi.org/10.14203/jissh.v3i1.50

Shah Alam S, Mohamed Sayuti N (2011) Applying the theory of planned behavior (TPB) in halal food purchasing. Int J Commer Manag 21(1):8–20. https://doi.org/10.1108/10569211111111676

Shariff SM, Lah NA (2014) Halal certification on chocolate products: a case study. Proc Social Behav Sci 121(September 2012):104–112. https://doi.org/10.1016/j.sbspro.2014.01.1112

Supaat SH, Nizam NZ (2019) Marketing potential of halal food products. Int J Recent Technol Eng

Taghipour A (2021) Effects of Thai Muslim purchasing behavior towards foreign brands usage of the halal sign of imported goods in Thailand. SSRN Electron J. https://doi.org/10.2139/ssrn.3824059

Teng PK, Jamaliah W, Jusoh W (2017). Why buying Halal labelled food? Understanding the spending behavior of non-Muslim consumers in Malaysia. *Int J Bus Manage* 1(2):78–85. https://doi.org/10.26666/rmp.ijbm.2017.2.12

Tjitroresmi E, Suhodo DS (eds) (2014) Peluang Usaha Produk Halal di Pasar Global. LIPI Press, In Peluang Usaha Produk Halal di Pasar Global

Vizano NA, Khamaludin K, Fahlevi M (2021) The effect of halal awareness on purchase intention of halal food: a case study in Indonesia. *J Asian Fin Econ Bus* 8(4):441–453. https://doi.org/10.13106/jafeb.2021.vol8.no4.0441

Weber M (2001) The protestant ethic and the spirit of capitalism (2001st ed.). Routledge

Wibowo MW, Ahmad FS (2016) Non-Muslim consumers' halal food product acceptance model. Proc Econ Fin. https://doi.org/10.1016/s2212-5671(16)30125-3

Yeo BL, Mohamed RHN, Muda M (2016) A study of Malaysian customers purchase motivation of halal cosmetics retail products: examining theory of consumption value and customer satisfaction. Proc Econ Fin. https://doi.org/10.1016/s2212-5671(16)30110-1

Yunus NSNM, Rashid WEW, Ariffin NM, Rashid NM (2014) Muslim's purchase intention towards non-Muslim's halal packaged food manufacturer. Proc Soc Behav Sci 130:145–154. https://doi.org/10.1016/j.sbspro.2014.04.018

What Drives Travelers to Recommend Halal Destinations? Reflections on Aceh, Indonesia

14

Cut Kesuma Pahlufi, Edwin Darmasetiawan, Muhammad Dharma Tuah Putra Nasution, and Ku Halim Ku Ariffin

Abstract

Halal tourism is not solely limited to the provision of halal food services and religious ritual facilities; it is necessary to shape the unique aspects of halal tourism in various destinations. Halal tourism may restrict the activities that travelers may engage in, notwithstanding the convenience of visiting several locations. Concerns over the restricted activities travelers may engage in during their visit to the destinations would elicit diverse perceptions. Some travelers may have a favorable perception of this place, while others may have a different one. Accordingly, the objective of this paper is to explore the perceived values of travelers and the kinds of perceptions that stimulate them to recommend halal tourism destinations. We hope that the paper will contribute to the enrichment of knowledge, especially in the field of halal tourism marketing.

Keywords

Halal tourism • Aceh • Intention to recommend • Perceived halal attributes • Perceived social values • Perceived epistemic values

C. K. Pahlufi
PT Awal Bintang Kesuma, Medan, Indonesia

University of Surrey, Guildford, UK

E. Darmasetiawan
PT. MNC Land, Tbk, Jakarta, Indonesia

M. D. T. P. Nasution (✉)
Department of Management, Universitas Pembangunan Panca Budi, Medan, North Sumatera, Indonesia
e-mail: dharmatuah@gmail.com

K. H. K. Ariffin
Faculty of Business and Communication, Universiti Malaysia Perlis, Arau, Malaysia

A. Rafiki (ed.), *Digitalization in Halal Management*, Contributions to Environmental Sciences & Innovative Business Technology,
https://doi.org/10.1007/978-981-99-5146-8_14

14.1 Introduction

Aceh Province in Indonesia is rich in natural tourism, cuisine tourism, and cultural tourism possibilities, although it is arguably best recognized as a sharia-compliant or halal tourism destination. Since the World Islamic Tourism Organization classified Aceh Province as a halal tourism destination in 2015, it has gained increased popularity among travelers, particularly those from outside Indonesia. Halal tourism is a part of the tourism sector in which services and products are regulated to meet halal standards (Mohsin et al. 2016). Halal tourism will give an opportunity to the segment of Muslim travelers who prefer to adhere to their religion's principles and practices while still traveling abroad (Henderson 2009). Even though many facts suggest that the concept of tourism is acceptable to Islam, this does not indicate that behaviors and activities that are contradictory to Islamic teachings could be tolerated in the realm of tourism (Al-Ansi et al. 2019; Battour et al., 2012, 2018). Hotels and restaurants that serve alcoholic beverages, for instance, are definitely against Islamic principles (Battour, 2011; Prebensen et al., 2013). In fact, the view of religious culture will become more positive if the identity of tourism is effectively applied and actualized (El-Gohary, 2016). Inversely, religious judgments toward tourism will become negative, despite the fact that the intention is to delight travelers. It is possible, however, that it will be rejected if it deviates from religious and cultural norms. Consequently, it is plausible that halal tourism will impose certain restrictions on travelers while maintaining their comfort. The concept of halal tourism in reality has garnered attention from several regions in Indonesia, including Lombok, West Sumatra, and Riau Islands, among others. However, it seems that halal tourism in other regions is limited to providing mostly halal food and beverage services and religious ritual facilities, but in Aceh, Islamic sharia principles are deeply embedded and govern many aspects of people's lives. For instance, male and female travelers should participate in activities separately, non-Muslim travelers should wear closed apparel while entering particular holy areas, and beach festivities should not be held at night to avoid breaking Sharia in Aceh.

Such a situation is considerable enough to indicate the evidence that the implementation of halal tourism in Aceh is more comprehensive and not partial when compared to other regions of Indonesia. In a way, the scope of tourism in Aceh is not limited to hotel or restaurant services. It is also undeniable that there are worries over the restricted spaces available for travelers to enjoy tourist attractions, which has the potential to generate divergent perceptions. Some travelers may react favorably to halal tourism destinations with Sharia practices, while others may have different perceptions or even have diverse experiences. One of the cultural elements that shapes a person's perception and behavior is their belief system (Zamani-Farahani and Musa 2012). This might be understood to mean that a person's view of others is affected by their religious beliefs. Misunderstandings may emerge when religiously diverse visitors interact with the locals they visit (Henderson 2003). This is especially apparent when non-Muslim travelers visit Muslim countries or when Muslim travelers visit non-Muslim countries, as in the case of South Korea (Han et al. 2019a, b). Cognitive and affective factors in common serve a significant role in analyzing travelers' perceptions, but how can one comprehend the perceptions without considering halal attributes in a discussion? Hence, perceived halal attributes (physical and non-physical) will be crucial factors for Muslim travelers when selecting to acquire tourism services (Battour et al. 2011; Eid and El-Gohary 2015a, b). We assume that a discussion on travelers' perceived social and epistemic values in the context of halal tourism destinations will be more comprehensive if it includes perceived halal attributes.

14.2 Literature Review

14.2.1 Islamic Tourism and Halal Tourism

In academic literature, the terms halal tourism and Islamic tourism are commonly used interchangeably (Aji et al. 2020; Battour and Ismail 2016; Boğan and Sarıışık 2019). The two terms have their own views. As per Aji et al. (2020), something that becomes halal may or may not have an Islamic name, and something that seems to have an Islamic name may or may not adopt the concept of halal or permissible in Islam. The finest example to describe this is "Mecca Bingo" in England. Even though the location utilizes the name "Mecca," which is linked with Islamic attributes and strongly associated with the Muslim holy city, gambling business operations are prohibited or haram in Islam (Wilson and Liu 2011). The concept of Islamic tourism relates to religious travel and pilgrimages associated with their deeds and faith (Vargas-Sánchez and Moral-Moral 2019). This sort of tourism is primarily focused on traveling to an object with dominant Islamic attributes and is conducted on purpose by Muslims in order to solidify their religious beliefs (Aji et al. 2020). Umrah and Hajj, for instance, constitute the finest examples of Islamic tourism (Eid and El-Gohary 2015a). Other scholars have determined that the term "halal tourism" is frequently utilized in Western countries and is preferred by a number of global destinations (Khan and Callanan 2017; Vargas-Sánchez and Moral-Moral 2019). Meanwhile, halal tourism adjusts the availability of tourism products and services in line with Islamic values (Mohsin et al. 2016). This type of tourism has provided opportunities for Muslim traveler segments that always respect the principles and practices set by their religion (Henderson 2009). The concept of halal tourism leads to travel for enjoyment purposes, taking time off or for social reasons, learning about other cultures, and admiring God's creation (Vargas-Sánchez and Moral-Moral 2019). Halal tourism consists mostly of the consumption practices of Muslim travelers in accordance with their way of life and religious provisions (Elaziz and Kurt 2017). Halal tourism might vary due to the variety of Muslim cultures throughout the globe (Stephenson 2014; Zamani-Farahani and Musa 2012). Included in the application of halal requirements in hotels and other tourism subsectors, this is one of the industries in which management plays a significant role in adjusting services to the needs of halal travelers.

Although the majority of hotel managers are confused by the concept and standardization included in categorizing a hotel as halal (Karim et al. 2017), in this context, Samori and Saturday (2014) argue that a hotel may not be able to meet all Sharia requirements in a short period of time. The alternative is to refer to these establishments as "Muslim Friendly Hotels," which offer basic services to halal travelers, particularly with respect to food and beverage, even though they have yet to implement all Islamic provisions (Samori and Sabtu 2014; Vargas-Sánchez and Moral-Moral 2019).

14.2.2 Travelers' Perceived Values

Prior research in the tourism-marketing field has included perceived value as a multidimensional concept (Bradley and Sparks 2012; Chang et al. 2014; Dean and Suhartanto 2019; Isa et al. 2018; Prebensen et al. 2014; Sanchez et al. 2006) with ties to a variety of future traveler behaviors (Al-Ansi and Han 2019; Jeaheng et al. 2019a; Rodrigo and Turnbull 2019). Travelers' behavior is not solely viewed from a cognitive-logical standpoint but also as an experience in which cognitive (quality and price) and affective (emotional and social value) factors serve as key aspects (Sanchez et al. 2006). In a sense, consumption value theory has given an insightful approach to identifying the diverse consumption values that travelers associate with their decisions in various consumption settings such as travel and tourism (Rodrigo and Turnbull 2019). According to consumption value theory, these values drive consumer concerns, which play various roles in different contexts (Sheth et al. 1991b). As noted by Sheth et al. (1991a, b), consumption choices are impacted by five values: functional, emotional, social, economic, epistemic, and conditional values, which encourage consumers to consume certain products or services (Sheth et al. 1991b). We highlight social and epistemic values in our paper.

14.2.3 Perceived Social Values

Social value is related to the utilities or benefits derived from acceptance in different social groups after consuming products and services (Sheth et al. 1991b). By definition, social value is the utility that is perceived and obtained from an alternative as a result of its relationship with one or more certain social groups (Sheth et al., 1991a). The term "value" refers to the ability of a product or service to relate to or associate with social groups (Rodrigo and Turnbull, 2019). Creating social value in tourism is typically possible by increasing interactions and relationships among travelers during the trip, selecting experienced travel operators, and gaining added value from adopting tourism travel packages, for instance, where perceived social value is associated with special prestige and recognition (Williams and Soutar 2009). Social value in a destination relates to social interactions and the willingness to satisfy or accept one or more certain social groups, such as ethnic, socioeconomic, and cultural (Phau et al. 2014). Aspects such as the friendliness of the local community, social acceptance, and their need to be respected by others emerge as significant social values for travelers (Rodrigo and Turnbull, 2019). The attitude and treatment of the local community will give travelers a positive impression and experience. In the context of halal tourism, Eid and El-Gohary (2015b) highlight that it is a reflection of the experience of travelers after using halal products and services. Rodrigo and Turnbull (2019) underline how perceived social values reflect major personal values, such as the need to be respected and build warm relationships with others.

14.2.4 Perceived Epistemic Values

Epistemic value is related to an individual's perceived utility, which is derived from alternative attempts to arouse curiosity, provide novelty, and satisfy the desire for knowledge (Sheth et al. 1991a). Ashton et al. (2020) note the perceived epistemic value in the tourism context is mostly presented with new forms of knowledge and experience. Phau et al. (2014) implicitly added that tourists who experience epistemic values in a destination, they will feel their curiosity fulfilled for a different or new experience. Ashton et al. (2020) assert that perceived epistemic value can be generated through the experience of participating in various activities related to the local natural or cultural environment. Rodrigo and Turnbull (2019) admit that epistemic value arises when they visit places that make them experience something new and different from their home country.

Even though travelers adhere to Islamic values, it doesn't mean they don't want to experience new things or activities. For example, when they try new types of local food or have different experiences related to customs, culture, or local wisdom that blends with Islamic values, Rodrigo and Turnbull (2019) further found that Muslim tourists are trying to get a unique holiday experience and reflect on the fact that holidays are considered a means of needing to obtain a memorable, unique experience. Explicitly explained by Chen et al. (2016), individual impressions (cognitive, affective, and conative) on a destination are relatively interrelated. In detail, when travelers have a new (epistemic) experience about a destination, it will produce an affective impression, and their intention is relatively related to actions based on tourists' cognition and affection (Chen et al. 2016), and positive impressions of tourists will arouse their intention to recommend a destination (Eid et al. 2019). Therefore, understanding the perceived value of travelers from a consumption value perspective, as suggested by Sheth et al. (1991a, b), will provide opportunities for academics and practitioners to gain a more comprehensive understanding related to travelers visiting halal destinations.

14.2.5 Perceived Halal Attributes

Placing Islamic attributes on tourism destinations has brought new directions in tourism development, especially regarding Islamic-oriented tourism standards

(Battour et al. 2010), such as the existence of alcohol and gambling-free zones and the availability of halal food. Islamic attributes are intended to evaluate the perceived value of Muslim travelers as an Islamic value (Eid and El-Gohary 2015a). Furthermore, Eid and El-Gohary (2015a) noted that, in addition to cognitive and affective values, Islamic attributes in a destination will represent the level of consumer experience assessment after they use products and services, or as a reflection of an experience. Battour et al. (2011) implicitly explain Islamic attributes of a destination specifically to build interest for the Muslim tourist segment to travel to that destination.

Scholars and other academics associate Islamic attributes with halal values (Isa et al. 2018), halal-friendly destinations (Al-Ansi and Han 2019; Jeaheng et al. 2019b), or attributes of halal (Battour et al. 2014, 2017). Halal attributes are divided into physical and non-physical attributes. Physical attributes include all halal food and beverages, tools, and equipment that better support Muslim worship activities (Battour et al. 2014; Boğan and Sarıışık 2019; Eid and El-Gohary 2015a). Non-physical attributes are intangible attributes that comply with Sharia provisions such as prayer rooms, separate swimming pools for Muslims, and separate spas and hair salons for men and women (Battour et al. 2017; Boğan and Sarıışık 2019; Eid and El-Gohary 2015a).

The perceived value of Muslim travelers has shown different results between visiting destinations in Muslim and non-Muslim countries (Al-Ansi and Han 2019; Battour et al. 2017; Rodrigo and Turnbull 2019). Battour et al. (2017) show that the value of Islamic environmental quality in Malaysia is significant for attracting Muslim travelers, especially those from Middle Eastern countries. Al-Ansi and Han (2019) outline the overall impact of Muslim travelers' perceptions, especially on the halal services of a destination. Rodrigo and Turnbull (2019) emphasized that halal travel service providers must meet the expectations of travelers regarding their religious activities. Aspects such as halal food, travel, hotels, and destinations will be highly valued, and they are part of the halal compliance expected by Muslim travelers. In addition to consumption value, this means that the value of Islamic attributes will directly and indirectly impact tourist behavior in the future, whether they intend to revisit or recommend halal tourism destinations (Han et al. 2019a, b).

14.2.6 The Significance of Driving Travelers to Recommend Halal Destinations

Behavioral intention is an important concept for understanding tourism destination selection and future motives and behaviors (Afshardoost and Sadegh 2020). Understanding behavioral intentions from a research perspective further contributes to the tourism knowledge base. The literature notes that behavioral intention refers to individual desires and actions in the future to participate, consume, and use or not use certain products or services (Heesup Han and Hyun 2017). In other words, intention describes the behavior that the individual anticipates or plans (Hoz-Correa and Muñoz-leiva 2018).

As stated by Oliver (2014), behavioral intention is an individual's interest in or willingness to engage in certain behaviors. Scholars and academics agree that this intention can be observed from the situation before making a visit, which means the intention to visit (Molinillo et al. 2018); or after making a visit, which is interpreted as an intention to revisit (Loi et al. 2017); or the intention to recommend (Al-Ansi et al. 2018; Prayag et al., 2017). Previous studies have also noted that behavioral intention is a functional construct that can be translated into the intention to revisit or recommend (Choo et al. 2016; Mason and Paggiaro 2012). Both are used to describe behavioral intentions (Han et al. 2019a, b). Intention to revisit refers to travelers' plans to return to the same location or destination, and it includes traveler post-consumption behavior (Cole and Scott 2004).

It's difficult to persuade new travelers to visit a destination; therefore, the likelihood that previous travelers will return is crucial. The likelihood that travelers will return to a place relies on the favorable impression created by the attributes of the destination they experienced during their first visit (Kim et al. 2012). Those who have a positive perception of the place are more likely to revisit it than those who have a negative perception (Loi et al. 2017). Consequently, taking use of the information provided by travelers who have a favorable view of the attributes of a place will result in future positive behavior at that destination.

Chen et al. (2016) implicitly explained that cognitive impressions arise when individuals know about a destination, whereas affective impressions occur when individuals experience an experience about that destination. Meanwhile, intention or conation refers to an action based on both cognition and individual affection (Chen et al. 2016). The intention to visit has also been discussed by Jeaheng et al. (2019a). They identified hotel attributes and evaluated how these features strengthened Muslim intentions to visit Thailand. Jaheng et al. (2019a) emphasized that what is most important in terms of predicting the affective and cognitive judgments of a destination is to identify the different attributes that ultimately influence behavioral intentions for Muslim customers.

In the context of halal-friendly tourism destinations, Han et al. (2019a, b) explained that behavioral intention refers to the plans or desires of Muslim travelers in the future regarding the intention to recommend halal-friendly tourism destinations. In a study of UAE visitors, Eid et al. (2019) found that travelers' impressions of a destination have a strong influence on their intention to recommend it. In line with this, Qu et al. (2011) explained that the overall impression (cognitive, affective, and unique) of a destination has a significant direct influence on the intention to return and recommend destinations to others. Han et al. (2019a, b) revealed that the affective impression of a destination has a significant and positive influence on the intention to revisit and recommend halal-friendly destinations. Following the line of thought of previous researchers (cf. Eid et al. 2019; Qu et al. 2011; Han et al. 2019a, b), we conclude that behavioral intention includes tourists' intention to recommend a destination as a way that seems to be able to evaluate the success of the tourism sector in the context of halal destinations such as Aceh. We evaluate the travelers' intentions to recommend destinations to others with regard to post-visit behavior. Most of the time, this kind of traveler behavior has a lot of commercial value for businesses that work in tourism, but it is often overlooked since the focus is more on the visit to the destination.

14.3 Discussion

We assume that the more favorable travelers' perceptions of halal attributes are, the more likely they will be to recommend halal destinations. Explanations related to halal destination attributes that may produce a positive experience for travelers, such as the freedom of travelers to enjoy visits within the limits that do not violate Sharia or the opportunity to interact directly with the cultural arts and local wisdom of the Muslim community in Aceh, have the possibility of generating traveler intentions to share memorable experiences with family or friends. Nonetheless, we predict that not all travelers will recognize that the halal practices in place in Aceh will result in an equally favorable experience. Similarly, in terms of social values, we expect that the more favorable travelers' perceptions of social values, the more likely they are to share their positive experiences with halal destinations with others. We predict that when travelers experience interacting with the local community, they are welcomed warmly and their arrival is appreciated. This makes it possible for travelers to believe that the destinations they are visiting are the right ones and are being considered by their friends and relatives for future visits. Travelers can have memorable social experiences when they are able to make themselves part of the local community of a destination and when their values and beliefs are respected. By respecting their values and beliefs and their dignity, this will create a positive experience for travelers, which will ultimately lead them to tell positive stories to other potential travelers. We also predict that epistemic value is reflected in unique, interesting, and special experiences for travelers when they are involved in various activities while visiting halal destinations. Travelers who feel epistemic value can experience new adventures, such as visiting Iboih beach destinations and enjoying a special food menu, art, and culture, or visiting the Acehnese people, who have integrated into Islamic culture.

14.4 Conclusion

In this section, the conclusions refer to the objectives of this paper. Therefore, the conclusion of this chapter underscores that in addition to perceptions of halal attributes and epistemic values, social values tend to bring a positive impression to travelers, so we assess this impression as being able to arouse traveler intentions to recommend halal tourism destinations in Aceh Province. We also conclude that when travelers gain positive and valuable experiences from halal attributes, it is possible for them to share interesting experiences with friends and relatives, including recommending a visit to Aceh. Likewise, with regard to social values, the experience of travelers interacting with local people who receive them kindly and whose presence is appreciated will lead them to believe that the destinations they visit are appropriate and attractive destinations, so that they can be considered destinations visited by travelers, their friends, and their relatives. Related to epistemic value, travelers may be able to gain new knowledge regarding Acehnese specialties, customs, and culture that are in harmony with Islamic provisions. This will be a one-of-a-kind experience for travelers, allowing them to share with others the uniqueness of halal tourisms in Aceh.

References

Afshardoost M, Sadegh M (2020) Destination image and tourist behavioural intentions: a meta-analysis. Tour Manage 81(December 2019):104154

Aji HM, Muslichah I, Seftyono C (2020) The determinants of Muslim travellers' intention to visit non-Islamic countries: a halal tourism implication. J Islamic Market

Al-Ansi A, Han H (2019) Role of halal-friendly destination performances, value, satisfaction, and trust in generating destination image and loyalty. J Dest Market Manage 13(December 2018):51–60

Al-Ansi A, Olya HG, Han H (2019) Effect of general risk on trust, satisfaction, and recommendation intention for halal food. Int J Hosp Manag 83:210–219

Ashton AS, Limisariyapong S, Islam R (2020) The development of value perception toward cultural tourism destination: a Northeast Thailand case study. ESTEEM J Soc Sci Human 4:243–252

Battour M, Battor M, Bhatti MA (2014) Islamic attributes of destination: construct development and measurement validation, and their impact on tourist satisfaction. Int J Tour Res 16(6):556–564

Battour M, Battor M, Ismail M (2012) The mediating role of tourist satisfaction: a study of Muslim tourists in Malaysia. J Travel Tour Mark 29(3):279–297

Battour M, Hakimian F, Ismail M, Boğan E (2018) The perception of non-Muslim tourists towards halal tourism. J Islamic Market 9(4):823–840

Battour M, Ismail MN (2016) Halal tourism: concepts, practises, challenges and future. Tour Manage Perspect 19:150–154

Battour M, Ismail MN, Battor M (2011) The impact of destination attributes on Muslim tourist's choice. Int J Tour Res 13(6):527–540

Battour M, Ismail MN, Battor M, Awais M (2017) Islamic tourism: an empirical examination of travel motivation and satisfaction in Malaysia. Curr Issue Tour 20(1):50–67

Battour MM, Ismail MN, Battor M (2010) Toward a halal tourism market. Tour Anal 15(4):461–470

Battour MMK (2011) The impact of Islamic attributes of destination on tourists' motivation, satisfaction and destination loyalty. University of Malaya

Boğan E, Sarıışık M (2019) Halal tourism: conceptual and practical challenges. J Islamic Market 1(1):87–96

Bradley GL, Sparks BA (2012) Antecedents and consequences of consumer value: a longitudinal study of timeshare owners. J Travel Res 51(2):191–204

Chang LL, Backman KF, Huang YC (2014) Creative tourism: a preliminary examination of creative tourists' motivation, experience, perceived value and revisit intention. Int J Cult Tour Hosp Res 8(4):401–419

Chen C, Lai YR, Petrick JF, Lin Y (2016) Tourism between divided nations: an examination of stereotyping on destination image. Tour Manage 55:25–36

Choo H, Ahn K, Petrick JF (2016) An integrated model of festival revisit intentions. Int J Contemp Hosp Manag 28(4):818–838
Cole ST, Scott D (2004) Examining the mediating role of experience quality in a model of tourist experiences. J Travel Tour Mark 16(1):79–90
Dean D, Suhartanto D (2019) The formation of visitor behavioral intention to creative tourism: the role of push–pull motivation. Asia Pac J Tour Res 24(5):393–403
Eid R, El-Gohary H (2015a) Muslim tourist perceived value in the hospitality and tourism industry. J Travel Res 54(6):774–787
Eid R, El-Gohary H (2015b) The role of Islamic religiosity on the relationship between perceived value and tourist satisfaction. Tour Manage 46:477–488
Elaziz MF, Kurt A (2017) Religiosity, consumerism and halal tourism: a study of seaside tourism organizations in Turkey. Tourism 65(1):115–128
El-Gohary H (2016) Halal tourism, is it really Halal? Tour Manage Perspect 19:124–130
Han H, Al-Ansi A, Koseoglu M, Lin P, Park J, Yu J, Kim W (2019a) Halal tourism: travel motivators and customer retention. J Travel Tour Mark 36(9):1012–1024
Han H, Al-Ansi A, Olya HGT, Kim W (2019b) Exploring halal-friendly destination attributes in South Korea: perceptions and behaviors of Muslim travelers toward a non-Muslim destination. Tour Manage 71:151–164
Han H, Hyun SS (2017) Impact of hotel-restaurant image and quality of physical-environment, service, and food on satisfaction and intention. Int J Hosp Manag 63:82–92
Henderson JC (2003) Managing tourism and Islam in peninsular Malaysia. Tour Manage 24(4):447–456
Henderson JC (2009) Islamic tourism reviewed. Tourism Recreat Res 34(2):207–211
Hoz-Correa A and Muñoz-leiva F (2018) The role of information sources and image on the intention to visit a medical tourism destination: a crosscultural analysis. J Travel Tourism Market 2018:1–16.
Isa SM, Chin PN, Mohammad NU (2018) Muslim tourist perceived value: a study on Malaysia Halal tourism. J Islamic Mark 9(2):402–420
Jeaheng Y, Al-Ansi A, Han H (2019a) Halal-friendly hotels: Impact of halal-friendly attributes on guest purchase behaviors in the Thailand hotel industry. J Travel Tour Mark 36(6):729–746
Jeaheng Y, Al-Ansi A, Han H (2019b) Impact of halal-friendly attributes on guest purchase behaviors in the Thailand hotel industry. J Travel Tour Mark 36(6):730–747
Karim MHA, Ahmad R, Zainol NA (2017) Length of working experience as a moderator in a hotel managers' behavioural intention to adopt Islamic quality standards for hotels in peninsular Malaysia. Journal of Global Business and Social Entrepreneurship 1(3):133–151
Khan F, Callanan M (2017) The "Halalification" of tourism. J Islamic Market 8(4):558–577
Kim K, Hallab Z, Kim JN (2012) The moderating effect of travel experience in a destination on the relationship between the destination image and the intention to revisit. J Hosp Market Manag 21(5):486–505
Loi LTI, So ASI, Lo IS, Fong LHN (2017) Does the quality of tourist shuttles influence revisit intention through destination image and satisfaction? The case of Macao. J Hosp Tour Manag 32:115–123
Mason MC, Paggiaro A (2012) Investigating the role of festivalscape in culinary tourism: the case of food and wine events. Tour Manage 33(6):1329–1336
Mohsin A, Ramli N, Alkhulayfi BA (2016) Halal tourism: emerging opportunities. Tour Manage Perspect 19:137–143
Molinillo S, Liebana-Cabanillas F, Anaya-Sanchez R, Dimitrios B (2018) DMO online platforms: image and intention to visit. Tour Manage 65:116–130
Oliver RL (2014) Satisfaction: a behavioral perspective on the consumer. Routledge
Phau I, Quintal V, Shanka T (2014) Examining a consumption values theory approach of young tourists toward destination choice intentions. Int J Cult Tour Hosp Res 8(2):125–139
Prayag G, Hosany S, Muskat B, Del Chiappa G (2017) Understanding the relationships between tourists ' emotional experiences, perceived overall image, satisfaction, and intention to recommend. J Travel Res 56(1):41–54

Prebensen NK, Woo E, Chen JS, Uysal M (2013) Motivation and involvement as antecedents of the perceived value of the destination experience. J Travel Res 52(2):253–264
Prebensen NK, Woo E, Uysal MS (2014) Experience value: antecedents and consequences. Curr Issue Tour 17(10):910–928
Qu H, Hyunjung L, Hyunjung H (2011) A model of destination branding: integrating the concepts of the branding and destination image. Tour Manage 32(3):465–476
Rodrigo P, Turnbull S (2019) Halal holidays: how is value perceived by Muslim tourists ? Int J Tour Res April 2019, 1–18.
Samori Z, Sabtu N (2014) Developing halal standard for Malaysian hotel industry: an exploratory study. Procedia Soc Behav Sci 121:144–157
Sanchez J, Callarisa L, Rodriguez RM, Moliner MA (2006) Perceived value of the purchase of a tourism product. Tour Manage 27(2):394–409
Sheth JN, Newman BI, Gross BL (1991a) Consumption values and market choices: theory and application. Cincinnati: Thomson SouthWestern.
Sheth JN, Newman BI, Gross BL (1991b) Why we buy what we buy: a theory of consumption values. J Bus Res 22(2):159–170
Stephenson ML (2014) Deciphering "Islamic hospitality": developments, challenges and opportunities. Tour Manage 40:155–164
Vargas-Sánchez A, Moral-Moral M (2019) Halal tourism: state of the art. Tour Rev 74(3):385–399
Williams P, Soutar GN (2009) Value, satisfaction and behavioral intentions in an adventure tourism context. Ann Tour Res 36(3):413–438
Wilson J, Liu J (2011) The challenges of islamic branding: navigating emotions and halal. J Islamic Mark 2(1):28–42
Zamani-Farahani H, Musa G (2012) The relationship between Islamic religiosity and residents' perceptions of socio-cultural impacts of tourism in Iran: case studies of Sare'in and Masooleh. Tour Manage 33(4):802–814

Social Media Marketing and Halal Brand Equity 15

Yossie Rossanty, Endang Sulistya Rini,
Beby Karina Fawzeea Sembiring, and Amlys Syahputra Silalahi

Abstract

Despite the growing body of studies on social media marketing, to date, halal brand equity has gotten little concern in the context of social media marketing. We assume that halal brand equity represents the value of a brand that is different from the common brand value. The establishment of brand equity in a halal brand begins with the individual's selective processing of the brand based on prior consumption experiences, which leads to the formation of attitudes and behaviors toward the brand. Accordingly, this study will attempt to explore the literature related to social media marketing and halal brand equity.

Keywords

Social media • Marketing • Halal brand equity

15.1 Introduction

A brand is an idea or an image that a company or product creates in the minds of its target audience (Wang and Yang 2010). The company's brand will expand if it is successful at building a positive image in customers' minds via strategic advertising. Many studies have already shown the powerful influence brand equity has on consumer behavior. Product characteristics are generally attributed as the source of consumer-based brand equity, the factor that makes buyers choose one brand over another. We argue that in some segments of the Islamic market, brand equity based

Y. Rossanty (✉) · E. S. Rini · B. K. F. Sembiring · A. S. Silalahi
Doctoral Program in Management Science, Faculty of Economics and Business, Universitas Sumatera Utara, Medan, North Sumatera, Indonesia
e-mail: yosunpab@gmail.com

Universitas Pembangunan Panca Budi, Medan, North Sumatera, Indonesia

A. Rafiki (ed.), *Digitalization in Halal Management*, Contributions to Environmental Sciences & Innovative Business Technology,
https://doi.org/10.1007/978-981-99-5146-8_15

on halal is created more by psychological characteristics and consumer behavior than by product features.

Halal, when applied to food, indicates that it is permissible for Muslims to consume (Kasim 2012). The halal version of a brand is a near-identical copy of the original, offering the same practical and aesthetic benefits as the original but with halal certification from an outside agency. The Cadbury chocolate offered in Malaysia is distinct from the Cadbury chocolate marketed in the UK due to their shared product features and brand image. Products sold in Malaysia must be labeled as halal if they are to be consumed by Muslims.

Several theoretical frameworks provide reasons for Muslim customers to seek for halal products. From a value viewpoint, the halal label benefits Muslim buyers and influences their purchase decisions (Jamal and Sharifuddin 2014). Bonding happens when Muslim customers believe a company shares their Islamic beliefs or worldview, which may lead to a preference for halal products (Hollenbeck and Kaikati 2012).

Both Muslim and non-Muslim scholars and professionals have taken an interest in the emerging field of halal marketing and branding (Sandikci 2011). Thus, halal branding is a kind of marketing that has gained a lot of scholarly interest in recent years (Borzooei and Asgari 2013). Brand-loyal customers are more likely to make snap judgments about what to buy.

When it comes to establishing and maintaining positive connections with target audiences, social media marketing has the greatest potential influence (Bruhn et al. 2012). Based on findings from their study, Wang et al. (2021) concluded that customers will use the information they find on social media as one of many decision criteria when making purchases, with product quality, price, promotion intensity, business reputation, and brand standing among the most important. Thus, social media has become a significant element in consumers' purchase choices (Bruhn et al. 2012).

Based on findings from their study, Wang et al. (2021) concluded that customers will use the information they find on social media as one of many decision criteria when making purchases, with product quality, price, promotion intensity, business reputation, and brand standing among the most important. Thus, social media has become a significant element in consumers' purchase choices (Bruhn et al. 2012). Therefore, it may increase its bottom line by establishing a halal brand equity (HBE) (Valette-Florence et al. 2011).

15.2 Literature Review

15.2.1 Social Media Marketing

The advent of social media has provided businesses with a new way to interact with their target demographic (Kelly et al. 2010). The power to shape the public's perception of a brand has shifted from marketers to the direct links established between content creators and consumers online (Tsai and Men 2013; Godey et al. 2016).

Users of social media were seen as entertainment seekers and experience consumers from a hedonic perspective (Manthiou et al. 2013). Multiple studies point to fun as a major factor in people using social media (Kaye 2007; Muntinga et al. 2011; Park et al. 2009). For instance, research by Shao (2009) indicates that artificial user motivation is high when amusement is included (User Generated Content). To a certain extent, entertainment drives social network engagement, as noted by Park et al. (2009). Muntinga et al. (2011) found that social media users consume content-related brand for pleasure relaxation and so on entertainment. Courtois et al. (2009) argue that relaxation and escape, which is reason for looking for entertainment, encourage loading content on social media.

Social media interactions in a manner fundamentally change communication among brands and customers (Gallaugher and Ransbotham 2010; Kaplan and Haenlein 2010). Daugherty et al. (2008) found that interaction social is an important motivator for making content artificial user, and social media could offer help to consumer as well as room for discussion and exchanging minds. According to Muntinga et al. (2011), social interaction describes users who contribute to social media platforms related to a brand to meet other like-minded people, interact, and engage in conversations about a certain product or brand. Zhu and Chen (2015) shared social media into two groups (based on profile and content), depending on the nature of connections and interactions kindly specifically, social media-based profile focus on members individually.

Information/topic related with members and goals mainly is for push social media users for connected with information/topics certain, social media-based profile push connection age, as a result of genuine curiosity in the person hiding behind the profile (e.g., on Facebook, Twitter, and WhatsApp). Conversely, material shared on social media platforms is centered on, well, content, conversation, and comments. The primary goal is to facilitate consumer demand for user-provided content profiles (such as those seen on sites like Flickr, Instagram, Pinterest, and YouTube).

According to Gallaugher and Ransbotham (2010), social media-based consumer discourse is more active than ever, and the authors conceive it in relation with framework work megaphone, magnet, and monitor (3-M). The megaphone stands for business-to-consumer connection, the magnetic net for consumer-to-business, and the monitor for consumer-to-consumer communication. The reason why brands need to provide original content that reflects the demographics of their audience, encourages participation in open dialogue, and provides useful resources is because these factors all contribute to higher levels of engagement.

Establishing a connection not only enhances trust and fosters a sense of affinity (Manthiou et al. 2013) but also plays a significant role in primary product search channels and the trending themes of conversation (Naaman et al. 2011). Consumers are increasingly often migrating to various sorts of social media for gain information and regard them as a trust to information sources, rather than company-sponsored communications via marketing conventional (Mangold and Faulds 2009; Volmer and Precourt 2008).

Information on social media has four sub-motives: control, knowledge, pre-purchase information, and inspiration, as stated by Muntinga et al. (2011). *Understanding one's social surroundings requires constant monitoring, explaining, and observation. For the purpose of learning more about a product or brand, "knowledge" refers to information acquired by customers via the experiences and insights of other consumers. Foreknowledge of a product's qualities is essential. Before making a purchase, it is a good idea to check out the feedback left by other customers on the brand's official community site. In conclusion, fresh brand-related ideas may be gleaned from consumer-related information as a source of inspiration.*

Brands may increase customer loyalty and brand awareness by allowing visitors to personalize and express themselves on their websites (Martin and Todorov 2010). *Customization, in the context of social media, refers to tailoring one's posts to a certain group of people. Zhu and Chen* (2015) *claim that there are two distinct sorts of posts: personalized messages and general announcements. Messages may be tailored to a single person or group (like a Facebook post), or they can be broadcast to anybody who might be interested (e.g., Twitter tweets).*

15.2.2 Social Media Marketing and Customer-Based Brand Equity

A customer's impression and attitude toward a brand may be affected by a variety of factors, such as the customer's familiarity with the brand, the brand's perceived quality, and the customer's desire for the brand. Changes to the equity draft brand are substantial as a result of the development process. Keller's (1993) brand equity models, which focus on the relationship between brand awareness and brand perception, are widely accepted. Strength of brand awareness is measured by a consumer's consistent capacity to recognize the brand under varying environmental and perceptual cues or "knots" (Rossiter and Percy 1987). The likelihood that a certain brand would come to mind when a given need arises is what we mean by "brand awareness" (Keller 1993). On the other hand, Keller defines a brand's image as the consumer's mental connection with the product. Consumers' confidence in a brand rises as a result of marketing actions conducted in social media contexts (Bruhn et al. 2012; Kim and Ko 2012). Action social media marketing is part of the mix promotion in the new paradigm of communication, say Mangold and Faulds (2009). Bruhn et al. (2012) found that both conventional and social channels of communication had a substantial influence on brand equity. The influence of conventional media is stronger on the conscience brand, while the communication via social media has a significant impact on the image brand.

Equity as a means of building a successful business has been the subject of much debate, as has the nature of an equity brand and the methods by which it may be established via various forms of advertising (Christodoulides and De Chernatony 2010). Big bachelors agree, in part, that more equity from more brands has a positive effect on brand choice, willingness to pay a premium price, and customer loyalty, but they vary on the relative relevance and assessment of these factors (Aaker 1991; Keller 1993, 2003; Keller and Lehmann 2006). With a focus on the effect of brands on the three most pressing customer responses, Kim and Ko (2012) found that social media settings facilitated by social media marketing had a positive correlation with consumers' propensity to make future purchases.

15.2.3 Social Media Marketing and Halal Brand Equity

What is permissible (halal) and what is not allowed (haram) for true Muslims is laid out in detail by Islamic Sharia, which is a scientific interpretation of the Quran (which Muslims believe is the word of Allah, revealed to the Prophet Muhammad) and Sunnah (Documents of the life of the Prophet Muhammad) (Kasim 2012). There are two requirements that religiously branded items must satisfy before they are considered appropriate for individual consumption among observant Muslims. Devout Muslims' purchasing habits are heavily influenced by Islamic Sharia, which also served as the intellectual impetus for the development of the halal market. First, the brand is not violating any Islamic taboos in the product area in which it operates. Many well-known brands that are widely eaten in non-Muslim cultures are automatically disqualified from the Muslim consumer's range of considerations due to the first criterion.

To continue, in order to be considered lawful under Islamic Sharia law, branded items must be manufactured and sold in a way that is consistent with Islamic Sharia principles (De Run et al. 2010). For instance, beef satisfies the first criterion since its consumption is permitted in Islam. Beef is permitted (halal) for Muslim consumers if it is produced in a thayyib (clean) manner and slaughtered in conformity with Islamic rules (Saeed et al. 2001). The primary challenge for companies selling halal products is to adhere to Islamic religious (Sharia law) guidelines throughout the production and distribution processes. Many multinational companies serving the halal market invest significantly in obtaining third-party halal certification to establish consumer trust in the religious permissibility. This certification becomes a hallmark of their brand, demonstrating institutional support for the significance of faith in consumption within Muslim-majority cultures. According to the theory of institutions, commercial enterprises face challenging institutional constraints (DiMaggio and Powell 1983; Lawrence and Suddaby 2006; Ab Talib et al. 2016; Yulianto et al. 2021) Most multinational corporations (MNCs) actively engage in conveying the halalness of their brands to develop Muslim customer trust via ads, public communications, and the exhibition of third-party halal certification badges. Once a product has been given the seal of approval by a national or international halal certification body, the certification insignia should be clearly displayed on the product's packaging or at the delivery location. Altering other parts of the marketing mix may help bring the multinational corporation and its goods to the attention of Muslim customers. Experts in the area of halal marketing agree that it is crucial for brands, particularly especially MNC companies, to cement their irrefutable halal status in the eyes of Muslim customers (Izberk-Bilgin 2012; Siala 2013).

In an analysis that mirrors Huntington's perspective on the consumer world, Alserhan (2010a, p. 35) highlights the difficulties multinational corporations (MNCs) confront when attempting to alter their public perception to become more acceptable in Islamic countries (Huntington 1996). MNC companies in the Islamic market have always been seen as either a glorification of Western culture or, worse, as infidel brands (Izberk-Bilgin 2012). While Western products often offer superior value compared to local alternatives, this frequently influences many Muslim customers to choose them. Nevertheless, when a company is perceived as not adhering to Islamic religious ideals or faces accusations of guilt by association, as seen in the case of several Danish businesses during the Prophet Muhammad cartoon dispute, it can impact consumer choices. (Abosag and Farah 2014), such preference may swiftly fade. MNC companies can only gain the trust of Muslim consumers concerning the religious permissibility of their goods by obtaining third-party halal certification from a respected national or international agency. Brands may gain the confidence of the vast majority of Muslim customers by displaying one of these certificates and emblems. To succeed in the halal market, a brand must follow Islamic principles and standards rather than those of non-Islamic marketplaces (Alserhan 2010a).

According to the notion of self-concept, Muslim consumers act in a manner that supports and strengthens their Islamic identity (Hollenbeck and Kaikati 2012; Strizhakova et al. 2008). Also, obtaining and using material objects is a major component of how religious identities are formed and shared. As a result, it is not shocking that Muslims have a hard time being true to themselves as Muslims while still adhering to halal standards of consumption (Izberk-Bilgin 2012). It is not an exaggeration to argue that coming out as Muslim and behaving as Muslim promotes halal consumption (Wilson and Liu 2010). Past studies on Muslim food preferences have shown, for instance, that consuming halal cuisine is a highly visible indicator of commitment to and pride in one's Muslim identity (Bonne et al. 2007; Mohd Suki and Mohd Suki 2015; Sandikci and Ger 2010). We define consumer halal choice behavior as individual consumption behavior largely focused on the search for and consumption of products that have a distinguishing halal status, and we argue that this activity is driven by identity adjustment concerns. Muslim identity is nurtured, preserved, and strengthened via the purchase and use of halal goods and services (Jamal and Sharifuddin 2014; Sandikci and Ger 2010). It follows that a person's religious identity is the primary motivation for consuming in accordance with the halal standard.

The term "halal" refers to companies that make a conscious effort to accommodate the specific religious needs of Muslim customers. Multinational corporations also have this problem with halal branding, when there is a lack of customer confidence. One way firms may reassure Muslim buyers and show they care about their needs is by getting halal certification from a reputable third party and displaying the label prominently on their packaging (Ishak et al. 2016).

For items with minimal symbolic value, for instance, past research has demonstrated that the appearance of religious symbols on product packaging favorably influences customer purchase intentions (Bakar et al. 2013). Consumers who are dedicated to a halal lifestyle are more likely to choose a halal brand when that lifestyle requires them to buy a branded product (such as tetra-packaged milk, processed meat items, cosmetic goods, medications, or auto loans), conforming to the letter and spirit of the law. Value accrued by a business as a result of a brand's influence on its target market is known as "brand equity" (Erdem et al. 1999, p. 302). In the case of financial services, for instance, the halal consumption behavior of Muslims mandates that they not engage in conventional banking due to the payment or receipt of interest, and this in turn creates a preference for brands that offer Sharia-compliant financial services, regardless of whether or not they offer such services. With sharia, the price goes up by a little amount (Taib et al. 2008).

Following from the above, we argue that CBHBE may be predicted in large part by consumers' halal choice behavior, since this turns abstract ideas into real-world behaviors. This conforms to the widely held idea that beliefs and behaviors must be consistent, just as we anticipate that attitudes will be reflected in acts that reveal them (El-Bassiouny, 2014; Schiffman et al. 2010).

Having a strong brand makes marketing efforts more successful, which in turn benefits the company. Increased brand loyalty, positive consumer attitudes, market share growth, and financial returns are the benefits (Aaker 1996). Customers' familiarity with a provider's product or service is the foundation of consumer-based brand equity, which is mostly attributable to but not limited to the provider's strategic marketing initiatives (Christodoulides and de Chernatony 2010). According to our definition, CBHBE occurs when Muslim buyers choose one brand over another that is essentially the same but is considered halal. Accordingly, businesses that cater to the halal market help consumers feel more connected to themselves, which is one of the three pillars of consumer-based brand equity (Chatzipanagiotou et al. 2016). Because to CBHBE, many multinational corporations (MNCs) in the Muslim world, notably those in the fast-moving consumer goods (FMCG), pharmaceutical, banking, and tourism industries, seek and achieve halal certification (Alserhan 2010a, b; Minkus-McKenna 2007). To boost one's sense of self-worth, the image alignment theory suggests that consumers should go beyond external validation and instead focus on satisfying their own criteria of excellence (Greenwald 1989; Hogg et al. 2000; Ng et al. 2014). As a result, SERB is said to be attained by patronage of products that are expressly in line with Islamic religious ideals. It is possible for such companies to fulfill religious and cultural significance for Muslims.

Brand equity refers to the value of a brand's name and logo in influencing consumer behavior (Torres and Tribo 2011). Awareness, associations, loyalty, image, and perceived quality are the five pillars that make up the multifaceted idea of brand equity (Chieng and Lee 2011). One factor that was taken into account for each brand in this analysis was HBE (Butt et al. 2017). Example: Consumers who are thinking about making a purchase may utilize the brand's halal certification as a deciding factor, dismissing any options that do not clearly state their status as halal. HBE adds value to the business by making marketing efforts more efficient. Increased brand loyalty, positive consumer attitudes, market share growth, and financial returns are the benefits (Preko et al. 2021). Several reports from completed research projects highlight the connection between HBE and consumer choice. According to studies done by Ansary and Nik Hashim (2018), brand equity significantly influences consumers' choices to make purchases. This lends credence to the claim made by Severi and Ling (2013), which states that consumers are more likely to make a purchase when their level of brand equity is high, as brand equity indicates preferences connected to buying or using intents. Brand awareness, brand perception, and brand loyalty are the pillars on which strong brand equity is built. Long-term advantages for the business may be gained by increasing the PDM process's credibility with customers if the product's brand equity is enhanced (Chang and Liu 2009). Consistent with these results, Butt et al. (2017) discovered that HBE is an excellent predictor of the buying habits of consumers. Consumers should easily be able to identify a brand by its halal logo or emblem. Although this is the lowest ranking in HBE's system, it is nonetheless important since it provides a glimpse into the brand's overall health (Wan Ismail 2020).

In the marketing realm, the ability to utilize social media in an engaging way with clients is seen as more vital than traditional technical expertise. It is common knowledge that the marketing idea benefits from the openness and rapidity of social media in terms of both feedback and participation (Schivinski and Dabrowski 2016). The primary factor in deciding whether or not to make an online purchase is the availability of relevant search results, backed up with simple access to relevant social media (Kim and Ko 2012). Typically, a buyer will start the process of acquiring an item by researching it online. Customers will take into account the acquired data when making a purchase decision. This data might be in the shape of product pricing, brands, quality, specs, reviews from past customers, and so on (Yogesh and Yesha 2014). Customers are really pampered by the many facilities supplied by social media, as well as the information included within, which may be seen as a breath of fresh air for producers, particularly those that sell products and services through social media. When consumers see positive posts about a business on social media, they are more likely to be ready to spend a premium price for it. The attitudes and acceptability of brand equity are affected by the primary reasons people use social media (Yazdanparast et al. 2016). Consumers' usage of social media for two-way communication extends to almost every facet of their life, including but not limited to their relationships, travel, and purchases. Companies are using ISM with great enthusiasm as a means to contact customers and build their brands (Zahoor et al. 2017). Manufacturers and consumers alike may utilize ISM, an interaction tool, to communicate with one another and exchange data in the form of text, photos, audio, and video. Since ISM is accessible to everyone, it may be used to expand the reach of a company's promotional network (Setiawati and Aini 2019). ISM is characterized by its user-friendliness, as it's widely recognized that contemporary consumers seek quick and straightforward solutions. Naturally, this is essential for ISM to appeal to potential buyers.

15.3 Conclusion

The primary purpose of this research was to create and evaluate a theoretical model of the factors in an individual's mental and behavioral makeup that are most predictive of CBHBE. The development and testing of a model to explain the precursors of CBHBE was a significant contribution of this work. By testing the hypothesis that consumer-based brand equity emerges mostly from psychological and behavioral traits of customers rather than product attributes, our study adds to the growing body of literature on international branding. Based on the results of this research, companies may increase their CBHBE by tailoring their marketing to the three aforementioned psychological and behavioral dimensions.

Ads that show Muslim customers actively searching out halal goods in stores are more likely to persuade these shoppers to continue their halal consumption habits and boost the persuasive power of such ads. By prominently displaying the halal certification insignia on product packaging and offering information about the brand's halal status, businesses may aid Muslim customers in their search for products that meet their dietary preferences.

It is also worth noting that halal certification is not the only technique to let consumers know that a product complies with Islamic dietary guidelines. Brand names with religious connotations; country of origin (e.g., food products from the Kingdom of Saudi Arabia or Iran may hardly raise questions about their halal status, while products from China or the West may receive significant consumer skepticism); the use of religious messages; and the sponsorship of religious events and festivals are all other ways that implicitly communicate halal.

Halal investigations and more specifically aspects related to branding, and business, are still in their early stages. However, the broader implications associated with halal and its potential to penetrate other disciplines should not be overlooked, nor should their approach remain limited. Halal is a paradigm that requires appreciation of: dual nature, situation specific and culture, figurative, esoteric and symbolic states; a strong ethical standpoint; a relationship away from process-driven materialistic and mechanical thinking; and strategic management practice.

As a result, it is suggested that the model be investigated further by including these components, as the country of origin impact may be important to the growth of CBHBE. Finally, it is vital to comprehend that halal brand equity may vary from product to product, culture to culture, and even among various sectors of the Muslim nation-state.

References

Aaker DA (1991) Managing brand equity. Free Press, New York, NY
Aaker DA (1996) Measuring brand equity across products and markets. Calif Manag Rev 38(3)
Ab Talib MS, Md Sawari SS, Abdul Hamid AB, Chin TA (2016) Emerging Halal food market: an institutional theory of halal certificate implementation. Manage Res Rev 39(9):987–997
Abosag I, Farah MF (2014) The influence of religiously motivated consumer boycotts on brand image, loyalty and product judgment. Eur J Mark 48(11/12):2262–2283
Alserhan BA (2010a) Islamic branding: a conceptualization of related terms. J Brand Manag 18(1):34–49
Alserhan BA (2010b) On Islamic branding: brands as good deeds. J Islamic Market 1(2):101–106
Ansary A, Nik Hashim NMH (2018) Brand image and equity: the mediating role of brand equity drivers and moderating effects of product type and word of mouth. RMS 12(4):969–1002
Bakar A, Lee R, Rungie C (2013) The effects of religious symbols in product packaging on Muslim consumer responses. Australas Mark J 21(3):198–204
Bonne K, Vermeir I, Bergeaud-Blackler F, Verbeke W (2007) Determinants of halal meat consumption in France. Br Food J 109(5):367–386
Borzooei M, Asgari M (2013) The halal brand personality and its effect on purchase intention. Interdiscip J Contemp Res Bus 5(3):481–491
Bruhn M, Schoenmueller V, Schafer DB (2012) Are social media replacing traditional media in terms of brand equity creation? Manage Res Rev 35(9):770–790
Butt MM, Rose S, Wilkins S, Haq JU (2017) MNCs and religious influences in global markets: drivers of consumer-based halal brand equity. Int Mark Rev 34:884–905
Chang HH, Liu YM (2009) The impact of brand equity on brand preference and purchase intentions in the service industries. Serv Ind J 29(12):1687–1706
Chatzipanagiotou K, Veloutsou C, Christodoulides G (2016) Decoding the complexity of the consumer-based brand equity process. J Bus Res 69(11):5479–5486
Chieng FYL, Lee GC (2011) Customer-based brand equity: a literature review. J Arts Sci Commerce 2(1):33–42
Christodoulides G, De Chernatony L (2010) Consumer-based brand equity conceptualization and measurement: a literature review. Int J Res Mark 52(1):43–66
Courtois C, Mechant P, De Marez L, Verleye G (2009) Gratifications and seeding behaviour of online adolescents. J Comput-Mediat Commun 15:109–137
Daugherty T, Eastin MS, Bright L (2008) Exploring consumer motivations for creating user-generated content. J Interact Advert 8(2):16–25
De Run E, Butt MM, Fam KS, Yin JH (2010) Attitudes towards offensive advertising: Malaysian Muslims' views. J Islam Market 1(1):25–36
DiMaggio PJ, Powell WW (1983) The iron cage revisited: institutional isomorphism and collective rationality in organizational fields. Am Sociol Rev 48(2):147–160
El-Bassiouny N (2014) The one-billion-plus marginalization: toward a scholarly understanding of Islamic consumers. J Bus Res 67(2):42–49
Erdem T, Swait J, Broniarczyk S, Chakravarti D, Kapferer J, Keane M, Roberts J, Benedict J, Steenkamp EM, Zettelmeyer F (1999) Brand equity, consumer learning and choice. Mark Lett 10(3):301–318
Gallaugher J, Ransbotham S (2010) Social media and customer dialog management at Starbucks. MIS Q Exec 9(4):197–212
Godey B, Manthiou A, Pederzoli D, Rokka J, Aiello G, Donvito R, Singh R (2016) Social media marketing efforts of luxury brands: influence on brand equity and consumer behavior. J Bus Res 69(12):5833–5841
Greenwald AG (1989) Why are attitudes important? In Pratkanis AR, Breckler SJ, Greenwald AG (eds) Attitude structure and function. Erlbaum, Hillsdale, NJ, pp 1–10
Hogg MK, Cox AJ, Keeling K (2000) The impact of self-monitoring on image congruence and product/brand evaluation. Eur J Mark 34(5/6):641–667

Hollenbeck CR, Kaikati AM (2012) Consumers' use of brands to reflect their actual and ideal selves on Facebook. Int J Res Mark 29(4):395–405

Huntington SP (1996) The West unique, not universal. Foreign Aff 28–46

Ishak S, Awang AH, Hussain MY, Ramli Z, Md Sum S, Saad S, Abd Manaf A (2016) A study on the mediating role of halal perception: determinants and consequence reflections. J Islam Market 7(3):288–302

Izberk-Bilgin E (2012) Infidel brands: unveiling alternative meanings of global brands at the nexus of globalization, consumer culture, and Islamism. J Consum Res 39(4):663–687

Jamal A, Sharifuddin J (2014) Perceived value and perceived usefulness of halal labeling: the role of religion and culture. J Bus Res 68(5):933–941

Kaplan AM, Haenlein M (2010) Users of the world, unite! The challenges and opportunities of Social Media. Bus Horiz 53:59–68

Kasim NAA (2012) Disclosure of Shariah compliance by Malaysian Takaful companies. J Islam Account Bus Res 3(1):20–38

Kaye BK (2007) Web site story: An exploratory study of blog use motivations. In: Tremayne M (ed) Blogging, citizenship and the future of media. Routledge, New York, NY, pp 127–148

Keller KL (1993) Conceptualizing, measuring, and managing customer-based brand equity. J Mark 57(1):1–22. https://doi.org/10.1177/002224299305700101

Keller KL (2003) Brand synthesis: the multidimensionality of brand knowledge. J Consum Res 29(4):595–600

Keller KL, Lehmann DR (2006) Brands and branding: research findings and future priorities. Mark Sci 25(6):740–759

Kelly L, Kerr G, Drennan J (2010) Avoidance of advertising in social networking sites: the teenage perspective. J Interact Advert 10(2):16–27

Kim AJ, Ko E (2012) Do social media marketing activities enhance customer equity? An empirical study of luxury fashion brand. J Bus Res 65:1480–1486

Lawrence TB, Suddaby R (2006) Institutions and institutional work. In: Clegg SR, Hardy C, Lawrence T, Nord WR (eds) The sage handbook of organization studies. Sage, London, pp 215–254

Mangold WG, Faulds DJ (2009) Social media: the new hybrid element of the promotion mix. Bus Horiz 52:357–365

Manthiou A, Chiang L, Tang L (2013) Identifying and responding to customer needs on Facebook Fan pages. Int J Technol Human Interact 9(3):36–52

Martin K, Todorov I (2010) How will digital platforms be harnessed in 2010, and how will they change the way people interact with brands? J Interact Advert 10(2):61–66

Minkus-McKenna D (2007) The pursuit of halal. Progress Groc 86(17):42

Mohd Suki N, Mohd Suki N (2015) Does religion influence consumers' green food consumption? Some insights from Malaysia. J Consum Mark 32(7):551–563

Muntinga DG, Moorman M, Smit EG (2011) Introducing COBRAs: Exploring motivations for brand-related social media use. Int J Advert 30(1):13–46

Naaman M, Becker H, Gravano L (2011) Hip and trendy: characterizing emerging trends on twitter. J Am Soc Inform Sci Technol 62(5):902–918

Ng PF, Butt MM, Khong KW, Ong FS (2014) Antecedents of green brand equity: an integrated approach. J Bus Ethics 121(2):203–215

Park N, Kee KF, Valenzuela S (2009) Being immersed in social networking environment: Facebook groups, uses and gratifications, and social outcomes. Cyberpsychol Behav 12(6):729–733

Preko A, Mohammed I, Allaberganov A (2021) Antecedents of brand equity on halal tourism destination. J Islamic Market 13(8):1685–1702

Rossiter JR, Percy L (1987) Advertising and promotion management. McGraw-Hill Book Company

Saeed M, Ahmed ZU, Mukhtar SM (2001) International marketing ethics from an Islamic perspective: a value-maximization approach. J Bus Ethics 32(2):127–142

Sandikci O (2011) Researching Islamic marketing: past and future perspectives. J Islam Market 2(3):246–258

Sandikci Ö, Ger G (2010) Veiling in style: how does a stigmatized practice become fashionable? J Consum Res 37(1):15–36
Schiffman LG, Kanuk LL, Wisenblit J (2010) Consumer behavior. Pearson Prentice Hall, Boston, MA
Schivinski B, Dabrowski D (2016) The effect of social media communication on consumer perceptions of brands. J Market Commun 22(2):189–214
Setiawati S, Aini W (2019) Increase adult learning motivation through promotion of their needs. KOLOKIUM: Jurnal Pendidikan Luar Sekolah 2(2):111–119
Severi E, Ling KC (2013) The mediating effects of brand association, brand loyalty, brand image and perceived quality on brand equity. Asian Soc Sci 9(3):125–137
Shao G (2009) Understanding the appeal of user-generated media: a uses and gratifications perspective. Internet Res 19(1):7–25
Siala H (2013) Religious influences on consumers' high-invement purchasing decisions. J Serv Mark 27(7):579–589
Strizhakova Y, Coulter RA, Price LL (2008) The meanings of branded products: a cross-national scale development and meaning assessment. Int J Res Mark 25(2):82–93
Taib FM, Ramayah T, Razak DA (2008) Factors influencing intention to use diminishing partnership home financing. Int J Islam Middle East Financ Manag 1(3):235–248
Torres A, Tribo JA (2011) Customer satisfaction and brand equity. J Bus Res 64(10):1089–1096
Tsai WHS, Men LR (2013) Motivations and antecedents of consumer engagement with brand pages on social networking sites. J Interact Advert 13(2):76–87
Valette-Florence P, Guizani H, Merunka D (2011) The impact of brand personality and sales promotions on brand equity. J Bus Res 64(1):24–28
Vollmer C, Precourt G (2008) Always on: advertising, marketing, and media in an era of consumer control. McGraw-Hill, New York
Wan Ismail WR et al (2020) Halal Malaysia brand equity mishap: false recognition of brand mere recognition using mixed method approach. J Islamic Mark 13(1):5–19
Wang X, Wang Y, Lin X, Abdullat A (2021) The dual concept of consumer value in social media brand community: A trust transfer perspective. Int J Inform Manag 59:102319
Wang X, Yang Z (2010) The effect of brand credibility on consumers' brand purchase intention in emerging economies: the moderating role of brand awareness and brand image. J Glob Mark 23(3):177–188
Wang X, Yu C, Wei Y (2012) Social media peer communication and impacts on purchase intentions: a consumer socialization framework. J Interact Mark 26(4):198–208
Wilson JAJ, Liu J (2010) Shaping the Halal into a brand? J Islam Mark 1(2):107–123
Yazdanparast A, Joseph M, Muniz F (2016) Consumer based brand equity in the 21st century: an examination of the role of social media marketing. Young Consumers 17(3):243–255
Yogesh F, Yesha M (2014) Effect of social media on purchase decision. Pac Bus Rev Int 6(11):45–51
Yulianto H, Sutomo Y, Palupiningtyas D, Nugraheni KS (2021) Investigating the role of interactive social media and halal brand equity on purchase decision-making: direct and indirect effects. Meas Bus Excellence
Zahoor SZ et al (2017) Impact of social media brand communications on consumer-based brand equity. Indian J Commerce Manage Stud 8(3):124–131
Zhu YQ, Chen HG (2015) Social media and human need satisfaction: implications for social media marketing. Bus Horiz 58:335–345

The Role of Interactive Social Media and Emotional Value on Halal Cosmetic Purchase Intention

16

Fitriani Tobing, Abdul Rahman Lubis, Permana Honneyta Lubis, Syafruddin Chan, and Said Musnadi

Abstract

This chapter aims to describe and discuss interactive social media and emotional values that play an important role in influencing the intention to purchase halal cosmetics. This study adopts a content analysis technique where various sources are referred to including research articles, official documents, and reports. The large number of cosmetic choices available in the market raises concerns for Muslim consumers because of the safety factor of the ingredients contained in cosmetics, whether they are safe from ingredients that are not prescribed by Islamic teachings. Cosmetics that have a halal logo can affect the emotional value of consumers because for them by buying halal cosmetics they feel that the products they consume have followed religious rules or the orders of Allah SWT. This perceived utility influences consumer buying behavior. The role of interaction on social media is also very important in providing information, reviews or reviews from consumers who have used halal cosmetics so that they are able to change consumer perceptions or beliefs to buy cosmetics that have a halal logo.

Keywords

Halal cosmetic • Interactive social media • Emotional value • Purchase intention

F. Tobing (✉)
Faculty of Economics and Business, Universitas Medan Area, Medan, Indonesia
e-mail: tobingfitri@gmail.com

A. R. Lubis · P. H. Lubis · S. Chan · S. Musnadi
Faculty of Economics and Business, Universitas Syiah Kuala, Banda Aceh, Indonesia

A. Rafiki (ed.), *Digitalization in Halal Management*, Contributions to Environmental Sciences & Innovative Business Technology,
https://doi.org/10.1007/978-981-99-5146-8_16

16.1 Introduction

The concept of halal is important for Muslims all around the world (Handriana et al., 2020; Isabelle Aoun Laurent Tournois 2015). Muslim consumers must obey the teachings of their religion by using halal products. Halal products have traditionally been focused on the food sector, but cosmetics and personal care products are becoming an important concern for Muslim consumers (Annabi and Ibidapo-Obe 2017). Currently, all countries both Muslim and non-Muslim countries, have realized the need for halal products and they are trying to provide halal products, as well as facilities and infrastructure (Septiarini et al. 2022).

Halal is a symbol that guarantees quality and is a lifestyle choice for a Muslim. Products that have a halal label have advantages that can benefit the producers who sell halal products as well as the consumers of those products. These products are of good quality, suitable for consumption and are in accordance with religious orders. The halal lifestyle has become a behavior that is carried out not only in Muslim-majority countries but also countries that have a Muslim minority population. A product being labeled halal gives a sense of security to Muslims because the process of making these products was carried out with Islamic religious orders. Muslims in the world are increasingly emphasizing religious sensitivity, including a sensitivity to cosmetic products.

Halal cosmetics prioritize safety, quality, and are free from animal testing, this provides a great opportunity for development because the market for halal products is not limited to Muslim consumers but also non-Muslims (Khalid et al., 2021). Halal cosmetics follow Islamic rules, including being free from pork, pork by-products, pork-based gelatin, alcohol, or other animal by-products (Mukhtar and Butt, 2012). According to Nordin et al., (2021) halal cosmetics have several criteria including not containing parts of animals that are prohibited, and products from animals that have not been slaughtered according to Islamic law. Halal products also do not contain alcohol, parts of the human body, do not threaten and endanger health, use permitted and clean tools, and avoid contact with substances that are not in accordance with sharia and Islamic law.

Halal cosmetics have reached the global cosmetics market segment (Anubha, 2021; Azam and Abdullah 2020; Handriana et al., 2020; Isabelle Aoun Laurent Tournois 2015; Widyanto and Sitohang, 2021). Indonesia with its large Muslim population is expected to become the country with the strongest halal economy in the world. According to Briliana and Mursito, (2017) there are 760 cosmetic companies spread throughout Indonesia. The number of players in the cosmetics industry shows that the business value is very promising. The increase in the halal cosmetics industry shows the great interest of consumers in buying halal cosmetics.

Intention is a strong indicator in identifying a person's actual buying behavior (Barber and Christopher Taylor, 2013). A consumer's behavioral intent is a reflection of their purchase intent. Purchase intent is a consumer's thoughts, desires, and interests about a product that leads to informed actions about the product. Purchase intent is the conscious consumer's plan to purchase a product. (Fishbein and Ajzen, 1977; Garg and Joshi, 2018; Handriana et al., 2020). In previous research, many factors were shown to influence purchase intention, and one of the strongest variables was emotional value (Ariffin et al., 2016; Firdaus et al., 2022; Hussin et al., 2016; Mohd Suki and Mohd Suki, 2015; Muhamed et al., 2019; Wang, 2010). In previous studies that examined the relationship between emotional value and purchase intention, the research results showed inconsistencies. Research (Hussin et al., 2016; Muhamed et al., 2019; Wang, 2010) identified that emotional value has a significant effect on intention. However, different results were found in research by (Ariffin et al., 2016; Firdaus et al., 2022; Mohd Suki and Mohd Suki, 2015), where emotional value did not have a significant effect on behavioral intention. A study by Yulianto et al., (2021) adds the interactive role of social media as a variable that reinforces the intention to purchase halal cosmetics. Social media plays an important role in influencing consumer attitudes toward halal brands. Consumers will be willing to pay higher prices for halal products if these products were previously advertised on social media. Social media platforms that are generally used by millennials in seeking information about halal cosmetics include Facebook, Instagram, blogs and websites, WhatsApp, Line, and Shopee (Ishak et al., 2019; Yulianto et al., 2021).

16.2 Literature Review

16.2.1 Emotional Value and Purchase Intention

Emotional value is part of the perceived value dimension, namely feelings or affective conditions that arise when using a product. The interactions that occur between consumers and producers can trigger and change consumer emotions in determining whether to buy or not to buy a product. Emotional value emerges when consumers have felt the use of a product. Emotional value is not only perceived based on the usefulness of a product but also the hedonic value because rational and emotional factors shape consumer needs. Emotional value plays an important role in determining human behavior along with functional and social values. Emotional value is thought to have a strong influence over a consumer's decision to buy a product. According to (Mohd Suki and Mohd Suki, 2015) emotional value indicates 'perceived utility' and evokes feelings when customers positively perceive the company's products or services, thereby influencing consumer purchasing decisions.

Information about halal labels can provoke consumer feelings or emotions toward halal products. By buying halal products, consumers feel that their need for personal values is fulfilled because the products they consume follow religious teachings, or the commands of Allah SWT, even though other similar products might be superior to halal products (Firdaus et al., 2022). Research conducted by Muhamed et al. (2019) shows a significant influential relationship between emotional value and the behavior of buying food supplies that have halal certificates in Kuala Lumpur, Selangor. This opinion is in line with research conducted by Joshi et al. (2021) which examined the intention to buy green among adolescent consumers in India which showed a significant effect. The emotional involvement that emerges becomes their driving force to make green purchases.

16.2.2 Role of Interactive Social Media Toward Emotional Value and Attitude to Purchase Intention

Purchase intention shown by consumers is the most important element used by marketing managers to predict future sales and determine how to act and influence consumer buying behavior. To understand how individuals make decisions, it is very important to start with understanding intentions first (Ali et al., 2018; Fauzi, 2022; Garg and Joshi, 2018; Handriana et al., 2020; Haque et al., 2018; Ishak et al., 2019; Khan et al., 2020; Ngah et al., 2021; Shahid et al., 2021; Widayanto and Sitohang, 2021; Widyanto and Sitohang, 2021). In most cases, people typically make decisions about which products to buy before actually purchasing said products. (Hutter et al., 2013).

Before deciding to buy a product, customers will do research first by collecting information related to price, brand, quality, specifications, and reviews from previous customers who have used the product (Yogesh and Yesha, 2014). The interactions they have on social media with the brands they like will affect their cognitive, affective, and behavioral involvement (Zhang et al., 2021). Interactive social media is a tool that helps consumers interact with producers and other consumers, as well as businesses through messages delivered in the form of information such as text, images, audio, and video (Yulianto et al., 2021). Research (Kim and Ko, 2012) explains that social media makes it easy for consumers to find information before deciding to make a purchase online. Every review displayed on social media becomes a reference for consumers to evaluate a product's quality before they finally decide to buy. Research conducted by Arsil et al. (2018) explains that information about products that have a halal label can affect consumers' emotions, because they feel that when using these products, they are indirectly obeying religious orders. Even if the quality and price of products that do not use the halal label are superior, they will still choose halal products because consumers prioritize emotional value over functional value (Handani, 2021).

16.2.3 Halal Cosmetic Development in Indonesia

The growing awareness of the Muslim population's need for halal products affects the demand for halal products. Indonesia has an enormous cosmetic industry potential, and it produces attractive and high-quality cosmetic products that meet the tastes and needs of the world market. The cosmetics industry in Indonesia can be used as a supporter of tourism and the creative economy with a variety of Indonesian cosmetics supported by the potential wealth of raw materials. The Ministry of Tourism and Creative Economy is taking part in the Proudly Made in Indonesia National Movement with the hopes that it can increase the consumption of creative economy production, help the wheels of the economy, and local MSMEs run as a means of marketing local Indonesian cosmetic products, and create the most job opportunities possible. Indonesia has the opportunity to develop into the world's leading producer of halal products. Opportunities for the development of the halal industry are backed by several supporting factors, including, (a) the high number of people who are Muslim; (b) high demand for and purchasing power of halal products in Indonesia; (c) the opening of export opportunities to both OIC and non-OIC countries; (d) import substitution policy; (e) the establishment of the National Sharia Economic and Finance Committee (KNEKS), and (f) the enactment of the Halal Product Guarantee Law and the establishment of the Halal Product Assurance Organizing Agency (BPJPH). According to Law no. 33 of 2014 concerning Guarantees for Halal Products; it mandates that those products circulating in Indonesia are guaranteed to be halal. Halal certification aims to provide convenience and certainty to the public regarding the availability of halal products and to provide added value for businesses that produce and sell halal products nationally and internationally.

Indonesian Muslim spending and global Muslim consumer ranking

Sector	2020 (US$ Billion)	2025 (US$ Billion)	2019–20 (Covid-19 Impact)	5 Years CAGR % (2020–25) (%)	Global Muslim consumer ranking
Food and beverage	$135	$204	−6.44	14.64	1
Fashion	$15.6	$23.28	−4.99	8.34	6
Pharmaceuticals	$5.13	$6.81	−4.29	5.83	4
Cosmetics	$4.19	$7.59	0.71	12.62	2
Travel and tourism	$3.37	$8.03	−4.65	18.96	5
Media and recreation	$20.73	$31.82	−70	8.95	2

Source Author's Compilation from Indonesia Halal Market Report 2021–2022

According to the 2021–2022 Indonesia Halal Economic Report, the Indonesian halal cosmetics industry is currently estimated to be generating USD 4.19 billion in revenue annually and is projected to grow by nearly 8% per year until 2025. This is caused by a high demand in domestic and export markets. Regulations on halal products provide great opportunities for the cosmetic industry. Apart from giving local cosmetic manufacturers a competitive advantage over foreign brands in the domestic market, cosmetics companies that have halal certificates can also expand their market share into overseas markets. The implementation of halal certifications for Indonesian cosmetic products has proven to be a game changer for the Indonesian cosmetics industry.

Cosmetics help a woman to always look her best, and the halal label listed on the cosmetic packaging provides assurance that the cosmetic product is safe and is in accordance with religious orders (Briliana and Mursito, 2017). For a Muslim, obeying Allah's decree is a form of obedience, even though obedience is only based on belief without knowing the wisdom of Allah's decree. Halal is an obligation as well as a necessity, and is a manifestation of human obedience in carrying out God's laws which is a manifestation of faith. Those who fulfill it mean that they have been obedient to Allah, so that for them there is a reward from Allah, and a violation of the rules of Shari'a is seen as rebellion and deserves punishment.

Everyone has a different level of religiosity. High and low levels of one's religiosity will also affect behavior in consuming a product (Delener, 1990). In halal cosmetics, the most important factor that can influence a person to be loyal is due to emotional attachment to the product as well as the quality of the product, and not because of one's religion (Suhartanto et al., 2020). This study is supported by Khan et al. (2020) which states that Generation Y in consuming halal cosmetics is more concerned with the safety of the raw materials contained within the cosmetics and the halal logo rather than religious beliefs. Today's consumers always want to evaluate their products before finally deciding to buy. Social media reviews and the like are the main reference for a consumer when deciding to purchase a product. This has resulted in a decreased dependence on information sourced from traditional advertising (Anubha and Shome 2021; Chakraborty, 2019; Chen and Chang, 2018). This change in the consumer paradigm is caused by the significant increase in the number of Internet users from year to year, thus forming a trend in consumer behavior that will only seek information through new mass communication channels such as social media (Men and Tsai, 2013). The current trend is that millennial consumers prefer to see reviews from social media such as Instagram, Facebook, and other social media before buying cosmetic products. Virtual communities in the digital world are one of the factors that influence consumer perceptions or beliefs to buy halal products (Sardana et al., 2021).

16.3 Method

This paper uses a content analysis in elaborating the role of interactive social media and emotional value on halal cosmetics purchase intention.

16.4 Discussion

According to the report by Markets (2021) Indonesia has become the world's largest halal economy market due to its large Muslim population. Domestic expenditure of this population on halal economic products and services reached $184 billion in 2020 and is expected to reach $281.6 billion in 2025 with an increase of CAGR of 14.96%. With a large Muslim population, Indonesia is expected to become the world's strongest halal economy. Indonesia must compete not only with other majority Muslim countries, but also with non-Muslim countries such as

Thailand, South Korea, Russia, Mexico, Japan, and Spain, which have begun to play an active role in the development of the halal industry.

There are several obstacles that pose challenges to the development of the halal industry in Indonesia, including the low number of MSME products (micro, small, and medium enterprises) that have halal certificates, the not yet optimal halal information system in providing Halal Product Assurance (JPH) services, and human resources whose experience with granting a halal certificate or logo is still limited. Every marketer must understand the consumers' needs and build consumer trust and confidence in halal cosmetic products. They can use social media platforms as a strategy to convince consumers to use halal products, because the ingredients used are only halal materials according to religious orders. Social media is the first step that can be used to create awareness of the halal brand which over time will develop into loyalty to the halal brand. The role of social media as a marketing communication tool is very important for companies to interact with customers before they decide to purchase halal cosmetic products.

16.5 Conclusion

The Qur'an is a guideline for Muslims in determining their actions in the world. All Muslims believe that using halal cosmetics is a form of submission to Allah SWT. Halal labels convey a sense of security and happiness in using products because the products they consume comply with religious precepts and teachings. Thus, Muslims pay close attention to a product's halal status, spend money on halal cosmetics, recommend halal cosmetics to others, and prioritize halal-labeled cosmetics. The use of social media as a marketing medium will create awareness of halal cosmetics, and eventually this will develop into consumer loyalty to halal cosmetics. There are significant differences that exist in each consumer group, e.g., young consumers are more interested in using cosmetic products that have further benefits, while older consumers tend to choose benefits that are more general and use cosmetic products that can cover deficiencies in their facial skin conditions. By understanding the differences that come from consumers, industry players can determine what strategies are suitable for each consumer group, then improve their brand performance and then drive the advancement of the cosmetics industry as a whole.

References

Ali A, Xiaoling G, Sherwani M, Ali A (2018) Antecedents of consumers' Halal brand purchase intention: an integrated approach. Manag Decis 56(4):715–735. https://doi.org/10.1108/MD-11-2016-0785

Annabi CA, Ibidapo-Obe OO (2017) Halal certification organizations in the United Kingdom. J Islamic Mark 8(1):107–126. https://doi.org/10.1108/jima-06-2015-0045

Anubha, Shome S (2021) Customer engagement and advertising effectiveness: a moderated mediating analysis. J Internet Commer. https://doi.org/10.1080/15332861.2021.1955324

Anubha (2021) Mediating role of attitude in halal cosmetics purchase intention: an ELM perspective. J Islamic Mark ahead-of-p(ahead-of-print). https://doi.org/10.1108/jima-04-2021-0112

Ariffin S, Yusof JM, Putit L, Shah MIA (2016) Factors influencing perceived quality and repurchase intention towards green products. Procedia Econ Finance. https://doi.org/10.1016/s2212-5671(16)30142-3

Arsil P, Tey YS, Brindal M, Phua CU, Liana D (2018) Personal values underlying halal food consumption: evidence from Indonesia and Malaysia. Br Food J. https://doi.org/10.1108/BFJ-09-2017-0519

Azam MSE, Abdullah MA (2020). Global Halal industry: realities and opportunities. Int J Islamic Bus Ethics. https://doi.org/10.30659/ijibe.5.1.47-59

Barber NA, Christopher Taylor D (2013) Experimental approach to assessing actual wine purchase behavior. Int J Wine Bus Res 25(3):203–226. https://doi.org/10.1108/IJWBR-2012-0013

Briliana V, Mursito N (2017) Exploring antecedents and consequences of Indonesian Muslim youths' attitude towards halal cosmetic products: a case study in Jakarta. Asia Pac Manag Rev 22(4):176–184. https://doi.org/10.1016/j.apmrv.2017.07.012

Chakraborty U (2019) Perceived credibility of online hotel reviews and its impact on hotel booking intentions. Int J Contemp Hosp Manag. https://doi.org/10.1108/IJCHM-11-2018-0928

Chen CC, Chang YC (2018) What drives purchase intention on Airbnb? perspectives of consumer reviews, information quality, and media richness. Telematics Inf. https://doi.org/10.1016/j.tele.2018.03.019

Delener N (1990) The effects of religious factors on perceived risk in durable goods purchase decisions. J Consum Mark. https://doi.org/10.1108/EUM0000000002580

Fauzi MA (2022) Consumer purchase of halal certified product: a quantitative systematic literature review. J Islamic Mark. https://doi.org/10.1108/jima-09-2021-0299

Firdaus FS, Ikhsan RB, Fernando Y (2022) Predicting purchase behaviour of Indonesian and French Muslim consumers: insights from a multi-group analysis. J Islamic Mark. https://doi.org/10.1108/JIMA-05-2021-0169

Fishbein M, Ajzen I (1977) Belief, attitude, intention, and behavior: an introduction to theory and research. Philos Rhetoric 10(2)

Garg P, Joshi R (2018) Purchase intention of "Halal" brands in India: the mediating effect of attitude. J Islamic Mark 9(3):683–694. https://doi.org/10.1108/JIMA-11-2017-0125

Handani N (2021) Product awareness of international muslim students in South Korea about the importance of halal food. Res Horiz. https://doi.org/10.54518/rh.1.1.2021.1-6

Handriana T, Yulianti P, Kurniawati M, Arina NA, Aisyah RA, Ayu Aryani MG, Wandira RK (2020) Purchase behavior of millennial female generation on Halal cosmetic products. J Islamic Mark 12(7):1295–13150. https://doi.org/10.1108/JIMA-11-2019-0235

Haque A, Anwar N, Tarofder AK, Ahmad NS, Sharif SR (2018) Muslim consumers' purchase behavior towards halal cosmetic products in Malaysia. Manage Sci Lett 8(12):1305–1318. https://doi.org/10.5267/j.msl.2018.9.009

Hussin S, Talhah HF, Rozekhi NA, Hashim R (2016) The perceived value of celebrity chef endorsement toward consumer purchase intention on food products. In: Regional conference on science, technology and social sciences (RCSTSS 2014). https://doi.org/10.1007/978-981-10-1458-1_83

Hutter K, Hautz J, Dennhardt S, Füller J (2013) The impact of user interactions in social media on brand awareness and purchase intention: The case of MINI on Facebook. J Product Brand Manage 22(5):342–351. https://doi.org/10.1108/JPBM-05-2013-0299

Ishak S, Che Omar AR, Khalid K, Intan IS, Hussain MY (2019) Cosmetics purchase behavior of educated millennial Muslim females. J Islamic Mark 11(5):1055–1071. https://doi.org/10.1108/JIMA-01-2019-0014

Joshi Y, Uniyal DP, Sangroya D (2021) Investigating consumers' green purchase intention: Examining the role of economic value, emotional value and perceived marketplace influence. J Clean Prod. https://doi.org/10.1016/j.jclepro.2021.129638

Khalid NR, Wel CAC, Mokhtaruddin SA (2021) Product positioning as a moderator for halal cosmetic purchase intention. Iran J Manage Stud 14(1):39–60. https://doi.org/10.22059/IJMS.2020.279978.673617

Khan N, Sarwar A, Tan BC (2020) Determinants of purchase intention of halal cosmetic products among Generation Y consumers. J Islamic Mark. https://doi.org/10.1108/JIMA-11-2019-0248

Kim AJ, Ko E (2012). Do social media marketing activities enhance customer equity? An empirical study of luxury fashion brand. J Bus Res. https://doi.org/10.1016/j.jbusres.2011.10.014

Markets IH (2021) Indonesia Halal Markets Report 2021/2022. 1–118

Men LR, Tsai WHS (2013) Beyond liking or following: Understanding public engagement on social networking sites in China. Pub Relat Rev. https://doi.org/10.1016/j.pubrev.2012.09.013

Mohd Suki N, Mohd Suki N (2015) Consumption values and consumer environmental concern regarding green products. Int J Sust Dev World. https://doi.org/10.1080/13504509.2015.1013074

Muhamed AA, Ab Rahman MN, Mohd Hamzah F, Che Mohd Zain CR, Zailani S (2019) The impact of consumption value on consumer behaviour: A case study of halal-certified food supplies. Br Food J. https://doi.org/10.1108/BFJ-10-2018-0692

Mukhtar A, Butt MM (2012) Intention to choose Halal products: the role of religiosity. J Islamic Mark 3(2):108–120. https://doi.org/10.1108/17590831211232519

Ngah AH, Gabarre S, Han H, Rahi S, Al-Gasawneh JA, Park SH (2021) Intention to purchase halal cosmetics: do males and females differ? a multigroup analysis. Cosmetics 8(1):1–14. https://doi.org/10.3390/cosmetics8010019

Nordin FNM, Aziz A, Zakaria Z, Wan Mohamed Radzi CWJ (2021) A systematic review on the skin whitening products and their ingredients for safety, health risk, and the halal status. J Cosmet Dermatol. https://doi.org/10.1111/jocd.13691

Sardana D, Cavusgil E, Gupta N (2021) The growing popularity of spiritual brands: what drives purchase intent? Int Bus Rev. https://doi.org/10.1016/j.ibusrev.2020.101777

Septiarini DF, Ratnasari RT, Salleh MCM, Herianingrum S, Sedianingsih (2022) Drivers of behavioral intention among non-Muslims toward halal cosmetics: evidence from Indonesia, Malaysia, and Singapore. J Islamic Acc Bus Res. https://doi.org/10.1108/JIABR-02-2021-0064

Shahid S, Parray MA, Thomas G, Farooqi R, Islam JU (2021) Determinants of Muslim consumers' halal cosmetics repurchase intention: an emerging market's perspective. J Islamic Mark. https://doi.org/10.1108/JIMA-08-2021-0265

Suhartanto D, Dean D, Sarah IS, Hapsari R, Amalia FA, Suhaeni T (2020) Does religiosity matter for customer loyalty? Evidence from halal cosmetics. J Islamic Mark. https://doi.org/10.1108/JIMA-03-2020-0069

Tournois IAL (2015) Building holistic brands an exploratory study of halal cosmetics. J Islamic Mark 6(1):1–18

Wang EST (2010) Impact of multiple perceived value on consumers' brand preference and purchase intention: a case of snack foods. J Food Prod Market. https://doi.org/10.1080/10454446.2010.509242

Widayanto H, Sitohang IA (2021) Muslim millennial's purchase intention of halal-certified cosmetics and pharmaceutical products: the mediating effect of attitude. J Islamic Mark 13:1373–1394

Widyanto HA, Sitohang IAT (2021) Muslim millennial's purchase intention of halal-certified cosmetics and pharmaceutical products: the mediating effect of attitude. J Islamic Mark 13(6):1373–1394. https://doi.org/10.1108/JIMA-04-2020-0117

Yogesh F, Yesha M (2014) Effect of social media on purchase decision. Pac Bus Rev Int

Yulianto H, Sutomo Y, Palupiningtyas D, Nugraheni KS (2021) Investigating the role of interactive social media and halal brand equity on purchase decision-making: direct and indirect effects. Measuring Bus Excellence. https://doi.org/10.1108/MBE-02-2021-0025

Zhang L, Zhao H, Cude B (2021) Luxury brands join hands: building interactive alliances on social media. J Res Interact Mark 15(4). https://doi.org/10.1108/JRIM-02-2020-0041

Development of Halal Travel and Tourism in Indonesia

17

Endang Sulistya Rini
and Muhammad Dharma Tuah Putra Nasution

Abstract

This chapter aims to elaborate and discuss on challenges and opportunities of halal travel and tourism in Indonesia. The study adopts content analysis technique where various sources are referred including reports, official documents and research articles. The development of halal tourism is positive but needs significant initiatives by the stakeholders. Other opportunities of halal travel and tourism also have been revealed to be used as valuable information for strategic planning and initiatives.

Keywords

Halal travel • Halal tourism • Indonesia

17.1 Introduction

Indonesia is classified as a country that has just been introduced by the concept of halal tourism. This type of tourism has appeared first in several countries with sizable Muslim populations such as Saudi Arabia, Palestine, Turkey, the United Arab Emirates, Egypt, and Malaysia. Halal tourism has also begun to be developed by the Ministry of Tourism and Creative Economy in a number of regions in Indonesia. Halal tourism itself is expected to be an alternative for Muslim and non-Muslim tourists to travel and fulfill their spiritual needs. The domestic tourism

E. S. Rini (✉)
Faculty of Economics and Business, Universitas Sumatera Utara, Medan, Indonesia
e-mail: endang.sulistya@usu.ac.id

M. D. T. P. Nasution
Faculty of Social Sciences, Universitas Pembangunan Panca Budi, Medan, Indonesia

A. Rafiki (ed.), *Digitalization in Halal Management*, Contributions to Environmental Sciences & Innovative Business Technology,
https://doi.org/10.1007/978-981-99-5146-8_17

sector is also expected to be able to attract more local and foreign tourists through this tourism sector.

Several countries are investing in the development of halal tourism infrastructure as a strategic sector to support economic growth. The world has been shocked by the emergence of COVID-19. An infectious disease spreads through saliva droplets or out of the nose when a person with the virus coughs or sneezes (WHO 2020). Due to the spread of COVID-19, economic conditions in some countries faced economic recessions, including Indonesia. The Indonesian government predicts that the growth of gross domestic product (GDP) will experience a slowdown around 2.3 and 0.4 this year, and this figure represents an optimistic scenario calculated by the government (Guild 2020). The Indonesian Chamber of Commerce and Industry said that the tourism sector and its derivatives were the sectors most affected by COVID-19 (Bayu 2020) and very sensitive to economic shocks. Indonesian government introduced the new normal concept to overcome the above problems, which can be understood as a new form of adjustment in the order of life. New normal is not interpreted as a state of surrender or resignation in fighting against COVID-19. On the contrary, a new normal is a solution for people to return to their activities and carry on life provided that they comply with health protocols. Through the new normal policy, which is designed in stages, economic activities will return, including the opening of tourist objects or destinations for tourists who want to take a vacation after self-isolating at home.

17.2 Literature Review

17.2.1 Halal Tourism

Conceptually, halal tourism is a concept of tourism services and products based on Islamic sharia. Several things are included in this type of tourism such as the availability of halal hotels, halal restaurants, halal resorts, and halal trips. This concept is not only used in Islamic countries but can also be applied in countries where the population is not predominantly Muslim.

In Islamic law, "halal" means things that can be permitted, and in the context of tourism, halal tourism applies rules related to Islamic law or values. The existence of this tour should be considered considering that there is still no internationally recognized standard for halal tourism. This tourism is also still related to the concept of Islamic tourism, halal-friendly tourism destinations, halal travel, Muslim-friendly travel destinations, and halal lifestyle.

17.2.2 Halal Tourism Development in Indonesia

The global society has a strong need for halal travel. This incredible potential for halal tourism necessitates the creation of halal tourism product marketing services that must adhere to Islamic sharia law. The factor causing the increasing interest in halal tourist visits is the very rapid Muslim population (Battour and Ismail 2016). In many nations, including Indonesia, the trend of halal tourism is emerging as a prospective leading industry. Indonesia has a lot of natural resources that can be utilized to draw more international tourists, which makes it a huge potential market for halal tourism. Additionally, the millennial generation is beginning to show interest in sharia-based tourism, which is drawing more and more businesspeople to Indonesia's tourism potential (Akar and Dalgic 2018).

This is in line with the level of awareness of the religiosity of society which has begun to grow, not only in Islamic countries or where the majority of the population is Muslim but even in secular countries. Based on the State of the Global Islamic Economy Report 2021, it was revealed that the total number of trips by world Muslim tourists was worth US$194 billion. In 2023, it is predicted to increase to US$274 billion. The high expected number of tourists has brought both extraordinary opportunities and challenges for Indonesia's halal travel and tourism sector as it strives to become the global hub for the halal industry.

In order to achieve this, halal tourism business actors first need to receive intensive guidance in order to ensure their compliance with the MUI (Indonesia Ulema Council) concerning Guidelines for Organizing Tourism based on Sharia Principles and Law and Halal Products (Fadhilah et al. 2020). The existence of government intervention in halal tourism provides guarantees for halal products and services in terms of international market competition. According to Samsul et al. (2022), government support through local regulations and MUI Fatwa has a positive impact on interest in halal tourism visits. Another study was found to have a different opinion stating that halal certification, which is one of the government's efforts to provide guarantees for halal products, does not affect the satisfaction of tourism visitors.

Nevertheless, the existence of halal certification makes its own emotional experience for behavioral intentions to travel halal. In this case, the government and halal tourism business actors need to work together to provide creative and attractive socialization for foreign tourists so that tourists are interested in halal tourism products that are halal certified (Ratnasari et al. 2020). It cannot be denied that Muslim consumers may be more easily attracted to halal products because they are in line with their beliefs. It is different for non-Muslim consumers as they tend to look for products and services not only with good quality but also products that suit their needs, especially for tourists who are very concerned about the health and a safe environment. Therefore, halal tourism should be able to offer an attractive platform by creating halal products that are healthy and satisfy visitors (Aji et al. 2021).

The fact that the tourist industry has contributed to the economy's growth is evidence that it may be regarded as Indonesia's most important industry. In today's Indonesia, tourism is not only focused on infrastructure and visitor numbers but also creates the idea of Muslim-friendly diversification. Halal tourism is the name given to this idea. According to data from Ministry of National Development Planning, Republic of Indonesia year 2016 shows that Indonesia saw a 2.9% rise in global tourism in 2015 and 12 million tourists in 2016. Indonesia's tourism industry is ranked 32 out of a total of 117 countries worldwide based on a report from the Travel and Tourism Development Index 2021 published in May 2022. This position is up 8 ranks compared to the achievements in 2019 (Anam 2022). In 2019, based on data from the Ministry of Tourism and Creative Economy, the amount of foreign exchange generated was IDR280 trillion and contributed to the gross domestic product (GDP) of 5.5% and absorbed a workforce of 13 million people. Additionally, according to the State of the Global Islamic Economy Report 2020–2021, Indonesia is ranked fourth globally in terms of Islamic economic indicator value, moving up one spot from the previous year's position (Septyaningsih 2021).

Previous research examined the marketing strategy for controlling halal tourism in Indonesia and its enabling variables. For example, research conducted by Wibowo and Mandusari (2018) revealed that the halal tourism management model was built based on good cooperation between business actors, government, academics, and all levels of society. It is very important to build priority policies that can increase public awareness of the importance of developing halal products in the tourism sector. This is done as an effort to improve the economy (Fadhilah et al. 2020). Positive perceptions of halal products and Islamic values influence tourist intentions and lead to more conducive halal tourism development (Aji et al. 2021). Factors of environmental value perceived by tourists, destination performance, sustainable development, and halal tourism image are proven to affect tourist satisfaction. However, these environmental and non-environmental factors have not been proven to be able to moderate tourism development satisfaction. This study adds that environmental management factors can be one of the priorities for sustainable halal tourism development (Zulvianti et al. 2022). Likewise, religious belief factors (Raya 2022); travel reputation and halal tourist destinations (Widjaja et al. 2020); sharia accommodation and activities (Usman et al. 2019); development of internet-based geographic information systems (GIS) (Afnarius et al. 2020); tourist attributes, accessibility, infrastructure and formal structures affect tourist visit intentions (Mafudi et al. 2021; Katuk et al. 2020; Wardi et al. 2018). However, the attributes of halal tourism that characterize halal tourism in Islamic countries cannot be used in non-Islamic countries because perceptions and cultural contexts may be different (Jia and Chaozhi 2020; Minardi et al. 2021).

Previous research has shown elements that encourage the growth of halal tourism. However, which criteria best represent Muslim visitors' buying intentions in the halal travel and tourism industry remains to be determined. This study confirms that there is a dearth of empirical research on the factors that influence Muslim travelers' and tourists' desire to travel halal. The Indonesian government, on the other hand, wants to establish its nation as the global epicenter of halal tourism. The willingness of Indonesia's Muslim population to adopt halal travel and tourist services must thus be determined.

Halal products can be a variety of goods and services provided by halal tourism actors such as accommodation, travel agencies, transportation, tourist destinations, and halal restaurants. This halal tourism product is based on acceptable Islamic values, has an ethical dimension, and has transcendental standards. Halal tourism products are designed in such a way as to provide comfort for tourists. With the existence of halal tourism products, a Muslim tourist does not need to worry about how his worship and others related to carrying out his sharia. All are provided in the religion of halal product types, including; (1) halal food, free of alcohol, pork, and the like; (2) worship facilities; (3) Ramadan services; (4) recreational facilities that maintain privacy (Pradana et al. 2021; Yan et al. 2017; Loussaief and Haque 2018). Another study explains that to increase the purchase intention of Muslim tourists, halal tourism products must have adequate standardization, for example, halal certification, environmentally friendly products, diversity of halal-certified food products, guaranteed travel products, use of sophisticated technology, and easy and adequate access to worship (Ratnasari et al. 2020; Vargas-Sánchez and Moral-Moral 2019).

A tourist trip is a trip taken to obtain pleasure and fulfill the desire to know something. The characteristics of a person or group traveling on a tour are that the trip is carried out by traveling around an area and returning to their place of origin, the traveler lives temporarily, planned trips, and there are elements of tourism products (Mansfeld 1992). Various factors motivate the intention to travel such as physical motivation aimed at relaxation and exercise, cultural motivation to know other cultures, social factors to visit others, and status factors related to self-development (education, business travel, and hobbies) (McIntosh et al. 1999; Qiu et al. 2018).

In general, tourist trips include expenditures on food, tourist lodging, travel transportation, and entertainment. Demographic, social, and cultural factors determine the pattern of tourist trips made (Cai et al. 1996). Halal tourism is rooted in Islamic law which requires a trip for a useful purpose, for example, the purpose of religious spiritual travel (hajj and umrah), education, treatment, and knowledge. This means that halal tourism must be truly halal and comply with every rule of Islamic law (El-Gohary 2016). The trend of halal tourism in Asian countries includes the entire provision of tourism products and services by Islamic law (Samori et al. 2016). Halal travel and tourism activities can help preserve national culture and the environment. The presence of tourists makes people more concerned about the preservation of tourist attractions including cultural arts, natural beauty, and historical relics (Jaelani 2017). However, several obstacles affect halal travel and tourism, such as the weather and climate of halal tourist destinations that tourists go to, access to travel to tourist destinations, and attractions that attract tourists (Becken and Wilson 2013). In the operation of the travel and tourism industry, there exist various subsectors, with travel agents assuming critical roles both in organizing and acting as intermediaries for halal tourism travel services. The travel agency is responsible for organizing travel from preparation for departure to equipment during the trip of a person or group who intends to organize a halal tourism trip (Chandra 2014).

In fact, some data show positive development of halal travel and tourism including in Indonesia. When searching for articles in a reputable database called Scopus, it found 371 articles which have been published on the topic of halal travel and

Table 17.1 Total number of documents by year and total number of documents by country

No	Year of publication/Total number documents	Country/Total number of documents
1	2023 (17)	Indonesia (141)
2	2022 (86)	Malaysia (101)
3	2021 (61)	Turkey (22)
4	2020 (63)	New Zealand (19)
5	2019 (48)	United Arab Emirates (17)
6	2018 (31)	UK (15)
7	2017 (23)	Australia (14)
8	2016 (18)	Spain (14)
9	2015 (3)	China (13)
10	2014 (8)	South Korea (13)

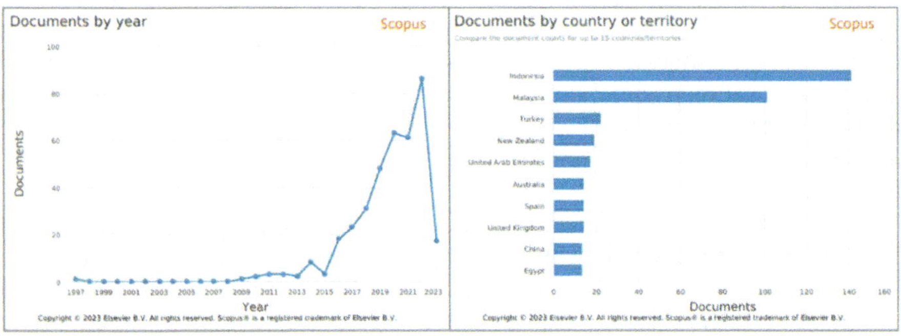

Fig. 17.1 Total number of documents by year and total number of documents by country

tourism. Table 17.1 and Fig. 17.1 show the researches that have been done from 1997 till 2023 on halal travel and tourism worldwide. Though the peak point of total research is in 2021, however, the increment actually starts in 2015. It can be assumed that potential of the halal travel and tourism industry has been noticed by researchers or practitioners since eight years ago.

Moreover, it seems not only Muslim countries that concern on the development of halal travel and tourism. For example, New Zealand, UK, Australia, Spain, China, and South Korea are the countries that conduct researches or studies on the topic of halal travel and tourism. Other potential countries like Finland, Sweden, and Thailand would see the potential of this industry. Though Indonesia and Malaysia lead the list (the dominant countries) that actively doing studies in the said topic, however, the quality of research in terms of citations or average citations per documents needs to be improved.

17.3 Method

This paper uses a content analysis in elaborating the development of halal tourism in Indonesia as well as revealing its challenges and opportunities.

17.4 Discussion

There are several other challenges that have contributed to the slow development of halal tourism in Indonesia. The challenges referred to include the following: First, there are currently no regulations that comprehensively regulate halal tourism in Indonesia. Therefore, the government is expected to immediately make regulatory standards related to hotels, spas, saunas and messages, tourist attractions, and travel agencies based on the DSN-MUI fatwa. Second, there is no good infrastructure for halal tourism in Indonesia. For example, currently, there are not many hotels that operate according to sharia principles as well as restaurants that are halal-certified. This is different if we compare it with Malaysia and Singapore, which already have a large number of sharia hotels and halal-certified restaurants from Indonesia. The lack of the two facilities above is certainly a challenge in itself for the development of halal tourism in Indonesia. Therefore, both the central and regional governments must work together to provide good facilities and infrastructure that are in accordance with sharia so that the interest of foreign Muslim tourists to come to Indonesia is even greater.

In the transition to the new normal, the use of digital technology is a strategic decision that can be relied on in supporting the promotion of halal tourism, especially in Indonesia. Start-up companies engaged in the information and communication technology sector must turn this challenge into a potential opportunity in creating applications that offer safety and comfort for Muslim tourists. Stephenson (2014) mentions that many interface applications can be operated easily via smartphones for halal service searches such as hotel search, vacation packages, travel guides, airport locations, and restaurant selections (for instance, Halal Trip and Muslim Pro). Responding to the danger of the spread of COVID-19, halal applications need to be equipped with sharia mobile banking service facilities to facilitate transactions. The government, through Bank Indonesia, also urges the public to support the non-cash movement as a preventive step in preventing COVID-19 in Indonesia (Ministry of Communication, 2020). This step is considered appropriate considering that in new normal conditions, people are still asked to limit physical contact.

An increase in Muslim tourists is an opportunity for the tourism sector to develop halal tourism. So that several countries began to take this opportunity by developing halal tourism, both countries with a Muslim majority and non-Muslims such as Japan, South Korea, Australia, and Thailand. It is expected that tourist attractions, hotels, restaurants, airlines, travel agents as well as all involved in tourism can be involved in halal tourism. Travel agents have halal tourism opportunities in various fields (Battour and Ismail 2016).

17.5 Conclusion

Halal tourism in Indonesia has great potential to develop. This is because Indonesia has a variety of beautiful tourist attractions and the majority of the population is Muslim. However, it is important to remember that Indonesia's potential will not be able to develop optimally if it is not managed properly. The government's full support, which is not half measures, is clearly needed. The various challenges mentioned above must be solved immediately so that halal tourism can develop and compete with other countries.

References

Afnarius S, Akbar F, Yuliani F (2020) Developing web-based and mobile-based GIS for places of worship information to support halal tourism: a case study in Bukittinggi, Indonesia. ISPRS Int J Geo-Inf 9(1):1–18. https://doi.org/10.3390/ijgi9010052

Aji HM, Muslichah I, Seftyono C (2021) The determinants of Muslim travellers' intention to visit non-Islamic countries: a halal tourism implication. J Islamic Mark 12(8):1553–1576. https://doi.org/10.1108/JIMA-03-2020-0075

Akar E, Dalgic T (2018) Understanding online consumers' purchase intentions: a contribution from social network theory. Behav Inf Technol 37(5):473–487. https://doi.org/10.1080/0144929X.2018.1456563

Anam K (2022) Pariwisata RI Naik Peringkat, Ranking 32 Dari 117 Negara. Cnbcindonesia.Com. https://www.cnbcindonesia.com/lifestyle/20220603154925-33-344181/pariwisata-ri-naikperingkat-ranking-32-dari-117-negara

Battour M, Ismail MN (2016) Halal tourism: concepts, practices, challenges and future. Tourism Manag Perspect 19:150–154. https://doi.org/10.1016/j.tmp.2015.12.008

Bayu DJ (2020) Kadin: sektor Pariwisata paling terdampak corona, Ribuan Hotel Tutup Artikel ini telah tayang di Katadata.co.id dengan judul "Kadin: Sektor Pariwisata Paling Terdampak Corona, Ribuan Hotel Tutup". https://katadata.co.id/berita/2020/04/19/kadin-sektor-par

Becken S, Wilson J (2013) The impacts of weather on tourist travel. Tour Geogr 15(4):620–639. https://doi.org/10.1080/14616688.2012.762541

Cai LA, Hong GS, Morrison AM (1996) Household expenditure patterns for tourism products and services. J Travel Tour Mark 4(4):15–40. https://doi.org/10.1300/J073v04n04_02

Chandra GR (2014) Halal tourism: a new goldmine for tourism. Int J Bus Manage Res (IJBMR) 4(6):45–62

El-Gohary H (2016) Halal tourism, is it really Halal? Tourism Manage Perspect 19:124–130. https://doi.org/10.1016/j.tmp.2015.12.013

Fadhilah FG, Widyastuti IP, Rizqullah MFS (2020) Wisata Halal Aceh Sebagai instrumen peningkatan perekonomian indonesia dalam menggandeng wisatawan mancanegara. Gorontalo Dev Rev 3(1):48–58

Guild J (2020) The political and institutional constraints on green finance in Indonesia. J Sustain Finance Investment 10(2):157–170

Jaelani A (2017) International review of management and marketing Halal tourism industry in Indonesia: potential and prospects. Int Rev Manage Mark 7(3):25–34. http://www.econjournals.com

Jia X, Chaozhi Z (2020) "Halal tourism": is it the same trend in non-Islamic destinations with Islamic destinations? Asia Pacific J Tourism Res 25(2):189–204. https://doi.org/10.1080/10941665.2019.1687535

Katuk N, Ku-Mahamud KR, Kayat K, Abdul Hamid MN, Zakaria NH, Purbasari A (2020) Halal certification for tourism marketing: the attributes and attitudes of food operators in Indonesia. J Islamic Mark 12(5):1043–1062. https://doi.org/10.1108/JIMA-03-2020-0068

Loussaief A, Haque A (2018) Determinants of tunisian consumer purchase intention Halal certified products: a qualitative study. Int Tourism Hospitality J 1(3):1–16. https://rpajournals.com/ithj

Mafudi M, Sugiarto S, Praptapa A (2021) Emerging opportunities for Halal Tourism in rural areas: insights from Indonesia. Fokus Bisnis Media Pengkajian Manajemen Dan Akuntansi 20(1):16–26. https://doi.org/10.32639/fokusbisnis.v10i1.726

Mansfeld Y (1992) From motivation to actual travel. Ann Tour Res 19(3):399–419. https://doi.org/10.1016/0160-7383(92)90127-B

McIntosh RW, Goeldner CR, Ritchie JRB (1999) Passenger transport. Tourism principles, practices, philosophies. Wiley

Minardi A, Astuti D, Suhadi (2021). Indonesia as the best Halal Tourism destination and its impacts to Muslim's travelers visit. Eur J Theol Philos 1(3):43–50. https://doi.org/10.24018/theology.2021.1.3.30

Pradana M, Huertas-García R, Marimon F (2021) Muslim tourists' purchase intention of halal food in Spain. Curr Issue Tour 24(13):1814–1818. https://doi.org/10.1080/13683500.2020.1797647

Qiu RTR, Masiero L, Li G (2018) The psychological process of travel destination choice. J Travel Tour Mark 35(6):691–705. https://doi.org/10.1080/10548408.2018.1435332

Ratnasari RT, Gunawan S, Mawardi I, Kirana KC (2020) Emotional experience on behavioral intention for halal tourism. J Islamic Mark 12(4):864–881. https://doi.org/10.1108/JIMA-12-2019-0256

Raya MKF (2022) Wali Pitu in Bali: inventing new halal tourism destinations in contemporary Indonesia. Contemp Islam 16(2):475–505. https://doi.org/10.1007/s11562-022-00503-8

Samori Z, Md Salleh NZ, Khalid MM (2016) Current trends on Halal tourism: cases on selected Asian countries. Tourism Manage Perspect 19:131–136. https://doi.org/10.1016/j.tmp.2015.12.011

Samsul S, Muslimin S, Jafar W (2022) Peluang dan tantangan industri Halal Indonesia menuju pusat industri Halal Dunia. Al-Azhar J Islamic Econ 4(1):12–24. https://doi.org/10.37146/ajie.v4i1.135

Septyaningsih I (2021) Kontribusi Sektor Halal Diharapkan Signifikan Bagi Ekonomi. Republika.co.id. https://www.republika.co.id/berita/qu5hr4380/kontribusi-sektor-halal-diharapkan-signifikan-bagiekonomi

Stephenson ML (2014) Deciphering "Islamic hospitality": developments, challenges and opportunities. Tour Manage 40:155–164. https://doi.org/10.1016/j.tourman.2013.05.002

Usman H, Sobari N, Sari LE (2019) Sharia motivation in Muslim tourism definition, is it matter? J Islamic Mark 10(3):709–723. https://doi.org/10.1108/JIMA-01-2018-0015

Vargas-Sánchez A, Moral-Moral M (2019) Halal tourism: state of the art. Tourism Rev 74(3):385–399. https://doi.org/10.1108/TR-01-2018-0015

Wardi Y, Abror A, Trinanda O (2018) Halal tourism: antecedent of tourist's satisfaction and word of mouth (WOM). Asia Pac J Tourism Res 23(5):463–472

Wibowo DE, Mandusari BD (2018) Pengaruh Labelisasi terhadap Keputusan Pembelian oleh Konsumen Muslim terhadap Produk Makanan di Kota Pekalongan. Indonesia J Halal 01(01)

Widjaja YI, Khalifa GSA, Abuelhassan AE (2020) The effect of Islamic attributes and destination affective image on the reputation of the Halal tourism destination of Jakarta. J Environ Manage Tourism 11(2):299–313. https://doi.org/10.14505/jemt.v11.2(42).08

World Health Organization. (2020). Coronavirus disease (COVID-19), 12 October 2020. https://apps.who.int/iris/bitstream/handle/10665/336034/nCoV-weekly-sitrep11Oct20-eng.pdf

Yan LS, Zahari NA, Ismail AS, Zain NM (2017) Halal tourism: a new world for tourism industry. Int J Asian Soc Sci 7(8):643–657. https://doi.org/10.18488/journal.1.2017.78.643.657

Zulvianti N, Aimon H, Abror A (2022) The influence of environmental and non-environmental factors on tourist satisfaction in Halal Tourism destinations in West Sumatra, Indonesia. Sustainability 14(15):9185. https://doi.org/10.3390/su14159185

Halal Lifestyle, Trends, and Branding of Muslim Societies in Indonesia

18

Ahmad Rafiki and Fahmi Natigor Nasution

Abstract

This chapter aims to elaborate and discuss on challenges and opportunities of the halal lifestyle, trends, and branding of Muslim societies in Indonesia. The study adopts a content analysis technique where various sources are referred including reports, official documents, and research articles. The halal lifestyle, trends, and branding have positive development but need significant initiatives by the stakeholders. Other opportunities for halal lifestyle, trends, and branding also have been revealed to be used as valuable information for strategic planning and initiatives.

Keywords

Halal branding • Halal trends • Halal lifestyle • Muslim societies • Indonesia

18.1 Introduction

Halal lifestyle in Indonesia the halal industry began to develop in Muslim-majority countries and demanded non-Muslim countries. The halal industry development is part of information technology, culture, and times. These new trends are easy to spread and attract the attention of the world community. Indonesia is one of the countries that followed and started implementing a halal lifestyle. Indonesia is the country with the fourth-largest Muslim population in the world. Based on World Population Review data, in 2020, the number of Muslims in Indonesia reached

A. Rafiki (✉)
Universitas Medan Area, Medan, Indonesia
e-mail: hahmadrafiki@gmail.com

F. N. Nasution
Universitas Sumatera Utara, Medan, Indonesia

A. Rafiki (ed.), *Digitalization in Halal Management*, Contributions to Environmental Sciences & Innovative Business Technology,
https://doi.org/10.1007/978-981-99-5146-8_18

229 million people or 87.2% of 273.5 million Indonesians. Meanwhile, based on Pew Research Center data, the Islamic population in the world reached 1.9 billion. By 2050, Muslims will reach 29.7% of the world's total population. As the country with the largest Muslim population, halal lifestyle consumption in Indonesia is the highest in Southeast Asia, reaching US$203.2 million.

According to the State of the Global Islamic Economy Report 2019/2020, the contribution of Muslims in the world to the halal lifestyle reached USD 2.2 trillion and Muslims spend USD 1369 trillion on food and beverage, followed by clothing products (fashion), amounting to USD 283 billion. Meanwhile, media and entertainment reached USD 220 billion, travel and tourism business amounted to USD 189 billion, and spending on pharmaceutical and cosmetic products amounted to USD 92 billion and USD 54 billion (Dinar Standard 2020). Meanwhile, based on the Global Islamic Economic Indicators report covering 81 countries incorporated in 2020, Malaysia, Saudi Arabia, UAE, Indonesia, and Jordan led the rankings. Indonesia and Saudi Arabia moved up the rankings, while Nigeria, Sri Lanka, and Singapore were in the top 15. Investment in companies relevant to the Islamic economy slowed in 2019/2020, falling by 13% to $11.8 billion. More than 54% of assets are in the halal product category. In comparison, Islamic finance and halal lifestyle attract 41.8% and 4% of an investment, respectively. Based on the data above, Muslim consumption of halal lifestyles is increasing from year to year. Industries that use the label "halal" are also growing and expanding in Indonesia.

At the same time, the halal lifestyle is not that simple because it must follow sharia principles in its implementation. Furthermore, it needs to be studied further about the halal lifestyle that is developing worldwide, including Muslims in Indonesia, is a necessity for Muslims' lives, or just brands to get traction from Muslim consumers, or indeed just an interesting "trend" to follow. The term "halal" is allowed under the provisions of Islamic sharia. The halal status can be attached to objects or deeds. For a Muslim, following Allah's command is a form of obedience as a servant.

Therefore, every Muslim is obliged to seek, use, and carry out something lawful as an embodiment of human obedience in carrying out the laws of Allah. However, it should be an obligation and a need for Muslims (Syamsuri et al. 2022). Lifestyle is a person's lifestyle shown in his life by manifesting in the activities carried out, desires, or interests owned, and the existence of opinions or opinions on something that happened (Asri and Ilyas 2022; Jailani and Adinugraha 2022). Lifestyle also means a lifestyle that describes individuals' activities, interests, and ideas that interact with the surrounding environment (Fatmi et al. 2020). Thus, lifestyle is a sign of the modern world. Lifestyle can help define attitudes, values and indicate wealth and social position in the contemporary world. Therefore, lifestyle is considered a vital project rather than leisure time (Nawawi et al. 2019).

18.2 Literature Review

18.2.1 Halal Needs

Indonesia is the center of the world's leading halal producer. A big vision that the government wants to realize. As a country that has the fourth-largest population in the world, namely 267 million people, and with the majority of the population being Muslims, Indonesia has strong potential to achieve this vision.

The government also continues to encourage the development of the halal industry in Indonesia in a sustainable manner. One of the commitments taken as a concrete step to achieve this vision is through accelerating the implementation of halal certification for micro and small business actors. The launch of the Sehati Program (Free Halal Certification) and the provision of halal facilitation for micro and small enterprises through several Ministries and Institutions is a strategy to accelerate this implementation. Thus, it is expected that 80 percent of food and beverage MSEs that have a business identification number (NIB), also have halal certification.

The micro and small enterprises receive facilitation in managing the halal certification process by the government through budget allocations prepared by ministries, institutions, regional governments as well as government-linked companies, and the private sector. Until 2021, there have been 17,863 certificates out of 263,190 products. This figure is of course still far from ideal if you look at the target of 10 million halal certifications.

18.2.2 The Importance of Halal Certification for Consumers

Halal certification is important in providing certainty and a sense of security for consumers. To find out whether a product is halal or not is not just looking at the final product. Consumers also want to know how halal principles are applied in the process of producing these products starting from the provision of ingredients, processing, and packaging, to presentation (Masruroh 2020; Budiandru et al. 2021).

Halal certification is needed because the issue of halal toyyiban is an integral part of the process. Halal is usually normative or lawful, while toyyiban ensures that products have quality. Halal is a process and not just an end product (Helmiati 2021). For example, consumers want to know the halal supply chain process for a product, starting from the farm to the products sold in stores. Halal certification provides guarantees and protection to consumers. Consumers have the right to consume, use, and utilize products and services that are guaranteed to be halal. Halal certification provides guarantees and protection for that (Indarti et al. 2020; Purwanto et al. 2021).

18.2.3 Halal Certification Acceleration Provides Benefits for Businesses

In addition to consumers, the acceleration of the implementation of halal certification will certainly have a positive impact on micro and small enterprises. It also provides the potential for additional income for business actors. With the current regulations, it will be more orderly and transparent. There are benefits of halal certification for business actors. From a business perspective, halal certification will also provide additional income or extended profit (Hasan 2021). In other words, from the business actor's point of view, the interest in taking care of halal certification is quite good. However, business actors still need to continue to be educated that this certification is not merely an administration, but also a standard that applies both nationally and globally (Sofiana et al. 2021; Nirwandar 2020). This will have an impact on a country's trade and economy.

18.2.4 Encourage MSMEs to Penetrate the Global Market

The various efforts made to make it easier for business actors to obtain halal product certification are closely related to Indonesia's vision as a center for halal producers. The more MSME products that enter the global halal product value chain, the greater the potential to achieve this vision. This is because the main requirement for MSMEs to penetrate the global halal market is a halal certification (Alamsyah et al. 2022).

It's simple, halal is not monopolistic. For example, Thailand is an exporter of halal industrial products, especially food, even though it has a small Muslim population. Korea is also an exporter of halal cosmetics. All of these countries a small Muslim population is actively exporting halal products. They are very enthusiastic about registering their products to have their products halal audited (Akram 2022). Halal products are their priority so that their standards are recognized globally so that their sales from exports increase. Therefore, Indonesia needs to encourage and rise the production of more halal products (Azam and Abdullah 2020; Rachman and Syamsuddin 2019).

18.2.5 Halal Product Industry Data Codification

Not only accelerating halal certification, strategies and other concrete steps taken by the government in the context of developing the halal product industry is the codification of halal product industry data. On November 30, 2021, the National Sharia Economic and Finance Committee (KNEKS) signed three cooperation agreements and a memorandum of understanding between KNEKS and several related parties, namely the signing of the KNEKS agreement with the Directorate General of Customs and Excise, the National Single Window Institution (LNSW), and the Halal Product Assurance Organizing Body (BPJPH) regarding the codification of halal product data and its integration into trade and economic data.

The halal products with export and import trade transactions and industrial data are well integrated. So far, Indonesia's exports, whose value is very large, have not been recorded, which has resulted in Indonesia's position losing to Malaysia. This is the importance of integrating an information system for recording halal products for export so that both the volume and the transaction value can be recorded (Butt 2020; Slamet et al. 2022).

In addition, this codification is also important in terms of tracking the product's raw material content. This is not only a form of producer transparency toward consumers but also a way to protect producers from losses. There used to be chocolate in Malaysia that suffered huge losses because when we traced it, it contained non-halal ingredients. After it was announced, the market size was down. Hence, there needs to be an integrated codification. In addition, there needs to be mutual recognition and agreement between exporters and importers so that there is codification and standardization between suppliers, especially for exports (Yuli and Wojtyla 2020; Mujahidin 2020).

Based on some findings from techniques in bibliometric analysis, the research on halal supply chains has been done by many researchers in various countries. There are 1839 documents sourced from a reputable database called Scopus by searching through keywords: "halal brands," "halal branding," "halal products," "halal industry." As depicted in Table 18.1, the highest hit on the number of documents is in the years 2020 and 2022 with a total number of documents 267 and 256, respectively. It is also shown that the number of documents increase incrementally since 2011, which considers the interest in studies on this topic to be perceived positively.

As depicted in Table above, shows that Universiti Teknologi MARA leads the list of affiliations as the most active institution that produced articles on the topic of halal branding, products, and industry. In the top ten lists, nine of them are Universities located in Malaysia. One of them is located in Indonesia (Universitas Gadjah Mada). This figure also indicated that the research on the topic of halal branding, products, and industry is mostly explored by researchers from Southeast Asia countries rather than Western countries.

Moreover, it shows that Rohman, A. leads the list of dominant authors with 48 documents that studied halal branding, products, and industry, then followed by other authors. The interest of researchers is increasing over the period which indicates the topic of halal branding, products, and industries could be extended to other subjects.

Meanwhile, it is also shown that the research on halal branding, products, and industry has been done mostly in the subject of business, management, and accounting with a percentage of 22.0%, followed by agricultural and biological sciences and social science with 12.5% and 10.1%, respectively. This indicates other subjects have an association with halal branding, products, and industry.

Furthermore, Malaysia and Indonesia are recognized as countries that are mostly involved in research of halal branding, products, and industry with a total

Table 18.1 Summary of top ten research output, affiliation, and document types

No	Year/ Number document	Author/Number of documents	Country/ Number of documents	Affiliation/Number of documents	Subject area / Number of documents
1	2023 (95)	Rohman, A. (49)	Malaysia (818)	Universiti Teknologi MARA (59)	Business, Management and Accounting (746)
2	2022 (251)	Ali, M.H. (16)	Indonesia (486)	Universiti Putra Malaysia (112)	Agricultural and Biological Sciences (430)
3	2021 (206)	Zailani, S. (16)	United Kingdom (95)	Universiti Kebangsaan Malaysia (98)	Social Sciences (351)
4	2020 (272)	Ngah, A.H. (15)	United States (68)	International Islamic University Malaysia (94)	Engineering (272)
5	2019 (193)	Erwanto, Y. (14)	India (49)	Universiti Malaya (87)	Economics, Econometrics and Finance (260)
6	2018 (171)	Jaswir, I. (14)	Pakistan (48)	Universitas Gadjah Mada (62)	Computer Science (254)
7	2017 (141)	Kamarulzaman, N.H.(14)	Thailand (48)	Universiti Utara Malaysia (59)	Environmental Science (164)
8	2016 (112)	Regenstein, J.M. (14)	China (44)	Universiti Sains Malaysia (56)	Arts and Humanities (159)
9	2015 (73)	Tieman, M. (14)	Australia (43)	Universiti Teknologi Malaysia (54)	Decision Sciences (132)
10	2014 (75)	Ali, M.E. (13)	Turkey (40)	Universiti Sains Islam Malaysia (47)	Biochemistry, Genetics and Molecular Biology (105)

of 1304 documents. Other non-Muslim populated countries placed in the 3rd and 4th rank in the list (UK and USA). This indicates that halal branding, products, and industry were studied by Muslim and non-Muslim communities.

18.3 Method

This paper uses content and bibliometric analyses in elaborating the development of halal lifestyles, trends, and branding of Muslim societies in Indonesia as well as revealing its challenges and opportunities.

18.4 Discussion

During the COVID-19 pandemic like this, it is only natural for people around the world to adopt a clean and healthy lifestyle. For Muslims, this lifestyle is familiar, and this lifestyle is known as the halal lifestyle. The halal lifestyle is the best life practice based on values originating from Islamic teachings in accordance with human nature. Even though it comes from Islamic values, a halal lifestyle can be applied by anyone because it is in accordance with human nature. For example, a clean and healthy lifestyle, eating in moderation and not excessively, always sharing with others, dressing and looking polite is a lifestyle that is in accordance with human nature. The halal lifestyle is not limited to Muslims, but can be applied by all people, even though they have different religions (Harjadi et al. 2021).

The halal lifestyle develops following the changing times which can be seen in the application of more modern Islamic values (Bux et al. 2022). The current generation of Muslims wants to be actualized by showing Muslim personalities today who are religious and contemporary. From these characteristics, it is not surprising that the halal lifestyle is a potential market (Annisa 2021). Countries in the world are developing the halal lifestyle industry are also called the halal industry because of its potential market (Zuraidah and Mu'is 2022).

Muslim countries respond to the market potential of the halal industry by providing products and services with Islamic values. This is supported by the rapid growth of the Muslim middle-class market in some countries. This middle class is more aware of Islamic values so it becomes the right target to develop the halal industry in the country. In this case, Indonesia through the Ministry of Tourism and Creative Economy focuses on developing the halal industry through the National Halal Industry Development Strategy, namely improving the quality of Creative Small and Medium Micro Enterprises (MSMEs) through business digitalization, business development and development of appropriate capital schemes. In addition, the Ministry of Tourism and Creative Economy plays a role in increasing the quantity and quality of creative economy human resources as part of the halal industry cluster that supports a halal lifestyle. In addition, regarding halal tourism, the Ministry of Tourism and Creative Economy also plays a role in developing the halal tourism ecosystem in Indonesia.

Based on our experience in developing the Islamic Creative Economy, the development of the halal industry, especially MSMEs, can be developed by paying attention to four main aspects, namely creative-innovative, based on sharia, having a social impact, and being scale-able. However, the COVID-19 pandemic has had an impact on the development strategy of the halal industry. Restrictions on people's activities at home certainly have an impact on overall economic activity which in turn also affects the halal industry itself. The decline in people's purchasing power in various halal industry clusters has occurred significantly, especially in the halal tourism cluster. Tourism activities stopped at the start of the pandemic and began to reopen by implementing various health protocols. Other halal industry clusters also have to adapt during the pandemic which ultimately affects the development of Indonesia's halal lifestyle.

In response to this, several strategies have been prepared to work on the potential of the halal industry even in the midst of the COVID-19 pandemic. Regarding the halal industry cluster which intersects with tourism and the creative economy, one of the development strategies undertaken is to encourage digital transformation and the application of health protocols so that people can return to being productive and safe at work (Lubis et al. 2022; Nafis 2019). The digital sector can become an enabler in the development of post-pandemic halal industry clusters. For example, in Muslim fashion, various products are now being displayed on e-commerce platforms, and currently, many virtual fashion shows are being held by utilizing digital technology (Voak and Fairman 2020).

For this reason, each halal industry cluster will be developed by utilizing digital media to get around the limitations caused by the COVID-19 pandemic. Then, as long as the pandemic occurs, the halal industry cluster which intersects with tourism and the creative economy is targeted to remain safe and productive in carrying out its activities. So that it can be part of the national economic recovery. Foresight in seeing opportunities amid a pandemic is very necessary. The creative sector that has the opportunity to develop is the application and game development sector as well as those related to digital media during a pandemic. As a result of the limited physical activity of the community, the level of consumption of applications, games, and other digital media has increased significantly.

18.5 Conclusion

The development of the halal lifestyle globally is indeed something new. Many Muslims are aware of it, but many have not. The halal lifestyle can develop quickly and readily accepted by Muslim communities. So many people think that it is just a trend. Based on the data above, the halal lifestyle is not just a trend or branding alone. However, Muslims need to carry out Islamic religious orders to consume and use halal and *tayyib* products. For Muslim communities, all activities must be of worship value and expect a reward from Allah SWT, including halal products. By the propositions in the Qur'an and Hadith, the Muslim community in the world must implement a halal lifestyle.

References

Akram HW (2022) Global halal business: a sectoral analysis. Int J Bus Globalisation 30(1)

Alamsyah A, Hakim N, Hendayani R (2022) Blockchain-based traceability system to support the Indonesian halal supply chain ecosystem. Economies 10(134)

Annisa AA (2021) Determining model of halal lifestyle: a study on the role of kopontren. Equilib J Ekon Syariah 9(1):153–178

Asri KH, Ilyas A (2022) Strengthening the Halal value chain ecosystem as the development of the Halal industry towards era 5.0. ALIF Sharia Economics J 1(1):37–47

Azam MSE, Abdullah MA (2020) Global Halal industry: realities and opportunities. Int J Islamic Bus Ethics 5(1)

Budiandru, Safuan, Alim MN (2021) Halal lifestyle, profitability, size, affects CSR as a learning process in Tawhidi epistimology. Influence Int J Sci Rev 3(2)

Butt MMZ (2020) Halal and Shariah Law. In: The Halal food handbook. Wiley, pp 27–38

Bux C, Varese E, Amicarelli V, Lombardi M (2022) Review Halal food sustainability between certification and blockchain: a review. Sustainability 14(2152)

Dinar Standard (2020) State of the global Islamic economy report 2019/2020: driving the Islamic economy revolution 4.0

Fatmi FO, Ahmad AN, Kartika B (2020) Determinants affecting purchase intention of Halal products: an article review. J Halal Prod Res 3(2):63–79

Harjadi D, Kartika NE, Gunawan WH (2021) Islamic marketing strategy in building Indonesia Halal business ecosystem. Int J Econ Bus Account Res (IJEBAR) 5(3)

Hasan Z (2021) Making Indonesia as integrated Halal zone through the implementation of Halal supply chain. J Islamic Econ Bus Res 1(1):1–14

Helmiati (2021) Muslim religiosity in a challenging secular state of Singapore. Qudus Int J Islamic Stud (QIJIS) 9(2):351–386

Indarti N, Budi ASL, Islam AM (2020) A systematic review of halal supply chain research: to where shall we go ? J Islamic Mark 12(9):1930–1949. https://doi.org/10.1108/JIMA-05-2020-0161

Jailani N, Adinugraha HH (2022) The effect of Halal lifetsyle on economic growth in Indonesia. J Econ Res Soc Sci 6(1)

Lubis FRA, Arsy M, Yunadita R (2022) Comparative study of the potential of the Halal industry in developed countries versus developing countries. ARBITRASE J Econ Acc 3(1):95–102

Masruroh N (2020) The competitiveness of Indonesia Halal food exports in global market competition industry. Economica J Ekon Islam 11(1):25–48

Mujahidin (2020) The potential of Halal industry in Indonesia to support economic growth. Al-Kharaj J Islamic Econ Bus 2(1):66–76

Nafis MC (2019) The concept of Halal and Thayyib and its implementation in Indonesia. J Halal Product Res 2(1)

Nawawi MSAM, Hussin MFA, Faid MS, Pauzi, Norhidayah, Man S, Sabri NM (2019) The emergence of halal food industry in non-Muslim countries: a case study of Thailand. J Islamic Mark 11(4):917–931. https://doi.org/10.1108/JIMA-05-2018-0082

Nirwandar S (2020) Halal gelatin and its business opportunity in Indonesia. Int J Halal Res 2(1):50–57

Purwanto H, Jati S, Rofiq A (2021) Policy analysis of enforcement of Halal product guarantee regulations through the regulatory impact analysis (RIA) approach. J Digit Mark Halal Ind 3(2):149–164

Rachman MA, Syamsuddin (2019) Halal industry in Indonesia: the role of Sharia financial instituion in driving industrial and Halal ecosystem. Al-Iqtishad J Islamic Econ 11(1)

Slamet, Abdullah I, Laila NQ (2022) The contestation of the meaning of Halal tourism. Heliyon 8(3)

Sofiana R, Utama S, Rohim A (2021) The problem of Halal certification regarding consumer protection in Malaysia and Indonesia. J Hum Rights Cult Legal Syst 1(3):180–193

Syamsuri, Nadhilah, Masrifah AR (2022) Halal industry products influence on current account in OIC economics in Asia on 2015–2019. JIEI J Ilmiah Ekon Islam 8(3):3401–3411

Voak A, Fairman B (2020) Anticipating human resource development challenges and opportunities in "Halal supply chains" and "Halal logistics" within ASEAN. Nusantara Halal J 1(1):1–9

Yuli SBC, Wojtyla E (2020) Challenges and strategies in developing human resources for the halal industry: evidence from Indonesia. J Innov Bus Econ 4(2):77–86

Zuraidah and Mu'is A (2022) Human resources in the development of the halal industry on industry revolution 4.0 during the Covid-19 pandemic. In: Reinforcement of the Halal industry for global integration revival, pp 108–114

The Challenges to Get Halal Certificate for Henna Cosmetic: An Analysis from Islamic Perspective and Medical Benefits

19

Muhammad Remy Othman, Sayidah Asma Basir, Norhidayah Pauzi and Zalina Zakaria

Abstract

The demand of the world's Muslim community for cosmetic products certified as halal by Halal Malaysia or Foreign Halal Certification Body is very high. This high demand is driven by the strength of the holistic halal ecosystem system that includes material review, management, and several other systems inter alia, the Halal Assurance System, Laboratory Analysis, and Halal Training. These systems have further strengthened consumers' confidence in Malaysia's halal-certified products. Despite that, in terms of legal provisions, Procedure 6(5) in the Malaysian Halal Certification Procedures Manual (Domestic) 2020 provides that products with negative implications for religion and society such as cigarettes, drugs, hair dyes, and nail dyes are not eligible to apply for the halal certificate. Henna or the scientific name *Lawsonia inermis* is common ingredient used in hair and nails dye products. Based on the provisions above, hair and nail dye products including those based on henna are not eligible to

M. R. Othman
Department of Al-Quran and Al-Hadith, Academy of Islamic Studies, Universiti of Malaya, Kuala Lumpur, Malaysia

S. A. Basir
Faculty of Law, Universiti Teknologi MARA (UiTM), Shah Alam, Malaysia

N. Pauzi
Department of Fiqh and Usul, Academy of Islamic Studies, University Malaya, Kuala Lumpur, Malaysia

Z. Zakaria (✉)
Department of Syariah and Law, Academy of Islamic Studies & Universiti Malaya Halal Research Center, University Malaya, Kuala Lumpur, Malaysia
e-mail: zalina_jsu@um.edu.my

A. Rafiki (ed.), *Digitalization in Halal Management*, Contributions to Environmental Sciences & Innovative Business Technology,
https://doi.org/10.1007/978-981-99-5146-8_19

apply for and obtain Malaysian halal status because they fall under the category of hair and nail dyes. The issues are what is the rule on the usage of henna, and to what extent do henna cosmetic products get Malaysian Halal Certification? The objective of this paper is to analyse the findings of benefit of henna from an Islamic and medical perspective. This paper uses qualitative methods to analyse the data. The findings will analyse the needs for re-evaluation by the Malaysian Halal Certification body for henna-based cosmetic products in order to be given halal certification. This is because the initial findings of the study show that the use of henna as ornamental colour is required by Shariah, especially for women. In fact, scientific evidence shows that henna has several benefits of its own, such as being rich in antioxidant compounds that can treat wounds. Therefore, if henna-based products are produced from natural henna, free from prohibited chemicals, safe to use, clean, and pure, it is submitted that the henna cosmetic products are eligible to be considered and be given Malaysian halal status.

Keywords

Consumer · Halal Malaysia · Henna cosmetic · Islamic perspective · Medical benefits

19.1 Introduction

Henna or the scientific name *Lawsonia inermis* L. is common ingredient used in hair and nails dye products. According to scientific evidence, henna leaves contain the colour hennotannic acid. When the skin is massaged, greased, or patched with henna, the hennotannic acid in the henna leaves causes the skin to turn red–orange. This is a result of how collagen and hennotannic acid react with skin cells and keratin in hair and nails (Sarang et al. 2017).[1]

Henna was mentioned in the hadith's popular discourse as an accessory that can be used to colour nails, beards, and grey hair, but it was not explicitly indicated how it can be utilised as medicine or as a remedy. Additionally, it discussed the legal perspective on how men and women use henna, as well as the many prohibitions relating to its use for transgender or gender-neutral purposes (Othman et al. 2021a, b, c)[2].

Many countries have granted halal status to henna-based cosmetic products. Garnier products such as henna-based hair dyes[3] have received halal status from

[1] Sarang et al. (2017).

[2] Othman et al. (2021a).

[3] Garnier Webiste, https://www.garnier.co.id/tips-dan-trick/pewarna-rambut/mau-mewarnai-rambut-yang-halal-kamu-bisa-ikuti-langkah-berikut-ini.

the Majelis Ulama Indonesia, which acts as halal authenticator. Additionally, Halal India Pvt Ltd granted halal status to henna-based products like Queen Nail Henna from Irresistible Cosmetics[4] and Iba Hair Colour by Iba Cosmetics.[5] Halal-certified henna products are also used in some hair and beauty salons in Brunei (Salleh and Aghwan 2019).[6]

However, Malaysian Halal Certification has not been granted to henna-based cosmetic products. In terms of legal provisions, Procedure 6(5) in the Malaysian Halal Certification Procedures Manual (Domestic) 2020[7] provides that products with negative implications for religion and society such as cigarettes, drugs, hair dyes, and nail dyes are not eligible to apply for the halal certificate. Based on the provisions above, hair and nail dye products including those based on henna are not eligible to apply for and obtain Malaysian halal status because they fall under the category of hair and nail dyes.

The issues are what is the rule on the usage of henna, and to what extent do henna cosmetic products get Malaysian Halal Certification? The issues are what is the rule on the usage of henna, and to what extent do henna cosmetic products get Malaysian Halal Certification? Therefore, the objective of this paper is to analyse the findings of benefit of henna from an Islamic and medical perspective.

19.2 Research Methodology

This paper uses qualitative methods to analyse the data based on the sub-topic which are fatwa text and decisions made by state fatwa committee, perspective of the *Fiqh al-Hadith*, and medical benefits of henna. The findings will analyse the needs for re-evaluation by the Malaysian Halal Certification body for henna-based cosmetic products.

19.2.1 Halal Certificate for Henna Cosmetic

Before analysing the demands for henna to be permitted to apply for halal status, some aspects should be addressed initially.

[4] Irresistable Cosmetics Website, Irri https://www.irresistiblecosmetics.co.za/collections/queen-nail-henna

[5] Iba Cosmetic Website, https://ibacosmetics.com/collections/hair-colors.

[6] Salleh and Aghwan (2019).

[7] Manual Prosedur Pensijilan Halal Malaysia (Domestik) (2020).

(1) **Henna in the Perspective of Hadith Nabawi**

According to the research of Othman et al. (2021a, b, c),[8] the henna leaves are *ma'aruf* leaves that are well known for being a natural colouring agent.[9] A total of 17 matn hadiths with a frequency of 32 were outlined from the hadith collection sourced from *al-Kutub al-Sittah* (*Ṣaḥīḥ al-Bukhārí*, *Sahíh Muslim*, *Sunan Abu Dāwūd*, *Jami` al-Tirmidhí*, *Sunan al-Nasa'í*, and *Sunan Ibn Mājah*) that mentions the use of henna either from the Prophet Muhammad PBUH himself or aṣ-saḥābah.

The research also discovered five groups of hadiths that mention applying henna. The following are the five groups of hadiths.

First: Hadiths that explain henna as a dye to darken grey hair. Among them are the following hadiths:

[1753] عَنْ أَبِي ذَرٍّ، عَنِ النَّبِيِّ قَالَ: "إِنَّ أَحْسَنَ مَا غُيِّرَ بِهِ الشَّيْبُ : الْحِنَّاءُ ، وَالْكَتَمُ

Translation: Narrated by Abu Dharr R.A, The Prophet PBUH said: "*Indeed the best of which grey hair may be changed is with Henna' and Katam*"

[6075] عَنْ مُحَمَّدِ بْنِ سِيرِينَ قَالَ : " سَأَلْتُ أَنَسَ بْنَ مَالِكٍ : أَخَضَبَ رَسُولُ اللَّهِ قَالَ :
إِنَّهُ لَمْ يَرَ مِنَ الشَّيْبِ إِلَّا قَلِيلًا

Translation: Muhammad Ibn Sirin reported: I asked Anas Ibn Malik if Allah's Messenger dyed his hair. He said: *"He had but little white hair."*

[6076] حَدَّثَنَا ثَابِتٌ قَالَ : "سُئِلَ أَنَسُ بْنُ مَالِكٍ عَنْ خِضَابِ النَّبِيِّ فَقَالَ : لَوْ شِئْتُ أَنْ أَعُدَّ شَمَطَاتٍ كُنَّ فِي رَأْسِهِ فَعَلْتُ ، وَقَالَ : لَمْ يَخْتَضِبْ، وَقَدِ اخْتَضَبَ أَبُو بَكْرٍ بِالْحِنَّاءِ وَالْكَتَمِ، وَاخْتَضَبَ عُمَرُ بِالْحِنَّاءِ بَحْتًا.

Translation: Thabit said: Anas Ibn Malik was asked about the Prophet PBUH dyeing his hair. He said: *"If I had wanted to count the number of white hairs on his head I could have done so."* And he said: *"He did not dye his hair, but Abu Bakr dyed his hair with henna and Katam and Umar dyed his hair with pure henna."*

[3623] عَنْ عُثْمَانَ بْنِ مَوْهَبٍ ، قَالَ : دَخَلْتُ عَلَى أُمِّ سَلَمَةَ ، قَالَ : " فَأَخْرَجَتْ إِلَيَّ شَعَرًا مِنْ شَعْرِ رَسُولِ اللَّهِ مَخْضُوبًا بِالْحِنَّاءِ وَالْكَتَمِ

Translation: From 'Uthman Ibn Mawhab he said: *"I entered upon Umm Salamah and she brought out for me a hair of the Messenger of Allah (ﷺ), which was dyed with henna and Katam."*

[8] Othman et al. (2021b).

[9] Abū Zakariyyā Yaḥyā ibn Sharaf al-Nawawī, Ṣāḥīḥ Muslim bi Sharḥ al-Nawawī, (Misr: al-Matba'ah al-Misriyyah bi al-Azhar, 1930), 15:95.

[4206] عَنْ أَبِي رِمْثَةَ ، قَالَ: انْطَلَقْتُ مَعَ أَبِي نَحْوَ النَّبِيِّ فَإِذَا هُوَ ذُو وَفْرَةٍ بِهَا رَدْعُ حِنَّاءٍ وَعَلَيْهِ بُرْدَانِ أَخْضَرَانِ

Translation: Narrated Abu Rimthah he said: I went with my father to the Prophet (ﷺ). *He had locks hanging down as far as the lobes of the ears stained with henna, and he was wearing two green garments.*

[5509] عَنْ جَابِرِ بْنِ عَبْدِ اللَّهِ قال : " أُتِيَ بِأَبِي قُحَافَةَ يَوْمَ فَتْحِ مَكَّةَ وَرَأْسُهُ وَلِحْيَتُهُ كَالثَّغَامَةِ بَيَاضًا ، فَقَالَ رَسُولُ اللَّهِ : " غَيِّرُوا هَذَا بِشَيْءٍ وَاجْتَنِبُوا السَّوَادَ

Translation: From Jabir Ibn 'Abdillah said: Abu Quhafah came to Rasulullah (ﷺ) on the day of the Fath Makkah in a situation where his head and beard were like plants with white flowers. Rasulullah (ﷺ) said: *"Change (the colour of hair and beard) this with something and avoid the colour black."*

[3462] إِنَّ أَبَا هُرَيْرَةَ رَضِيَ اللَّهُ عَنْهُ، قَالَ : إِنَّ رَسُولَ اللَّهِ قَالَ : " إِنَّ الْيَهُودَ وَالنَّصَارَى لَا يَصْبُغُونَ فَخَالِفُوهُمْ

Translation: Abu Hurairah Narrated: Allah's Messenger (ﷺ) said: *"The Jews and the Christians do not dye (their grey hair), so you shall do the opposite of what they do."*

[4208] عَنْ أَبِي رِمْثَةَ قَالَ :" أَتَيْتُ النَّبِيَّ أَنَا وَأَبِي فَقَالَ لِأَبِيهِ : مَنْ هَذَا؟ قَالَ : ابْنِي، قَالَ : لَا تَجْنِي عَلَيْهِ وَكَانَ قَدْ لَطَّخَ لِحْيَتَهُ بِالْحِنَّاءِ

Translation: Narrated by Abu Rimthah: I and my father came to the Prophet (ﷺ), the Prophet said to my father: Who is this? He replied: *"He is my son." He said: "Do not commit a crime on him. He had stained his beard with henna."*

[4164] عَنْ عَلِيِّ بْنِ الْمُبَارَكِ ، قَالَ : حَدَّثَتْنِي كَرِيمَةُ بِنْتُ هَمَّامٍ، "أَنَّ امْرَأَةً أَتَتْ عَائِشَةَ رَضِيَ اللَّهُ عَنْ خِضَابِ الْحِنَّاءِ ، فَقَالَتْ : لَا بَأْسَ بِهِ وَلَكِنْ أَكْرَهُهُ، كَانَ حَبِيبِي رَسُولُ اللَّهِ يَكْرَهُ رِيحَهُ

Translation: From 'Alī Ibn al-Mubārak said: Karimah, daughter of Hammam, told that a woman came to Aishah (Allah be pleased with her) and asked her about dyeing with henna. She replied: "*There is no harm, but I do not like it. My beloved, the Messenger of Allah, disliked its odour.*"

[4211] عَنِ ابْنِ عَبَّاسٍ ، قَالَ : " مَرَّ عَلَى النَّبِيِّ رَجُلٌ قَدْ خَضَّبَ بِالْحِنَّاءِ ، فَقَالَ : مَا أَحْسَنَ هَذَا؟ قَالَ : فَمَرَّ آخَرُ قَدْ خَضَّبَ بِالْحِنَّاءِ وَالْكَتَمِ، فَقَالَ : هَذَا أَحْسَنُ مِنْ هَذَا، قَالَ : فَمَرَّ آخَرُ قَدْ خَضَّبَ بِالصُّفْرَةِ، فَقَالَ : هَذَا أَحْسَنُ مِنْ هَذَا كُلِّهِ

Translation: Ibn 'Abbās RA Narrated: "*When a man who had dyed himself with henna passed by the Prophet (ﷺ), he said: How fine this is. When another man who had dyed himself with henna and katam passed by, he said: This is better than that. Then another man who had dyed himself with al-Sufrah, passed by, he said: This is better than all that.*"

[4165] عَنْ عَائِشَةَ رَضِيَ اللَّهُ عَنْهَا، أَنَّ هِنْدَ بِنْتَ عُتْبَةَ قَالَتْ : يَا نَبِيَّ اللَّهِ بَايِعْنِي، قَالَ: "لَا أُبَايِعُكِ حَتَّى تُغَيِّرِي كَفَّيْكِ كَأَنَّهُمَا كَفَّا سَبُعٍ

Translation: Narrated Aisha: When Hind, daughter of 'Utbah, said: Prophet of Allah, accept my allegiance, he replied; "*I shall not accept your allegiance till you make a difference to the palms of your hands; for they look like the paws of a beast of prey.*"

Second: Henna also acts as an ornamental specifically for women either for their nails or hair. Among them are the following hadiths:

[4166] عَنْ عَائِشَةَ رَضِيَ اللَّهُ عَنْهَا، قَالَتْ : " أَوْمَتِ امْرَأَةٌ مِنْ وَرَاءِ سِتْرٍ بِيَدِهَا كِتَابٌ إِلَى رَسُولِ اللَّهِ فَقَبَضَ النَّبِيُّ يَدَهُ، فَقَالَ : مَا أَدْرِي أَيَدُ رَجُلٍ أَمْ يَدُ امْرَأَةٍ؟ قَالَتْ : بَلِ امْرَأَةٌ، قَالَ : لَوْ كُنْتِ امْرَأَةً لَغَيَّرْتِ أَظْفَارَكِ يَعْنِي بِالْحِنَّاءِ

Translation: Narrated Ā'ishah: A woman made a sign from behind a curtain to indicate that she had a letter for the Messenger of Allah (ﷺ). The Prophet (ﷺ) closed his hand, saying: "I do not know this is a man's or a woman's hand." She said: "*No, a woman.*" *He said: "If you were a woman, you would make a difference to your nails, meaning with henna.*"[10]

[2305] ثُمَّ قَالَتْ عِنْدَ ذَلِكَ أُمُّ سَلَمَةَ دَخَلَ عَلَيَّ رَسُولُ اللَّهِ حِينَ تُوُفِّيَ أَبُو سَلَمَةَ وَقَدْ جَعَلْتُ عَلَى عَيْنِي صَبْرًا، فَقَالَ : " مَا هَذَا يَا أُمَّ سَلَمَةَ ؟ " فَقُلْتُ : إِنَّمَا هُوَ صَبْرٌ يَا رَسُولَ اللَّهِ لَيْسَ فِيهِ طِيبٌ، قَالَ:" إِنَّهُ يَشُبُّ الْوَجْهَ، فَلَا تَجْعَلِيهِ إِلَّا بِاللَّيْلِ وَتَنْزَعِينَهُ بِالنَّهَارِ، وَلَا تَمْتَشِطِي بِالطِّيبِ وَلَا بِالْحِنَّاءِ فَإِنَّهُ خِضَابٌ "، قَالَتْ : قُلْتُ : بِأَيِّ شَيْءٍ أَمْتَشِطُ يَا رَسُولَ اللَّهِ؟ قَالَ : " بِالسِّدْرِ تُغَلِّفِينَ بِهِ رَأْسَكِ

Translation: Then Ummu Salamah said; The Messenger of Allah came to visit me when Abu Salamah died, and I had put sabr in my eye. He asked: "*What is this, Ummu Salamah?*" I replied: "*It is only the juice of sabr and contains no perfume.*" He said: "*It gives the face a glow, so apply it only at night and remove it in daytime, and do not comb yourself with scent or henna, for it is a dye.*" I asked: "*What should I use when I comb myself, Messenger of Allah?*" He said: "*Use sidr leaves and smear your head copiously with them.*"

Third: The prohibition of wearing henna for men as an ornamental that resembles women (*tasyabbuh*). The hadith is as follows:

[10] Sulayman Ibn al-'Asy'ath al-Sajistani, *Sunan Abu Dawud*, Kitab al-Tarajjul, Bab fi al-Khadab li al-Nisa', Riyad: Maktabah al-Ma'arif li al-Nashr wa Tawzi', 744, no. Hadith 4166.

[4928] عَنْ أَبِي هُرَيْرَةَ ، أَنّ النَّبِيَّ " أُتِيَ بِمُخَنَّثٍ قَدْ خَضَّبَ يَدَيْهِ وَرِجْلَيْهِ بِالْحِنَّاءِ ، فَقَالَ النَّبِيُّ : مَا بَالُ هَذَا؟ فَقِيلَ : يَا رَسُولَ اللَّهِ، يَتَشَبَّهُ بِالنِّسَاءِ، فَأَمَرَ بِهِ فَنُفِيَ إِلَى النَّقِيعِ ، فَقَالُوا : يَا رَسُولَ اللَّهِ، أَلَا نَقْتُلُهُ؟ فَقَالَ : إِنِّي نُهِيتُ عَنْ قَتْلِ الْمُصَلِّينَ " قَالَ أَبُو أُسَامَةَ : وَالنَّقِيعُ نَاحِيَةٌ عَنِ الْمَدِينَةِ وَلَيْسَ بِالْبَقِيعِ

Translation: Narrated Abu Hurayrah an effeminate man (mukhannath) who had dyed his hands and feet with henna was brought to the Prophet (ﷺ). He asked: What is the matter with this man? He was told: *"Messenger of Allah! He imitates the look of women."* So he issued an order regarding him and he was banished to an-Naqi'. The people said: *"Messenger of Allah! Should we not kill him?"* He said: *"I have been prohibited from killing people who pray."*

Fourth: Henna functions as a material for wound treatments. The hadiths are as follows:

[2054]عَنْ عَلِيِّ بْنِ عُبَيْدِ اللَّهِ ، عَنْ جَدَّتِهِ سَلْمَى ، وَكَانَتْ تَخْدُمُ النَّبِيَّ صَلَّى اللَّهُ عَلَيْهِ وَسَلَّمَ قَالَتْ : "مَا كَانَ يَكُونُ بِرَسُولِ اللَّهِ قَرْحَةٌ وَلَا نَكْبَةٌ إِلَّا أَمَرَنِي رَسُولُ اللَّهِ أَنْ أَضَعَ عَلَيْهَا الْحِنَّاءَ"

Translation: Ali bin 'Ubaidullah narrated that his grandmother Salma, who used to serve the The Prophet (ﷺ) said: *"When there was a wound or cut on the Messenger of Allah (ﷺ) he would order me to put Henna on it."*

[3502] حَدَّثَنِي مَوْلَايَ عُبَيْدُ اللَّهِ، حَدَّثَتْنِي جَدَّتِي سَلْمَى أُمُّ رَافِعٍ مَوْلَاةُ رَسُولِ اللَّهِ صَلَّى اللَّهُ عَلَيْهِ وَسَلَّمَ، قَالَتْ: " كَانَ لَا يُصِيبُ النَّبِيَّ صَلَّى اللَّهُ عَلَيْهِ وَسَلَّمَ قَرْحَةٌ، وَلَا شَوْكَةٌ، إِلَّا وَضَعَ عَلَيْهِ الْحِنَّاءَ."

Translation: Salma Ummu Rafi, the server of the Messenger of Allah (ﷺ), said: *"When the Prophet (ﷺ) suffered any injury or thorn-prick, he would apply henna to it."*

Fifth: Henna functions to treat leg pain. The hadith is as follows:

[3858] حَدَّثَنَا فَائِدٌ مَوْلَى عُبَيْدِ اللَّهِ بْنِ عَلِيِّ بْنِ أَبِي رَافِعٍ، عَنْ مَوْلَاهُ عُبَيْدِ اللَّهِ بْنِ عَلِيِّ بْنِ أَبِي رَافِعٍ، عَنْ جَدَّتِهِ سَلْمَى خَادِمِ رَسُولِ اللَّهِ قَالَتْ: "مَا كَانَ أَحَدٌ يَشْتَكِي إِلَى رَسُولِ اللَّهِ وَجَعًا فِي رَأْسِهِ إِلَّا قَالَ : احْتَجِمْ، وَلَا وَجَعًا فِي رِجْلَيْهِ إِلَّا قَالَ : اخْضِبْهُمَا

Translation: It has been told to us Fā'id who served 'Ubaidillah Ibn 'Alī Ibn Abī Rāfi', from his master, 'Ubaidillah Ibn 'Alī Ibn Abī Rafi', from Salmah, the maid-servant of the Messenger of Allah (ﷺ) Narrated: *"No one complained to the Messenger of Allah (ﷺ) of a headache but he told him to get himself cupped, or of a pain in his legs but he told him to dye them with henna."*

To facilitate the classification of research into subthemes, researchers are arranged in a table format and will be discussed further in the discussion section. The list of hadiths on the application of henna from the *al-Kutub al-Sittah* source is presented in Table 19.1.

Table 19.1 List of hadiths on the use of henna in the source of *al-Kutub al-Sittah*

Group	List of hadith	Source of hadith					
1	Henna as a dye to darken grey hair	B	M	AD	T	N	IM
	إِنَّ أَحْسَنَ مَا غُيِّرَ بِهِ الشَّيْبُ : الْحِنَّاءُ ، وَالْكَتَمُ	–	–	(1)	(1)	(6)	(1)
	إِنَّهُ لَمْ يَرَ مِنَ الشَّيْبِ إِلَّا قَلِيلًا	–	(1)	–	–	–	–
	وَقَدِ اخْتَضَبَ أَبُو بَكْرٍ بِالْحِنَّاءِ وَالْكَتَمِ، وَاخْتَضَبَ عُمَرُ بِالْحِنَّاءِ بَحْتًا	–	(1)	–	–	–	–
	فَأَخْرَجَتْ إِلَيَّ شَعَرًا مِنْ شَعْرِ رَسُولِ اللَّهِ مَخْضُوبًا بِالْحِنَّاءِ وَالْكَتَمِ	–	–	–	–	–	(1)
	انْطَلَقْتُ مَعَ أَبِي نَحْوَ النَّبِيِّ فَإِذَا هُوَ ذُو وَفْرَةٍ بِهَا رَدْعُ حِنَّاءٍ وَعَلَيْهِ بُرْدَانِ أَخْضَرَانِ	–	–	(1)	–	–	–
	" أُتِيَ بِأَبِي قُحَافَةَ يَوْمَ فَتْحِ مَكَّةَ وَرَأْسُهُ وَلِحْيَتُهُ كَالثَّغَامَةِ بَيَاضًا ، فَقَالَ رَسُولُ اللَّهِ : " غَيِّرُوا هَذَا بِشَيْءٍ وَاجْتَنِبُوا السَّوَادَ	–	(1)	(1)	–	(1)	–
	إِنَّ الْيَهُودَ وَالنَّصَارَى لَا يَصْبُغُونَ فَخَالِفُوهُمْ	(1)	–	–	–	–	–
	لَا تَجْنِي عَلَيْهِ وَكَانَ قَدْ لَطَّخَ لِحْيَتَهُ بِالْحِنَّاءِ	–	–	(1)	–	(1)	–
	"أَنَّ امْرَأَةً أَتَتْ عَائِشَةَ رضِيَ اللَّهُ عَنْ خِضَابِ الْحِنَّاءِ ، فَقَالَتْ : لَا بَأْسَ بِهِ وَلَكِنْ أَكْرَهُهُ، كَانَ حَبِيبِي رَسُولُ اللَّهِ يَكْرَهُ رِيحَهُ	–	–	(1)	–	(1)	–
	" مَرَّ عَلَى النَّبِيِّ رَجُلٌ قَدْ خَضَّبَ بِالْحِنَّاءِ ، فَقَالَ : مَا أَحْسَنَ هَذَا؟ قَالَ : فَمَرَّ آخَرُ قَدْ خَضَّبَ بِالْحِنَّاءِ وَالْكَتَمِ، فَقَالَ : هَذَا أَحْسَنُ مِنْ هَذَا.....	–	–	(1)	–	–	(1)
	...لَا أُبَايِعُكِ حَتَّى تُغَيِّرِي كَفَّيْكِ كَأَنَّهُمَا كَفَّا سَبُعٍ	–	–	(1)	–	–	–
2	Henna also acts as an ornamental specifically for women either for their nails or hair						
	... لَوْ كُنْتِ امْرَأَةً لَغَيَّرْتِ أَظْفَارَكِ يَعْنِي بِالْحِنَّاء	–	–	(1)	–	(1)	–
	...وَلَا تَمْتَشِطِي بِالطِّيبِ وَلَا بِالْحِنَّاءِ فَإِنَّهُ خِضَابٌ	–	–	(1)	–	(1)	–
3	The prohibition of wearing henna for men as an ornamental that resembles women (*tasyabbuh*)						
	..أُتِيَ بِمُخَنَّثٍ قَدْ خَضَّبَ يَدَيْهِ وَرِجْلَيْهِ بِالْحِنَّاءِ ، فَقَالَ النَّبِيُّ : مَا بَالُ هَذَا؟ فَقِيلَ : يَا رَسُولَ اللَّهِ، يَتَشَبَّهُ بِالنِّسَاءِ.....	–	–	(1)	–	–	–
4	Henna as a material for wound treatments						
	...مَا كَانَ يَكُونُ بِرَسُولِ اللَّهِ قَرْحَةٌ وَلَا نَكْبَةٌ إِلَّا أَمَرَنِي رَسُولُ اللَّهِ أَنْ أَضَعَ عَلَيْهَا الْحِنَّاءَ	–	–	–	(1)	–	–
	...كَانَ لَا يُصِيبُ النَّبِيَّ صَلَّى اللَّهُ عَلَيْهِ وَسَلَّمَ قَرْحَةٌ، وَلَا شَوْكَةٌ، إِلَّا وَضَعَ عَلَيْهِ الْحِنَّاءَ	–	–	–	–	–	(1)
5	Henna to treat leg pain						
	...اخْتَجِمْ، وَلَا وَجَعًا فِي رِجْلَيْهِ إِلَّا قَالَ : اخْضِبْهُمَا	–	–	(1)	–	–	–
	Total	1	3	11	2	11	4
		32					

Abbreviation **B**: *Ṣaḥīḥ al-Bukhārí*, **M**: *Ṣāḥīḥ Muslim*, **AD**: *Sunan Abū Dāwūd*, **T**: *Jāmi` al-Tirmidhī*, **N**: *Sunan al-Nasā'ī*, **IM**: *Sunan Ibn Mājah*
Source Author's Analysis

(2) **Fatwa Text and Decisions Made by State Fatwa Committee**

In the management of fatwas in Malaysia, there are 14 mufti positions representing 13 states and federal (Putrajaya, Kuala Lumpur, and Labuan). In addition to the state-level fatwa committee, there is a level committee that has a specific role related to a fatwa or Islamic law, which is the Muzakarah Committee of the National Council for Islamic Religious Affairs Malaysia under the National Council for Islamic Religious Affairs (Majlis Kebangsaan Bagi Hal Ehwal Ugama Islam Malaysia (MKI). It serves as a national-level institution in Malaysia in order to coordinate and standardise fatwa decisions throughout the country. However, any decision decided at this stage will be re-discussed at each state fatwa institution whether there is need of fatwa or otherwise.[11]

The fatwa gazetted in each state is binding on Muslims who are in their respective states and applies together with the Shariah enactment for each state and can even be one of the sources of reference for judges in deciding cases in their respective state courts.[12] Meanwhile, the fatwa decided by the Muzakarah of the National Fatwa Committee managed by the Department of Islamic Development Malaysia or popularly known as JAKIM, at the federal level (under the Prime Minister's Department) which is not gazetted is regarded as a persuasive legal opinion but not binding in terms of Malaysian law.[13] As for the fatwa which is not a gazetted has no legal effect and is not regarded it as a source of law. However, that fatwa is considered an official fatwa, following the fact that a fatwa when formally applied by letter or inquiry will be considered as a government document and subjected to the regulations relating to it.[14]

Fatwas on henna are obtained from the esmaf.islam.gov.my website and the Official Portal of the Pahang State Mufti Department. The review collected the committee's decision on the ruling or fatwa from the website using the keyword "*inai.*" The selected legal decisions and fatwas are listed in Table 19.2.

Based on the above list, 5 out of 7 fatwas collected were decided by the state fatwa committee, namely the results of the fatwas of Kelantan (H01), Negeri Sembilan (H03), Sabah (H04), Sarawak (H05), and Pahang (H06).[15] While (H02)[16] is decided in the Muzakarah Fatwa Committee of the MKI, the State of Perak (H07)[17] adopts the opinion issued by the Muzakarah of the Fatwa Committee of the MKI with the recommendation that it is permissible for any party to leave a dubious act to an act that is not dubious. According to the fatwa, there are several aspects of the equation in the decision issued in all states:

[11] Rosele and Abdullah (2021).

[12] Portal Rasmi Fatwa Malaysia (2015).

[13] Aminudin (2015).

[14] Rosele and Abdullah (2020).

[15] Garis Panduan Pemakaian Inai Bercorak Menurut Pandangan Islam (2023).

[16] *Ibid.*

[17] Pemakaian Inai (2023).

Table 19.2 List of the legal decisions and fatwas that were chosen

Code	Fatwa	Information	Status Gazette
H01	Guidelines for the use of patterned inai according to islamic views	Jamaah Ulama' meeting of the Kelantan Islamic Religious and Malay Customs Council which convened on 02 May 2018	This fatwa is not gazetted
H02	The ruling of applying henna according to Shari'a	Muzakarah Fatwa Committee of the National Council for Islamic Religious Affairs malay-sia (MKI Muzakarah Committee) which convened in 2009	Yes gazetted
H03	Application of henna pat-terned according to Shari'a	Muzakarah of the Negeri Sembilan Fatwa Committee No. 04/2016-1437H con-vened on 28 April 2016	This fatwa is not gazetted
H04	Application of henna pat-terned according to Syari'a	Muzakarah of the Sabah State Fatwa Council Committee, bill. 2/2009 was convened on 09 December 2009	This fatwa is not gazetted
H05	Application of henna pat-terned according to Shari'a	Eighteenth Sarawak State Fatwa Board Meeting, which convened on 16 Muharam 1432H/ 22 December 2010M	This fatwa is not gazetted
H06	The Law of using patterned inai according to Shari'a	3rd/2014 Pahang State Islamic Law Consultative Committee Meeting, on 27th August 2014	Yes gazetted
H07	The ruling of applying henna is patterned according to Shari'a	The Perak State Mufti Department adopted the opinion issued by the 88th Muzakarah of the Fatwa Committee of the National Council for Islamic Religious Affairs Malaysia which convened on 2 October 2009	Yes gazetted

Source Author's Analysis

i. It is permissible for married or unmarried women to wear henna on both hands and feet as long as it does not cause slander.
ii. The rate of henna wear on both hands and feet is limited to the wrist and ankle only.
iii. The henna colour used is not as permanent as a tattoo and does not contain any dubious elements.

iv. Regarding the patterns or carvings used, patterns or carvings of leaves and plants are allowed. While animal patterns or carvings, emblems of deities or teachings or any depictions that contradict Shari'a are prohibited.

However, there are difference in fatwas on the ruling of henna use among men. Fatwas issued by H02, H04, H05, H06, and H07 prohibit men from wearing henna except with a slight rate on the fingers of the hand to express marriage as well as for medical purposes, in contrast to the fatwa decided by H03, which clearly prohibits the application of henna on men except for medical purposes only.

Irsyad al-Fatwa 157, at the Federal Territory Mufti Office, also tends not to allow men to wear henna on both hands and feet for the purpose of marriage. This is because henna is included in women's jewellery. However, for the purpose of medicating to cure certain diseases or to impose henna on the hair, it is permissible because there is a hadith that allows it.[18]

According to Prof. Dato' Arif Perkasa and Dr. Mohd Asri bin Zainul Abidin, women are encouraged to dye their fingers, especially during the marriage, while men are encouraged to dye their hair, especially those that are already Gray. The wearing of henna is one form of jewellery for women. So do men dye hair. It is encouraged if it is something that is not considered strange in its society. If otherwise, it should be avoided.[19]

Thus, it can be understood that henna is one of the ornaments for women. According to the opinion of the Shafi'i school, it is haram for men to wear henna as it is considered to resemble[20] women especially ornate for the purpose of marriage whether it is henna on the little finger, ring finger, and ghost finger on one hand only or both hands.[21] This is because there is a hadith that prohibits men from resembling women. The Prophet PBUH said from Abu Hurayrah:

[4928] عَنْ أَبِي هُرَيْرَةَ ، أَنّ النَّبِيّ " أُتِيَ بِمُخَنَّثٍ قَدْ خَضَّبَ يَدَيْهِ وَرِجْلَيْهِ بِالْحِنَّاءِ ، فَقَالَ النَّبِيُّ : مَا بَالُ هَذَا؟ فَقِيلَ : يَا رَسُولَ اللَّهِ، يَتَشَبَّهُ بِالنِّسَاءِ، فَأَمَرَ بِهِ فَنُفِيَ إِلَى النَّقِيعِ ، فَقَالُوا : يَا رَسُولَ اللَّهِ، أَلا نَقْتُلُهُ؟ فَقَالَ : إِنِّي نُهِيتُ عَنْ قَتْلِ الْمُصَلِّينَ " قَالَ أَبُو أُسَامَةَ : وَالنَّقِيعُ نَاحِيَةٌ عَنْ الْمَدِينَةِ وَلَيْسَ بِالْبَقِيعِ

Translation: Narrated Abu Hurayrah an effeminate man (mukhannath) who had dyed his hands and feet with henna was brought to the Prophet (ﷺ). He asked: What is the matter with this man? He was told: *"Messenger of Allah! He imitates the look of women."* So he issued an order regarding him and he was banished to an-Naqi'. The people said: *"Messenger of Allah! Should we not kill him?"* He said: *"I have been prohibited from killing people who pray."*

---

[18] Irsyad al-Fatwa ke-157: Hukum Lelaki Memakai Inai (2023).

[19] Apakah hukum memakai pewarna kuku selain daripada inai? (2023).

[20] استعمال الحناء والتزين بالكحل...رؤية شرعية (2023).

[21] استخدام الحناء للرجال (2023).

Imam al-Nawawi Rahimahullah mentions in *al-Majmu'*[22]:

أَمَّا خِضَابُ الْيَدَيْنِ وَالرِّجْلَيْنِ بِالْحِنَّاءِ فَمُسْتَحَبٌّ لِلْمُتَزَوِّجَةِ مِنْ النِّسَاءِ: لِلْأَحَادِيثِ الْمَشْهُورَةِ
فِيهِ وَهُوَ حَرَامٌ عَلَى الرِّجَالِ إلَّا لِحَاجَةِ التَّدَاوِي

Translation: As for colouring both hands and feet with henna, it is encouraged for those who marry among women based on the famous hadiths of it and it is haram for men except for medicine.

According to *al-Mawsuah al-Fiqhiyyah al-Kuwaitiyyah*, the *jurists* agreed that men are allowed to dye their hair and beard to exchange grey hair for henna and the like. They are allowed to wear henna throughout the body except on the hands and feet. They are not allowed to use henna on both hands and feet unless there is a Sharia excuse. This is because, colouring henna on both hands and feet resembles a woman, and it is prohibited. This is because colouring this on them, it is like women, which equals that of women is haram in Sharia.[23]

It can be concluded here that the ruling of applying henna to women is permissible for Sharia. On the other hand, men are prohibited from wearing henna because they resemble women, and this is not permissible in religion unless there is a Sharia excuse. However, based on the fatwas decided by some states above, it is permissible for men to wear henna on the bride's finger during the wedding ceremony. This decision is different from the opinion of the school of Islamic thought which prohibits wearing henna among men for being guilty of the Shari'a. Hence, it is encouraged for them to avoid doing so as it includes emulating a good example from the Prophet PBUH.

(3) **Medicinal Benefit of Henna**

The henna plant contains basic compounds from quinone types such as *lawsone (2-hydroxynaphthoquinone*) that gives a red–orange colour (Alem et al., 2020).[24] A study of Egyptian civilization has found the use of henna leaves (henna paste) for colouring hair, cheeks, lined their eyes, and stained their nails.[25] A review of the literature reveals numerous uses for the henna compound lawsone, including UV radiation protection for the skin, textile colouring, vulcanised rubber with an anti-ageing additive, and steel corrosion prevention (Jordao et al. 2015).[26] According to SEM analysis, henna dyestuffs may be able to repair cuticle damage and give dyed hair cuticles a smooth, moisture-rich appearance.[27]

---

22 *Ibid.*

23 *Ibid.*

24 Alem et al. (2020).

25 Chaudhri and Jain (2009).

26 Jordao et al. (2015).

27 Rippon (2013).

Furthermore, *lawsone* has a low allergic potential. In most cases, allergic reactions are caused by synthetic colouring additives added to henna mixtures rather than the henna itself.[28] The European Union's Scientific Committee on Consumer Safety (SCCS) concluded that henna was indeed slightly and transiently irritating to the eyes, but it can be used safely as a hair dye when the *lawsone* content is less than 1.4%.[29]

Through phytochemical experiments, Othman et al. (2020)[30] also discovered *isoplumbagin, esculetin, fraxetin, gallic acid* compounds, and other henna-derived compounds. These compounds play an important role in wound healing.

From the findings of the High Performance Liquid Chromatography-Quadrupole Time-Of-Flight Mass Spectrometry (HPLC-QTOFMS) on Ethanol: Water (80:20). According to Othman et al. (2020) the *Lawsonia Inermis* leaves extract has confirmed the presence of the first class group from the aromatic group which is the *phenolic, flavonoid, tannin, coumarin*, and *quinone* compound group. These compounds show that the presence of the aromatic group plays a role in treatment of wounds (Neha et al. 2019)[31] that act as an antioxidant agent (Othman et al. 2021a, b, c).[32]

The *isoplumbagin* compound has been reported as having potential in anti-inflammatory and antimicrobial activities (Ponugoti, 2018).[33]

According to Liang et al. (2017), *esculetin* and *fraxetin* compounds have the potential to be used as an anticoagulant/anti-clumps (warfarin medication) agent to stop blood clots that cause illnesses like stroke.[34]

19.3 Discussion

The findings of the analysis were conducted on three sub-themes, including henna from a hadith perspective and henna application, a comprehensive collection of national and state fatwas, and henna's medical benefits.

In general, the advantages of henna can be categorised into five groups or clusters based on 17 matn hadiths (refer Table 19.1). A total of 11 *Muslim* hadiths from **group one**, or 64% of them, mention henna as a permitted colouring agent

---

[28] Demirci et al. (2012).

[29] Scientific Committee on Consumer Safety (SCCS) Opinion on Lawsonia inermis (Henna) COLIPA n C169-SCCS/1511/13 Corrigendum, 12 November 2021. Available online: https://ec.europa.eu/health/scientific_committees/consumer_safety/docs/sccs_o_140.pdf (accessed on 14 February 2023).

[30] Othman et al. (2020).

[31] Neha et al. (2019).

[32] Othman et al. (2021a).

[33] Ponugoti (2018).

[34] Liang et al. (2017).

to darken grey hair dan bread. Henna from **group two** is grouped together specifically as a female accessory. Henna's colour is also acceptable for women to wear a nice with, according to **Hadith No. 4166** from *Sunan Abū Dāwūd*, which Imam Muḥammad Nāṣir al-Dīn al-Albanī deemed to be *hassan*. **Group three** apart from that, there is a prohibition on the use of henna colour for a man's nails that attempt to resemble a woman's such as **Hadith No. 4928** from (*Sunan Abū Dāwūd*) narrates of a *mukhannath* (khunsa). **Group four** includes two hadiths, **Hadith No. 2054** and **Hadith No. 3502**, which describe henna as a treatment material for wounds caused by thorns, swords, knives, and similar objects. Henna as a treatment for leg pain is in **group five**. In accordance with **Hadith No. 3858** (*Sunan Abū Dāwūd*), henna was necessary as a remedy for leg pain.

After analysing the groups of hadiths related to the use of henna, henna is a permitted colouring agent to darken grey hair dan bread for men. This is also discussed in the book *Tahdhib al-Sunan*, Ibn Qayyim mentions that the hadiths about colouring the hair and beard do not contradict each other. This is because the Prophet SAW had no objection to the practice of colouring hair and beards except for colouring them black. The permission (obligation) to colour with black colour is for a specific purpose only.[35] Furthermore, al-Nawawi states that it is permissible to colour (change the colour of) grey hair for men and women to yellow (using *sufrah*) or red (using henna or katam).[36]

Further regarding the application (dyeing) of henna on both hands and feet, this is required for women only where henna acts as one of the ornaments for women. As mentioned in **group two** regarding the application (dyeing) of henna's colour, it is acceptable for women to wear. The Prophet SAW once said:

[4166] عَنْ عَائِشَةَ رَضِيَ اللَّهُ عَنْهَا، قَالَتْ : " أَوْمَتِ امْرَأَةٌ مِنْ وَرَاءِ سِتْرٍ بِيَدِهَا كِتَابٌ
إِلَى رَسُولِ اللَّهِ فَقَبَضَ النَّبِيُّ يَدَهُ، فَقَالَ : مَا أَدْرِي أَيَدُ رَجُلٍ أَمْ يَدُ امْرَأَةٍ؟ قَالَتْ : بَلِ
امْرَأَةٌ، قَالَ : لَوْ كُنْتِ امْرَأَةً لَغَيَّرْتِ أَظْفَارَكِ يَعْنِي بِالْحِنَّاءِ

Translation: Narrated Ā'ishah: A woman made a sign from behind a curtain to indicate that she had a letter for the Messenger of Allah (ﷺ). The Prophet (ﷺ) closed his hand, saying: "I do not know this is a man's or a woman's hand." She said: "*No, a woman.*" *He said: "If you were a woman, you would make a difference to your nails, meaning with henna.*"[37]

---

[35] Safa al-Dawi Ahmad al-'Adawi, *Ihda' al-Dibajah 'Ala Sunan Ibn Majah*, 5: 45.

[36] *Ibid.* 44.

[37] Sulayman Ibn al-'Asy'ath al-Sajistani, *Sunan Abu Dawud*, Kitab al-Tarajjul, Bab fi al-Khadab li al-Nisa', Riyad: Maktabah al-Ma'arif li al-Nashr wa Tawzi', 744, no. Hadis 4166. **This hadith is considered hasan by al-Albani**. Ahmad Ibn Syu'aib Ibn 'Ali, *Sunan al-Nasa'i*, Kitab al-Zinah min al-Sunan, Bab al-Khadab li al-Nisa', Riyad: Maktabah al-Ma'arif li al-Nashr wa Tawzi', 771, no. Hadis 5089.

Only one hadith mentions the prohibition of wearing henna for men as an ornament that resembles a woman (*tasyabbuh) as follow:*

[4928] عَنْ أَبِي هُرَيْرَةَ ، أَنّ النَّبِيّ " أُتِيَ بِمُخَنَّثٍ قَدْ خَضَّبَ يَدَيْهِ وَرِجْلَيْهِ بِالْحِنَّاءِ ، فَقَالَ النَّبِيُّ : مَا بَالُ هَذَا؟ فَقِيلَ : يَا رَسُولَ اللَّهِ، يَتَشَبَّهُ بِالنِّسَاءِ، فَأَمَرَ بِهِ فَنُفِيَ إِلَى النَّقِيعِ ، فَقَالُوا : يَا رَسُولَ اللَّهِ، أَلَا نَقْتُلُهُ؟ فَقَالَ : إِنِّي نُهِيتُ عَنْ قَتْلِ الْمُصَلِّينَ " قَالَ أَبُو أُسَامَةَ : وَالنَّقِيعُ نَاحِيَةٌ عَنْ الْمَدِينَةِ وَلَيْسَ بِالْبَقِيعِ

Translation: Narrated Abu Hurayrah an effeminate man (mukhannath) who had dyed his hands and feet with henna was brought to the Prophet (ﷺ). He asked: What is the matter with this man? He was told: *"Messenger of Allah! He imitates the look of women."* So he issued an order regarding him and he was banished to an-Naqi'. The people said: *"Messenger of Allah! Should we not kill him?"* He said: *"I have been prohibited from killing people who pray."*[38]

Based on findings from the collection of fatwa text and decisions made by the state fatwa committee, there are several aspects of the equation in the decision issued in all states:

i. It is permissible for married or unmarried women to wear henna on both hands and feet as long as it does not cause slander.
ii. The rate of henna wear on both hands and feet is limited to the wrist and ankle only.
iii. The henna colour used is not as permanent as a tattoo and does not contain any dubious elements.
iv. Regarding the patterns or carvings used, patterns or carvings of leaves and plants are allowed. While animal patterns or carvings, emblems of deities or teachings or any depictions that contradict Shari'a are prohibited.

19.4 Conclusion

Many countries such as India, Indonesia, and Brunei have given henna-based cosmetic products halal status, for example, due to the fact that some colouring products such Garnier products have achieved halal logo status by Majelis Ulama Indonesia (one of Recognised Foreign Halal Certification Body (FHCB) that meet Malaysia's procedures and guidelines).[39] The Malaysian Halal Certification Procedures Manual (Domestic) 2020's Procedure 6(5) could be revised in the future.

[38] Sulayman Ibn al-'Asy'ath al-Sajistani, *Sunan Abu Dawud*, Kitab al-Adab, Bab fi al-Hukm fi al-Mukhannathin, Riyad: Maktabah al-Ma'arif li al-Nashr wa Tawzi', 891, no. Hadis 4928.

[39] Jabatan Kemajuan Islam Malaysia (JAKIM) (2023).

According to the research presented, in the *fiqh al-hadith* discussion towards the aspects of the use of henna, the initial findings of the study show that the use of henna is required by Shariah, especially for women. A total of 11 matn hadiths from the sources *Sunan Abū Dāwūd* and also *Sunan al-Nasā'ī*, followed by 4 matn hadiths from the sources *Sunan Ibn Mājah*. Source *Jāmi` al-Tirmidhī*, *Ṣaḥīḥ Muslim* and *Ṣaḥīḥ al-Bukhārí* that recorded 2 matn hadiths, 3 matn hadiths, and 1 matn hadith, respectively.

Henna can be used in five different ways. First, depending on the situation, such as jihad or when the *Fath Makkah*, there is some flexibility and obligation to colour the grey hair and men's beard refer **no. Hadith 1753,**[40] **6075,**[41] **6076,**[42] **and the other eight hadiths in** Table 19.1, but avoid using black colours refer **no. Hadith 5509**.[43] Second, only women are allowed to have their fingernails and feet painted the natural henna colour (refer **no. Hadith 4166**).[44] Third, **no. Hadith 4928,**[45] it is forbidden for men to use henna as an accessory that resembles women (*tasyabbuh*). In the fourth and fifth places, henna is used to cure wounds (**no. Hadith 2054**)[46] and relieve leg pain (**no.Hadith 3858**).[47]

These findings are also parallel to the point of view from the 88th Muzakarah of the Fatwa Committee of the National Council for the Malaysian Islamic Religious Affairs that convened in 2009 that had discussed the Law on the Use of Patterned Henna Based on the Law that states that:

1. It is permissible to use henna on both hands and feet for married or single women as long as it does not raise defamation.
2. It is permissible to use henna for the purpose of medicine on both hands and feet for men; it is not permissible except for a couple of fingers for the groom during wedding ceremonies.
3. The proportion of use of henna on both hands and feet is limited to the wrists and ankle only; meanwhile, the colours used should not be permanent like tattoos and contain any suspicious elements.

[40] Muhammad bin 'Īsā bin Sāwrah al-Tirmidhī (1996).

[41] Abu al-Husayn Muslim bin al-Hajjaj al-Qushairi al-Naysaburi (2000).

[42] Abu al-Husayn Muslim bin al-Hajjaj al-Qushairi al-Naysaburi (2000).

[43] Abu al-Husayn Muslim bin al-Hajjaj al-Qushairi al-Naysaburi (2000).

[44] Sulaymān bin al-'Asyath al-Sajistānī, *Sunan Abū Dāwūd*, (Riyadh: Maktabah al-Ma'arif li al-Nashr wa Tawzi', 1988) , 744 (Kitab al-Tarajjul, Bab fi al-Khidab li al-Nisa', no. Hadith 4166).

[45] Sulaymān bin al-'Asyath al-Sajistānī, *Sunan Abū Dāwūd*, (Riyadh: Maktabah al-Ma'arif li al-Nashr wa Tawzi', 1988) 891 (Kitab al-Adab, Bab fi al-Hukm fi al-Mukhannathin, no. Hadith 4928).

[46] Muhammad bin 'Īsā bin Sāwrah al-Tirmidhī (1996).

[47] Sulayman bin al-'Asy'ath al-Sajistani, *Sunan Abū Dāwūd*, (Riyadh: Maktabah al-Ma'arif li al-Nashr wa Tawzi', 1988) 693 (Kitab al-Tibb, Bab fi al-Hijamah, no. Hadith 3858).

4. It is allowed to have patterns or leaves and plant engravings, while animal patterns, symbols of deities, or the teachings of whatever image that are against the Syariah are absolutely prohibited.

In fact, scientific evidence shows that henna has several benefits of its own, such as being rich in antioxidant compounds that can treat wounds, antioxidant agent, antibacteria, anti-inflammatory, and antimicrobial. Therefore, if henna-based products are produced from natural henna, free from prohibited chemicals, safe to use, clean, and pure, it is submitted that the henna cosmetic products are eligible to be considered and be given Malaysian halal status.

References

Abu al-Husayn Muslim bin al-Hajjaj al-Qushairi al-Naysaburi (2000) "*Ṣaḥīḥ Muslim*," in al-Mawsu'at al-Hadith al-Sharif al-Kutub al-Sittah, ed. Salih bin 'Abd al-Aziz Al al-Shaykh, Kitab al-Fada'il, Bab Shaibah SAW, Riyadh: Dar al-Salam li Nashr wa al-Tawzi'

Alem FZ, Gita SA, Cougnaud L, Afnar C, Nounah I, Youssef B, de Oliveira AL, Charrouf Z, Rhourri-Frih B (2020) Lawsone quantification in Lawsonia inermis L. by HPLC-MS: How does the temperature and pluviometry affect lawsone concentration? Ind Crops Prod 158:112960

Chaudhri SK, Jain NK (2009) History of cosmetics. Asian J Pharm (AJP) 3(3):164–167

Demirci GT, Altunay IK, Atis G, Kucukunal A (2012) Allergic contact dermatitis mimicking angioedema due to paraphenylendiamine hypersensitivity: a case report. Cutan Ocul Toxicol 31(3):250–252

Jordao AK, Vargas MD, Pinto AC, da Silva FDC, Ferreira VF (2015) Lawsone in organic synthesis. RSC Adv 5(83):67909–67943

Liang C, Ju W, Pei S, Tang Y, Xiao Y (2017) pharmacological activities and synthesis of esculetin and its derivatives: a mini-review. Molecules 22:387

Manual Prosedur Pensijilan Halal Malaysia (Domestik) 2020

Muhammad bin 'Īsā bin Sāwrah al-Tirmidhī, *Sunan al-Tirmidhī*, Kitab al-Libas, Bab Ma Ja'a fi al-Khidab, Riyadh: Maktabah al-Ma'arif li al-Nashr wa Tawzi' 1996

Neha K, Haider MR, Pathak A, Yar MS (2019) Medicinal prospects of antioxidants: a review. Eur J Med Chem 178:687–704

Othman MR, Othman R, Ismail AA, Hazni H, Ahmad K, Razzak MA, Yusoff ZM, Awang K (2020) High-performance liquid chromatography quadrupole time-of-flight mass spectrometry (HPLC-QTOFMS) analysis on the ethanol: water (80: 20) extract of Lawsonia inermis leaves. Sains Malaysiana 49(7):1597–1613

Othman MR, Ahmad K, Yusoff ZM, Awang K, Md Ariffin MF, Othman R, Linn LI (2021a) (inai) dalam perspektif hadith : aplikasi dalam etnobotani Melayu dan kehadiran aktiviti antioksida. Islamiyyat J Antarabangsa Pengajian Islam Int J Islamic Stud 43(1):97–111

Othman MR, Ahmad K, Yusoff ZM, Awang K, Ariffin MFM, Othman R (2021b) Lawsonia inermis Linn.(Inai) dalam Perspektif Hadith: Aplikasi dalam Etnobotani Melayu dan Kehadiran Aktiviti Antioksida. Islamiyyat Int J Islamic Stud 43(1):97–111

Othman MR, Ahmad K, Yusoff ZM, Awang K, Ariffin MFM, Othman R (2021c) Lawsonia inermis Linn.(Inai) dalam Perspektif Hadith: Aplikasi dalam Etnobotani Melayu dan Kehadiran Aktiviti Antioksida. Islamiyyat Int J Islamic Stud 43(1):97–111

Ponugoti M (2018) A pharmacological and toxicological review of *Lawsonia inermis*. Int J Pharm Sci Res 9(3):902–915

Rosele MI, Abdullah LH (2020) Analisis Terhadap Keputusan Hukum Dan Fatwa Zakat Syarikat Di Malaysia

Rosele MI, Abdullah LH (2021) Analisis Terhadap Keputusan Hukum Dan Fatwa Zakat Syarikat Di Malaysia. J Pengurusan dan Penyelidikan Fatwa, 23(1):223, Jan 2021

Rippon DMLJA (2013) The coloration of human hair. The coloration of wool and other keratin fibres. Wiley, Bradford, UK, pp 358–389

Salleh ZHAM, Aghwan ZA (2019) Halal beauty salon services in Brunei Darussalam. In: Prosiding (Proceeding) 4th international seminar halalan thayyiban products and services, UNISSA Press, Universiti Islam Sultan, pp 410–420

Sarang H, Rajani P, Vasanthakumari MM, Kumara PM, Siva R, Ravikanth G, Uma Shaanker R (2017) An endophytic fungus, Gibberella moniliformis from Lawsonia inermis L. produces lawsone, an orange-red pigment. Antonie Van Leeuwenhoek 110(7):853–862

Interview

Aminudin M (2015) (Ketua Penolong Pengarah Kanan, Cawangan Penyelarasan Dan Kajian Fatwa, Bahagian Pengurusan Fatwa, Jabatan kemajuan Islam Malaysia), dalam temu bual dengan penulis, 11 April 2015

Website

"استخدام الحناء للرجال" (2013) Islam Online, accessed on 7 Feb 2023. https://fiqh.islamonline.net

"استعمال الحناء والتزين بالكحل...رؤية شرعية", *islamweb.net*, accessed on 5 Feb 2023, https://www.islamweb.net/ar/fatwa/12480/?searchKey=uY6PBZvC79aBU7vrB18S&wheretosearch=0&order=&RecID=8&srchwords=%C7%E1%CD%E4%C7%C1&R1=0&R2=0&hIndex=

Apakah hukum memakai pewarna kuku selain daripada inai? (2023) Official *website Jabatan Mufti Negeri Perlis*, accessed on 5 Feb 2023. https://muftiperlis.gov.my/index.php/minda-mufti/733-apakah-hukum-memakai-pewarna-kuku-selain-daripada-inai

Garis Panduan Pemakaian Inai Bercorak Menurut Pandangan Islam (2023) e-SMAF Fatwa Webite, Jabatan Kemajuan Islam Malaysia (JAKIM), accessed on 4 February 2023, http://e-smaf.islam.gov.my/e-smaf/fatwa/fatwa/find/3

Garnier Webiste, https://www.garnier.co.id/tips-dan-trick/pewarna-rambut/mau-mewarnai-rambut-yang-halal-kamu-bisa-ikuti-langkah-berikut-ini

Iba Cosmetic Website. https://ibacosmetics.com/collections/hair-colors

Irsyad al-Fatwa ke-157: Hukum Lelaki Memakai Inai (2023) Official website *Pejabat Mufti Wilayah Persekutuan*, accessed on 4 Feb 2023, https://muftiwp.gov.my/artikel/irsyad-fatwa/irsyad-fatwa-umum/1043-irsyad-al-fatwa-ke-157-hukum-lelaki-memakai-inai

Irresistable Cosmetics Website, Irri. https://www.irresistiblecosmetics.co.za/collections/queen-nail-henna

Jabatan Kemajuan Islam Malaysia (JAKIM) (2023) Senarai Badan Luar Negara Diiktiraf (2020), 24. Malaysia Halal Official Portal website, Accessed on 12 Feb 2023. https://www.halal.gov.my/v4/ckfinder/userfiles/files/cb2/CB%20LIST%201ST%20DEC%202020.pdf

Pemakaian Inai (2023) Official Website of the Perak Mufti Department, Accessed on 4 Feb 2023. http://mufti.perak.gov.my/artikel-khas/517-pemakaian-inai?start=3

Portal Rasmi Fatwa Malaysia (2015) "Senarai Akta/Enakmen Mufti dan Fatwa Negeri-negeri Seluruh Malaysia," laman sesawang *Portal Rasmi Fatwa Malaysia*, dicapai 26 Nov 2015. http://www.e-fatwa.gov.my/sites/default/files/enakmen_perlis.pdf

Prospects of Halal Supply Chain in Muslim Emerging Countries

20

Pipit Buana Sari and Yossie Rossanty

Abstract

This chapter aims to elaborate and discuss on prospects of the halal supply chain in some emerging countries. The study adopts content and bibliometric analyses where various sources are referred including reports, official documents, and research articles. The halal supply chain has positive development as well as prospects. There are potentials and relationships of halal supply chain with other sectors/industries that can be expanded. Significant initiative by the stakeholders is vital to ensure the continuous development of the halal supply chain. Other opportunities for a halal supply chain also have been revealed to be used as valuable information for strategic planning and initiatives.

Keywords

Halal supply chain • Halal concept • Emerging countries • Muslim

20.1 Introduction

Currently, Halal products are important because they are not only comply with religious law, but halal products are also healthy for consumption, bearing in mind the way of preparation and processing that pays attention to various rules that are in accordance with goodness. Halal products can be consumed by many parties, so this could cause halal products to be well-known globally.

The supply chain or halal logistics will develop into consumer demands, especially for Muslim consumers. Indonesia needs to immediately develop a halal supply chain that has developed in other countries, including Malaysia. The main

P. B. Sari (✉) · Y. Rossanty
Universitas Pembangunan Panca Budi, Medan, Indonesia
e-mail: pipitbsari@gmail.com

A. Rafiki (ed.), *Digitalization in Halal Management*, Contributions to Environmental Sciences & Innovative Business Technology,
https://doi.org/10.1007/978-981-99-5146-8_20

reason for the development of the halal supply chain in Indonesia is supported by the Muslim population (the largest Muslim population in the world). The percentage of Indonesian Muslims is around 12.5% of the world's population or as much as 88% of around 205 million Indonesians.

In addition, Indonesia must develop a halal supply chain because it will become one of the important determinants of product competitiveness in globalization, including in the ASEAN Economic Community (AEC). The world's Muslim population of around 1.75 billion or 24% of the world's population shows the potential for an important halal supply chain market as a consideration for the development of a halal supply chain in Indonesia.

The halal supply chain can be an opportunity for Indonesia to increase the competitiveness of its products. This opportunity is not only for producers but also for Indonesian logistics service providers. However, to turn this into an opportunity, the implementation of a halal supply chain requires coordination and collaboration between related parties, government, producers, logistics service providers, retailers, and related certification bodies.

20.2 Literature Review

20.2.1 Halal Supply Chain

In order to assure food safety in the halal supply chain, its process must use resources efficiently from raw materials to the finished product. It should be referred to the Islamic law or Sharia which that means divinely proclaimed and mandated legal code that governs Muslim behavior. To ensure that halal components are not mix up with haram things, halal supply chain management is crucial.

The right commitment must be made at the outset of creating an integrated halal supply chain. Top management must commit to build halal ecosystem through a halal policy, which serves as the cornerstone for supply chain organization (Indarti et al. 2020). The halal policy must strategically align with the company's strategy. Among the topics covered by the halal policies are (1) corporate accountability for maintaining halal integrity, (2) the scope of halal certification inside the business, (3) guarantees to customers, and (4) guarantee procedures (mechanisms control).

The halal supply chain model, which consists of logistics control, supply chain resources, business processes and network structure, and halal performance, is determined by the objectives of the supply chain. The orientation of supply chain goals can be broken down into two categories: (1) externally which known as customer service goals, like minimizing supply chain lead time and (2) internally which known as logistics goals, such condensing halal freight flows. The supply chain procedure establishes the degree of segregation necessary for terminal transportation, storage, and handling by land, air, and sea (Omar and Jaafar 2011).

Standard operation procedures (SOP) for the production process of halal products must be followed in order to ensure halalness at each stage of production in a company. The SOP considered necessary for handling goods that have been polluted by, or created purposefully or unintentionally from, filthy and/or unclean ingredients. Those halal products that do not fulfill the halal criteria, should be tracked back in time (traceability) precisely (Zulfakar et al. 2018).

Internal audits must be undertake periodically to assess the implementation of halal assurance system as part of integrated management. The appointed team has set a deadline for the immediate follow-up on the recommendation of the internal audit towards the company's halal assurance system. For cases that need to be followed up by top management, evaluation of the halal assurance system's application in businesses must also be done periodically in the form of a management review (Mohamed et al. 2020; Ngah et al. 2014).

Issues on Halal Supply Chain

The halal supply chain is a newly explored subject which then found by scholars that it has several debatable issues such as the complexity of handling supply chain management, a debatable question on halal status as well as questions related to halal integrity (Shafie and Othman 2006; Christopher and Holweg 2011; Tieman et al. 2012; Zulfakar et al. 2014; Bahrudin et al. 2011). Despite the issues mentioned above, the rise of Muslim consumers' trends on food ingredients is obvious (AbdulTalib and Abd-Razak 2013; Mohayidin and Kamarulzaman 2014), which demand the halal supply chain to be a solution for achieving halal product integrity (Kamaruddin et al. 2012).

In logisticians' perspectives, the halal supply chain is something potential to be engaged. Thus, they attempt to find ways to get investment to have it. However, some investors realize the additional tools, equipments, or facilities to support the halal supply chain in terms halal-only warehouse, halal audit team, and samak (ritual cleansing) services (Jaafar et al. 2011; Ab Talib et al. 2013). Nevertheless, some customers are willing to pay an additional fee for halal logistics (Kamaruddin et al. 2012; Tieman et al. 2013). With these arguments, halal supply chain management becomes a potential option that could be adopted by various companies or even a country. An innovative initiative needs to consider ensuring the halal supply chain is acceptable to be adopted by the logisticians and practitioners (Alqudsi 2014).

It is found that there is a lack of empirical studies on halal food supply chain integrity and assurance. In implementing both aspects of integrity and assurance, any firm should consider managing knowledge, commitment, and trust among the employees (Nik Muhammad et al. 2009). A better understanding of the halal concept application in society is also required.

Accurate halal supply chain management is a necessity to ensure the halal product of the food, beverage, pharmaceutical, and cosmetic industries. In the case of collagen and gelatin, for example, to obtain a halal final product, every link in the halal supply chain starts from the process of selecting and slaughtering animals, handling after slaughter, collecting skin and bones, the process of producing collagen and gelatin, the production process of the final product with collagen and gelatin ingredients, and each storage and transportation at each stage, until the final product is ready for consumption, must be guaranteed to be halal. To guarantee the halal final product that is ready for consumption, each link in the halal supply chain must implement a halal assurance system (Sumarliah et al. 2021).

The halal assurance system (HAS) is an integrated management system that is developed, implemented, and maintained by the company to manage materials, production processes, products, human resources, and procedures in order to maintain the continuity of the halal production process. As a prerequisite, the HAS must start with a written commitment from every company in each halal supply chain to implement a halal assurance system in their company (Hashim and Shariff 2016). Then the HAS must be supported by HR (halal management team/internal halal auditors) who are responsible for planning, implementing, evaluating, and continuously improving the halal assurance system (Manzouri et al. 2013).

20.3 Methods

This study adopts a content analysis with additional few items on bibliometric analysis. Bibliometric analysis manifests interconnections among the articles, whereby the data was retrieved from Scopus, a reputable and comprehensive database. The primary reason for choosing Scopus is its availability, nearly 60% larger than the Web of Science (WoS) database (Zhao and Strotmann 2015).

20.4 Discussion

Based on some findings from techniques in bibliometric analysis, the research on the halal supply chain has been done by many researchers in various countries. There are 300 documents sourced from a reputable database called Scopus. Below is the summary of elements related to research on the halal supply chain:

As depicted in Table 20.1, the highest hit on the number of documents is in the years 2020 and 2022 with a total of 52 documents. It showed a significant increment from the year 2015 to 2022. This means the interest in doing research on the halal supply chain is increasing. Though the halal supply chain is a newly explored subject, however, researchers are looking forward to finding more elaborations and studies on it.

Moreover, Malaysia is known as a country that is mostly involved in research on the halal supply chain with more than 169 documents compared to the other nine countries. This was followed by countries such as Indonesia, UK, and India in the 2nd, 3rd, and 4th places. It seems non-Muslim countries are engaged in research on the halal supply chain. Meanwhile as depicted in Figure 3 of Ali et al. (2022), Haleem et al. (2021), there are three authors that actively produced documents on the halal supply chain (Ali, M.H., Haleem, A., Tieman, M.) 13 documents, respectively. They also attempted to widespread their research results to other channels or media so that other groups could understand the concept of a halal supply chain.

Table 20.1 A summary of total number documents over authors, country, affiliations, and document type

No	Year/ Number document	Author/ Number of documents	Country/ Number of documents	Affiliation/ Number of documents	Document type/ Number of documents
1	2022 (53)	Ali, M.H. (13)	Malaysia (169)	Universiti Teknologi MARA (34)	Article (186)
2	2021 (43)	Haleem, A. (13)	Indonesia (71)	Universiti Kebangsaan Malaysia (21)	Conference paper (62)
3	2020 (52)	Tieman, M. (13)	UK (24)	Universiti Utara Malaysia (18)	Review (20)
4	2019 (28)	Khan, S. (12)	India (17)	Universiti Putra Malaysia (18)	Book chapter (17)
5	2018 (36)	Khan, M.I. (11)	Australia (10)	Universiti Kuala Lumpur (16)	Conference review (7)
6	2017 (26)	Jaafar, H.S. (8)	China (8)	Universiti Teknologi Malaysia (14)	Book (1)
7	2016 (18)	Tan, K.H. (8)	Brunei Darussalam (6)	Universiti Malaya (14)	Editorial (1)
8	2015 (6)	Rahman, N.A.A. (7)	Netherlands (6)	Jamia Millia Islamia (13)	Erratum (1)
9	2014 (6)	Zailani, S. (7)	South Korea (6)	Universiti Tun Hussein Onn Malaysia (11)	Note (1)
10	2013 (6)	Ab Talib, M.S. (6)	Thailand (6)	Universiti Malaysia Terengganu (9)	Retracted (1)

Meanwhile, the research on the halal supply chain has been done mostly on the subject of business, management, and accounting with a percentage of 26.8% which means that the topic of the halal supply chain was discussed mainly by researchers in the field of the halal supply chain. It also indicates that there is room for a halal supply chain to be explored and associated with other aspects or subjects such as engineering or sciences.

Halal Value Chain versus Halal Supply Chain

Halal value chain or can be called halal value chain is an industry-integrated effort starting from input, production, distribution, marketing, and consumption. In producing halal products, the input of raw materials must be considered, as well as the processing technology used. This integration effort is a form of developing the Islamic economy through strengthening which focuses on six potential industrial clusters or sectors, including food beverages, halal tourism, Muslim fashion, halal media-recreation, halal pharmaceuticals-cosmetics, and halal energy.

Halal value chain is the government's strategy for a country to develop the Islamic economy, particularly the halal industry. One way is to fulfill the demand for the domestic halal market and produce a variety of products. Fulfilling market demand will also be more effective due to the development of high domestic demand for halal products. This can also contribute to strengthening the balance of payments.

The halal supply chain must be regulated which requires consideration of both product attributes and market demands. The large and cold of chilled and frozen items in unitized and dry environments are thought to be more susceptible. Comparatively to less sensitive categories, the level of segregation is stricter for sensitive categories. It is crucial to consider an analysis on market situations, for example, some Muslim countries have tight procedures on logistical segregation than in non-Muslim ones. The rationale behind this is to lessen challenges in non-Muslim marketplaces while still fulfilling unique demands in Muslim markets based on Islamic schools of thought, regional fatwas, and local customs (such as ritual cleansing).

20.5 Conclusion

Several policies are needed so that the halal supply chain is maintained, for instance in a food crisis. *First*, the government needs to strengthen the halal product development model in more detail, so that it does not cause public anxiety when buying basic necessities. For example, before Eid season, the government needs to carry out a massive halal audit in the livestock sector. Include if there is a DNA manipulation on the animals or other artificial manipulation. All of this information must be completely recorded and recapitulated.

Second, encouraging the financial sector to provide Islamic financing for the procurement of halal food and medical needs, either in the form of opening new markets or individual financing. This is the time for all Islamic financial institutions to prove themselves as institutions that consistently uphold *maqasid shari'ah* in the midst of a crisis. *Third*, encourage every producer to increase the added value of their products and strengthen the safety of the products they produce. This step will increase consumer confidence amidst the many suspicions against manufacturers. Consumer confidence is very important because it can stimulate producers to continue to produce all the demands of society.

Fourth, the government is of course absorbing a lot of input and information by considering many parties. However, the government should not hesitate in protecting all consumers of halal products. The government's hesitation in dealing with food problems will have a negative impact psychologically on society. In addition, the key to success in strengthening the halal supply chain is consistently continuing to educate businesses and consumers regarding the importance of maintaining a halal supply chain.

One of the factors why the halal industry is not growing faster than expected, even though there is an increase in percentages, however no consensus on standards in all sectors. After all conditions return to normal, there will be a new balance for the halal industry with standards and market mechanisms that are more acceptable to the public and sufficient for the needs of all Indonesians and the world.

References

Ab Talib MS, Rubin L, Zhengyi VK (2013) Qualitative research on critical issues in halal logistics. J Emerg Economies Islamic Res 1(2):1–20

Abdul-Talib AN, Abd-Razak IS (2013) Cultivating export market-oriented behavior in halal marketing: addressing the issues and challenges in going global. J Islamic Mark 4(2):187–197

Ali MH, Iranmanesh M, Tan KH, Zailani S, Omar NA. (2022) Impact of supply chain integration on halal food supply chain integrity and food quality performance. J Islamic Mark 13(7):1515–1534

Alqudsi SG (2014) Awareness and demand for 100% halal supply chain meat products. Procedia Soc Behav Sci 130:167–178

Bahrudin SSM, Illyas MI, Desa MI (2011) Tracking and tracing technology for halal product integrity over the supply chain. In: Proceedings of the 2011 international conference on electrical engineering and informatics. IEEE, pp 1–7

Christopher M, Holweg M (2011) "Supply Chain 2.0": Managing supply chains in the era of turbulence. Int J Phys Distrib Logistics Manage 41(1):63–82

Haleem A, Khan MI, Khan S (2021) Conceptualising a framework linking halal supply chain management with sustainability: an India centric study. J Islamic Mark:12(8):1535–1552

Hashim HIC, Shariff SMM (2016) Halal supply chain management training: issues and challenges. Procedia Econ Finance 37:33–38

Indarti N, Lukito-Budi AS, Islam AM (2020) A systematic review of halal supply chain research: to where shall we go? J Islamic Mark 12(9):1930–1949. https://doi.org/10.1108/JIMA-05-2020-0161

Jaafar M, Abdul-Aziz AR, Maideen SA, Mohd SZ (2011) Entrepreneurship in the tourism industry: issues in developing countries. Int J Hosp Manag 30(4):827–835

Kamaruddin R, Iberahim H, Shabudin A (2012) Willingness to pay for halal logistics: the lifestyle choice. Procedia Soc Behav Sci 50:722–729

Manzouri M, Nizam Ab Rahman M, Saibani N, Rosmawati Che Mohd Zain C (2013) Lean supply chain practices in the Halal food. Int J Lean Six Sigma 4(4):389-408. https://doi.org/10.1108/IJLSS-10-2012-0011

Mohamed YH, Abdul Rahim AR, Ma'aram A (2020) The effect of halal supply chain management on halal integrity assurance for the food industry in Malaysia. J Islamic Mark 12(9):1734–1750. https://doi.org/10.1108/JIMA-12-2018-0240

Mohayidin MG, Kamarulzaman NH (2014) Consumers' preferences toward attributes of manufactured halal food products. J Int Food Agribusiness Mark 26(2):125–139

Ngah AH, Zainuddin Y, Thurasamy R (2014) Adoption of halal supply chain among Malaysian halal manufacturers: an exploratory study. Procedia Soc Behav Sci 129:388–395

Nik Muhammad NM, Md Isa F, Kifli BC (2009) Positioning Malaysia as halal-hub: integration role of supply chain strategy and halal assurance system. Asian Soc Sci 5(7):44–52

Omar EN, Jaafar HS (2011) Halal supply chain in the food industry-A conceptual model. In: 2011 IEEE symposium on business, engineering and industrial applications (ISBEIA). IEEE, pp 384–389

Shafie S, Othman MN (2006) Halal certification: an international marketing issues and challenges. In: Proceeding at the international IFSAM VIIIth world congress, vol 28. University of Malaya Press, Kuala Lumpur, p 30
Sumarliah E, Li T, Wang B, Moosa A, Sackey I (2021) The impact of customer halal supply chain knowledge on customer halal fashion purchase intention. Inf Resour Manag J (IRMJ) 34(3):79–100
Tieman M, Che Ghazali M (2013) Principles in halal purchasing. J Islamic Mark 4(3):281–293
Tieman M, Van Der Vorst JGAJ, Ghazali MC (2012) Principles in halal supply chain management. J Islamic Mark 3(3):217–243
Zhao D, Strotmann A (2015) Analysis and visualisation of citation networks. Synthesis Lect Inf Concepts Retrieval Serv 7(1):1–207. https://doi.org/10.2200/S00624ED1V01Y201501ICR039
Zulfakar MH, Anuar MM, Ab Talib MS (2014) Conceptual framework on halal food supply chain integrity enhancement. Procedia Soc Behav Sci 121:58–67
Zulfakar MH, Chan C, Jie F (2018) Institutional forces on Australian halal meat supply chain (AHMSC) operations. J Islamic Mark 9(1):80–98. https://doi.org/10.1108/JIMA-01-2016-0005

The manufacturer's authorised representative in the EU is Springer Nature Customer Service Centre GmbH, Europaplatz 3, 69115 Heidelberg, Germany. If you have any concerns regarding our products, please contact ProductSafety@springernature.com

Printed and bound by CPI Group (UK) Ltd, Croydon, CR0 4YY
30/09/2026
02225319-0008